# Scottish Collieries

## An Inventory of the Scottish Coal Industry in the Nationalised Era

Miles K Oglethorpe

## Published by

The Royal Commission on the Ancient and Historical Monuments of Scotland

in partnership with

The Scottish Mining Museum Trust

June 2006

## ISBN

10 digit – 1902419472

13 digit – 9781902419473

## Jacket front

Lady Victoria Colliery, Newtongrange, Midlothian, 1966. John Keggie, SC367264

## Jacket back

Rothes Colliery, Fife, late 1950s. SC446437

'Bait Time' at Highhouse Colliery, Ayrshire. Harrison Collection, SC446730

A new NCB lorry advertising coal as a 'progressive fuel' in the late 1950s. SMM:1998.616, SC706322

Scotland's last steam-powered coal train leaves Bedlay Colliery near Glenboig in Lanarkshire, December 1981. East Dunbartonshire Information & Archives P5250, SC706487

## End papers

(front) Killoch Colliery, near Ochiltree, Ayrshire, by Egon Riss. SC706731

(back) Bilston Glen Colliery, Loanhead, Midlothian, by Egon Riss. SC706505

## The Royal Commission on the Ancient and Historical Monuments of Scotland

The Royal Commission on the Ancient and Historical Monuments of Scotland (RCAHMS) collects, records and interprets information on the architectural, industrial, archaeological and maritime heritage of Scotland. Whether you are working, teaching, studying or simply exploring your local heritage, RCAHMS resources are available to assist your research. You can use our online databases and mapping services to view over 60,000 digital images and to search for information on more than 250,000 buildings or sites, over a million aerial photographs and some 2.5 million other photographs, drawings and manuscripts. You can also visit our search room to consult original archive material (Monday to Friday, 9.30am – 4.30pm).

RCAHMS
John Sinclair House
16 Bernard Terrace
Edinburgh
EH8 9NX

t 0131 662 1456
f 0131 662 1477
info@rcahms.gov.uk
www.rcahms.gov.uk
Registered charity SC026749

## The Scottish Mining Museum

The Scottish Mining Museum is based in Lady Victoria Colliery, which is situated nine miles to the south of Edinburgh in Midlothian. It is one of Europe's best-preserved 19th century coal-mining complexes, having closed in 1981 after over a century of production. It houses a nationally significant collection of mining equipment and materials, and a large library and archive relating to the coal industry in Scotland, the UK and beyond. It is now a five-star visitor attraction, a key stopping point on the Lothian tourist trail, and is one of Scotland's most important industrial museums.

The Scottish Mining Museum
Lady Victoria Colliery
Newtongrange
Midlothian
EH22 4QN
t 0131 663 7519
www.scottishminingmuseum.com/

### Supporting Organisations

The Coal Authority
The National Union of Mineworkers (Scotland)
The Scottish Coal Industry Special Welfare Fund
The Mining Institute of Scotland Trust
The Friends of The Scottish Mining Museum

# Contents

# Dedication

To 'The Two Georges' and their fellow volunteers

# Forewords

**Kathleen Dalyell**, MA, FRSAS, FSA Scot,
Chairman of RCAHMS, 1999–2004

The Royal Commission on the Ancient and Historical Monuments of Scotland (RCAHMS) was established in 1908 to survey and record the built heritage of Scotland. Its original remit explicitly excluded any structure dating from after 1707, and it was therefore not until the 1950s that industrial heritage was gradually embraced within mainstream recording programmes. During the 1960s, industrial monuments were increasingly included within routine survey work and for the first time, featured in official publications, such as the county inventories for Stirlingshire and Peeblesshire. Thereafter, industry became a much more significant element within the Commission's recording programmes, culminating in the publication of Geoffrey Hay and Geoffrey Stell's *Monuments of Industry – An Illustrated Historical Record in 1986.*

At this time, RCAHMS confirmed its commitment to the recording of industrial heritage by taking on the staff and archive of the Scottish Industrial Archaeology Survey, a research and recording unit based at the University of Strathclyde and founded and directed by Professor John R Hume in 1977, with the support of the Department of the Environment. On arrival at the Commission, the two new staff, Graham Douglas and Miles Oglethorpe, continued to carry out thematic and threat-based industrial surveys and also published books based on accumulated research and survey, such as *Brick, Tile and Fireclay Industries in Scotland* in 1993. By this time, however, very little of Scotland's once hugely important coal industry survived intact, and in order to ensure the survival of accurate information on the coal-mining sites throughout the country, an alternative strategy was necessary.

The result has been a recording exercise that has involved the collation of information from a variety of people and organisations, perhaps the most important of which has been the Scottish Mining Museum, based at Lady Victoria Colliery in Newtongrange, Midlothian. Whilst a few coal-mining sites have survived with some of their surface buildings recognisably intact, most have been razed, and all evidence of mining activity totally destroyed. The challenge has therefore been to gather together enough information to ensure that the mines and miners responsible for producing the coal on which almost every economic activity depended are not forgotten.

For me, this is of great personal importance. In 1963, when I married Tam Dalyell, Member of Parliament for West Lothian, there were six collieries in his parliamentary constituency: Easton (Bathgate), Kinneil (Bo'ness), Polkemmet and Whitrigg (Whitburn), Riddochill (Blackburn), and Woodend (Armadale). Since the closure of Polkemmet mine in unsatisfactory circumstances during the Miners' Strike of 1984, there are now none, and all of Scotland's remaining collieries have since closed. In 1963, I took about 20 of my 6th-year pupils from James Gillespie's High School for Girls down Polkemmet, with the enthusiastic support of the National Coal Board and the National Union of Mineworkers. I was conscious that the girls had previously no idea of the conditions in which coal was won. Having descended in the cage deep into the pit and walked through air thick with coal dust for over a mile in narrow low tunnels that dripped with water, we reached the coal face where miners were using their picks to get at the coals that the Anderton Shearer could not reach, some lying on their backs to do so. My pupils very quickly realised that the winning of the coal was too often at the price of fatal diseases such as pneumoconiosis and silicosis, and that mining also brought about a host of chronic illnesses and ailments such as bronchitis, arthritis and white finger. They also witnessed an extraordinary and unique working environment in which teamwork and camaraderie were a vital part of everyday life.

This is reflected throughout the central coalfields of Scotland, where the mining communities have set a pattern of values that have permeated the society in which they have lived and worked. Against this background, the physical evidence of the coal mines continues to disappear from our landscapes, with the dismantling of the spoil heaps and the demolition of winding towers and associated surface buildings and structures. At the same time, many of those who worked in the industry are gradually slipping away, taking with them unique knowledge of the working conditions in the mines and life in their communities.

With this in mind, the main purpose of this book is therefore to provide an inventory of the Scottish coal industry in the second half of the 20th century. Although it is no longer possible to take parties of school children down a working coal mine, the gazetteer can at least help to provide a reliable picture for future generations of the nature, extent and importance of the coal industry in Scotland. It should also be a useful skeleton onto which future more-detailed research on aspects of the coal industry might be based. Most important of all, we hope that it will help to ensure that the work, traditions and achievements of Scotland's mineworkers are properly recognised and remembered by future generations.

**Rhona Brankin**, MSP, Trustee of the Scottish Mining Museum Trust, 2002–04

Within only a few decades, the place of coal in the everyday lives of people in Scotland has been transformed from a position of dominance to one of near total obscurity. Yet, in 1947, almost all the energy on which the Scottish economy depended was provided by coal, and in addition to the 140,000 people who found themselves in the pay of the new National Coal Board, tens of thousands of other workers were engaged in activities directly and indirectly supporting the coal industry.

Inevitably, the rapidity of the decline of coal, and the intensity and hardship that this process generated, has resulted in the perpetuation of negative sentiments associated with the industry. The final decades were fraught with struggle and bitter disillusionment, culminating in the strike of 1984, after which the pace of decline accelerated. Even today, the physical remains of the industry are continuing to disappear, sometimes being cleared away for development projects, or as part of land improvement and reclamation schemes. As time passes, it is becoming difficult to recognise mining areas, and for many people, the massive contribution of coal miners to the economic, political and cultural wealth of Scotland remains unrecognised.

Against this background, it is especially encouraging to see the publication of this book, which is the result of many years of partnership between the Scottish Mining Museum and the Royal Commission on the Ancient and Historical Monuments of Scotland (RCAHMS). Together, they have attempted to ensure that accurate information on the nature, location and extent of Scotland's collieries has not been lost with the industry itself. This has been possible in particular because of the wealth of material in the collections of the Scottish Mining Museum (at Lady Victoria Colliery) and because of the dedicated work of its staff and volunteers. Indeed, following the disappearance of deep coal mining in 2002, Lady Victoria Colliery is now the only coherent surviving colliery complex in Scotland, and remains one of the most important mining monuments in Europe.

This book draws together the rich archive holdings of the Scottish Mining Museum and the survey and recording expertise of RCAHMS. Whilst it is intended to be of general interest, it is also designed to be a significant educational resource, especially for schools and colleges in former mining areas. As memories fade and the physical remains of the industry continue to disappear, this book provides accurate information on the mines themselves, and should itself be a useful platform for further research on coal mining in Scotland. In particular, it is hoped that some of the inevitable gaps in its coverage can be filled, and that a continuous process of enhancing information on the coalfields can be encouraged. Ultimately, therefore, the aim of this book is to ensure that the coal industry, the people who worked in it, and its associated communities are not overlooked, and that the central part coal has played in the evolution of modern Scotland is not forgotten.

# Acknowledgements

In the late 1980s and early 1990s, the once great British coal industry entered a new and near-terminal phase of decline in a process that culminated in its privatisation. During this period, the three Royal Commissions responsible for recording the built heritage in England, Scotland and Wales (RCHME, RCAHMS and RCAHMW) became aware that a hugely important industry was rapidly disappearing, and that efforts should be made to ensure that records of coal mines and their associated landscapes and communities were safeguarded. The Royal Commissions therefore embarked upon survey programmes designed to record the last surviving mines prior to closure, and also attempted to co-ordinate efforts to save coal-mining records that would have otherwise been lost in the closure process. Both RCHME (Thornes, 1994) and RCAHMW (Hughes *et al*, 1995) subsequently published books on the English and Welsh coal industries in the mid-1990s, and it was always the intention that RCAHMS would follow suit in Scotland.

Much of the initial impetus for this project was therefore provided through collaboration with colleagues at RCHME, notably Janet Atterbury, Keith Falconer and Robin Thornes, and in Wales through working alongside Stephen Hughes, Brian Malaws and Peter White of RCAHMW, and Peter Wakelin, then of Cadw. In Scotland, the initiative also benefited from the experience and support of John R Hume (then at Historic Scotland), Richard Gillanders of the British Geological Survey, Peter Anderson of the National Archives of Scotland, and Colin McLean, then the Director of the Scottish Mining Museum.

From the outset, the intention was that the Scottish book would be based around a gazetteer of the industry, and would concentrate on the recent era, from nationalisation in 1947 to privatisation in the mid-1990s. This was a time of rapid change which commenced with the total dominance of coal as the nation's principal source of energy, combined with immense confidence in a newly nationalised industry. There followed a period of breathtaking change as other power sources such as nuclear and natural gas grew in significance. Eventually, accelerating decline within the coal industry culminated in the loss of Scotland's last deep mine in 2002. Just as significant was the propensity for the physical remains of the industry to be rapidly and comprehensively cleared away, and it soon became apparent that what had been one of Scotland's most important industries was being almost entirely erased from the landscape.

The principal aim of the project was therefore to ensure that sufficient records survived to enable future recognition of the coal industry's immense historical contribution to Scottish life. As a start, the intention was to create a minimum record by compiling an illustrated gazetteer which, at the very least, could provide basic information on the name and location of collieries, the physical remains of most of which have disappeared entirely. Ultimately, this task took considerably longer than expected, partly because the work had to be accommodated within the broader Industrial Survey programme at RCAHMS, but also because of the wealth of record material encountered, notably through the National Coal Board (NCB), and latterly the Scottish Mining Museum.

RCAHMS is therefore greatly indebted to several people who worked for the NCB (latterly the British Coal Corporation). Bob Howden (Health & Safety Officer) helped to provide access to surviving pits, and mine managers and deputies who provided assistance both on the ground and under it included William Kerr and Tony Simpson. Perhaps the greatest quantity of information emanated from the NCB/British Coal Newbattle Archive, which was situated in a leased building within the Scottish Mining Museum's Lady Victoria Colliery. Prior to privatisation, abandonment plans were sifted and information co-ordinated within the archive, and of the staff employed on this project, John McDowell and Jimmy Kinnell were especially helpful, in particular providing access to the NCB's shaft register for Scotland. In later years, Colin Flaws of Scottish Coal was especially helpful, and the recent addition of relevant information by Alan Thomson has been especially significant. However, perhaps the greatest assistance was given by British Coal's Newbattle archivist, Sharon Urquhart, who not only provided access to unique information, but also supplied a desk and a continuous supply of tea and coffee. Amongst the records to emerge was a colliery questionnaire yielding information on every colliery taken into state ownership in 1947, and a wonderful collection of photographs.

The assistance received from NCB staff was all the more valuable because it was given against a background of great personal uncertainty, and in most cases, inevitable redundancy. It also contrasts with other parts of the UK, where an embattled situation made co-operation with heritage bodies more difficult. Help was, however, forthcoming from the NCB's archives in Mansfield, where Brian Thornton provided access to a wealth of material, including 'pit-profile wallets' and photographs mostly related to the activities of the Miners' Welfare Fund. Much of this material, and a great deal of other coal-related records, found its way to the Public Record Office (PRO) in Kew, London. The PRO, which has since become TNA (The National Archives) deserves great credit for the way in which it makes its records available to the public. In this case, the batch ordering system made a huge difference, given the large quantity of material that was of interest. Ultimately, the PRO's Image Library yielded some unique and important photographs, and the assistance of David Mole, Hugh Alexander and Paul Johnson in locating and copying them is greatly appreciated.

Back in Scotland, RCAHMS embarked on a search for record material throughout the coalfields. Help was encountered in particular from South Lanarkshire, North Lanarkshire, Ayrshire, East Dunbartonshire, Glasgow City, Clackmannanshire, Falkirk, Stirlingshire, West Lothian, East Lothian and Fife Council archives, museums and libraries. Enthusiastic and especially helpful individuals included Kevan Brown at Methil Heritage Centre, Sheena Andrew at Carnegie Library in Ayr, Don Martin, Richard Stenlake, and Guthrie Hutton, whose excellent county-based illustrated books on the coal industry have been both an inspiration and a source of essential information. Meanwhile, the Fife Pits website produced and maintained by Michael and Colin Martin and Chris Sparling (www.users.zetnet.co.uk/mmartin/fifepits/) proved to be an extraordinary source of information. Also in Fife, Wullie Braid's 'Cardenden Mining Archives' provided significant help, and the late Bien Bernard's infectious enthusiasm and generosity was greatly appreciated.

In the search for images of coal mines, it was particularly good to be able to rely on John R Hume, who, in addition to advice and encouragement, kindly provided access to the extensive photographic collection that he has been generating since the 1960s. However, perhaps the most extraordinary concentration of photographic material has emanated from Mauchline in

Ayrshire, where Terry Harrison and John McKinnon had collected, salvaged and created fantastic photograph collections of Ayrshire pits. We are particularly grateful to Georgina McKinnon for allowing us to use images from her late husband's collection.

Nevertheless, the greatest body of record material relating to the Scottish coal industry can be found at the Scottish Mining Museum's library and archive at Lady Victoria Colliery in Netwongrange, near Edinburgh which also holds coal-related records on behalf of the National Archives of Scotland. From the outset, the staff and volunteers at the museum provided unlimited access to their collections, and also continued to collect valuable archive material. As well as the then director, Colin McLean, Mike Ashworth and Rosemary Everett were especially helpful. Since then, the chairman of trustees, Donald Mockett, the current director, Fergus Waters, and colleagues Kirsty Lingstadt and Julia Stephen have continued to provide valuable support, and this book is being published in partnership with the museum.

Collating the gazetteer and providing background information and images would not have been possible without the volunteers at the Scottish Mining Museum. In particular, George Archibald and the late George Gillespie, otherwise known as 'The Two Georges', were immensely helpful, contributing important facts and context, and weeding out some of the author's more serious errors and misunderstandings. Perhaps most important was their assistance with image research, George Gillespie's encyclopaedic knowledge of individual collieries being particularly important in the identification and selection of photographs. Fellow volunteer, Campbell Drysdale, himself formerly a photographer for the NCB, also managed to conjure up images of collieries from nowhere, some of which were his own work. Furthermore, much of the information on the coal industry currently available in the museum's library has been collated by volunteers over many years, and it is pleasing to note that the volunteers are continuing the good work with great enthusiasm.

At RCAHMS, many colleagues have contributed to the completion of this book. In the Survey and Graphics Department, Alan Leith, John Borland, and Heather Stoddart were all involved in recording Lady Victoria Colliery, Kevin McLeod generated the book's distribution maps, and Oliver Brookes has been responsible for its production and design. Other contributions included those of Graham Douglas, who assisted with the recording of Cardowan, Monktonhall, Seafield and Bilston Glen collieries. Most staff in the photography department have participated in some way at some stage. In the field, Jim Mackie, Angus

Lamb and John Keggie have taken photographs at various collieries, and Robert Adam has photographed them from the air. Whilst in the office, Anne Martin has overseen a huge amount of image scanning by staff, who have included Tahra Duncan, Derek Smart, Stephen Thompson and Claire Brockley.

In order to ensure that the data included within the gazetteer corresponds with that in the National Monuments Record of Scotland database, several NMRS colleagues, particularly Miriam McDonald, Clare Sorensen, Jessica Taylor and David Easton, have also greatly assisted with map and aerial photography research, and with the assimilation of material into the collections. In many instances, this process was greatly helped by the work of the RCAHMS archaeology field teams in the Central Scotland Forest and elsewhere, such as at Muirkirk in Ayrshire and Wilsontown in Lanarkshire. Many tasks were made considerably easier by the provision in recent years of enhanced computer facilities, and the hard work of Kate Byrne and Jo Mc-Coy. Also essential has been the enhancement of the NMRS Oracle database by John McLeod and Christina Allan, and the development of Geographic Information Systems designed by Mark Gillick. Finally, encouragement and editorial guidance from Geoffrey Stell, Graham Ritchie, Lyn Taylor, Jack Stevenson, Roger Mercer, Diana Murray and Rebecca Bailey have been an enormous help, as has the input from Commissioners including Kathleen Dalyell, John R Hume, Roland Paxton, Anne Riches and Gordon Masterton.

RCAHMS and the Scottish Mining Museum are also greatly indebted to a number of organisations without whom the production of this book would have been impossible. Financial assistance was provided by The Coal Authority, the National Union of Mineworkers, The Mining Institute of Scotland Trust, The Scottish Coal Industry Special Welfare Fund, and the Friends of the Scottish Mining Museum.

# Chapter 1: Introduction

Coal has had a profound impact on both the people and the landscape of central Scotland. At its peak in the early 20th century, the Scottish coal industry directly employed almost 150,000 people. It had become a massive industry which fuelled a complex economy at the heart of the British Empire, and which exported coal throughout the world. It was also supported by a network of industries supplying a variety of products and services, notably engines for driving winding, haulage, ventilation, hydraulic and pneumatic machinery, locomotives, hutches and wagons. In addition, large-scale deep mining required huge quantities of materials, including steel for rails and mine roadway supports, and thousands of mostly wooden pit props. Collieries were connected to their markets by a dense network of railways, most of which were built to link the mines to an expanding national railway system. Also significant was the evolution of specialist professional bodies such as The Mining Institute of Scotland, and the Association of Mining, Electrical and Mechanical Engineers. This often resulted in the pioneering work of mine managers and deputies, surveyors and engineers, being recorded in published proceedings and journals. Working collieries and associated mining communities could be seen throughout the central belt of Scotland, and were a distinctive feature of the landscape (see Figure 1.1). Coal mining had become a fundamental part of Scottish life – so familiar that it was almost invisible, despite the bings.

As the 19th and 20th centuries progressed, coal miners, coal owners and mining communities were immensely influential elements within Scottish society. In particular, the shared experience of underground working engendered interdependence and shared values. Scottish miners, such as James Keir Hardie, helped to found the Labour Party, and trade union activity in subsequent decades assisted the struggle to attain better and safer working conditions in an industry which was notorious for its appalling record of disasters, fatalities and injuries.

Much of the current culture of teamwork and health and safety in the workplace emanated from the mines, becoming an essential element of working life in most industries and services. The industry established training centres at which education in safety and technical subjects was provided, and the organisation of regular mines rescue and first aid competitions helped to nurture the concept of collective responsibility. Coal was itself a vital part of everyday life, providing fuel for homes, gasworks, iron and steel works, electricity power stations, railway locomotives, ships and thousands of mills and factories. Less obvious, but also hugely important, were a number of industries utilising the by-products of coal, such as tar and chemicals, and heavy clay industries including brick, refractory and sanitary ware manufacture.

Although the early decades of the 20th century witnessed relative decline and stagnation, an optimistic mood prevailed after World War II. In 1947, nationalisation took the coal industry into the care of the state, and there was a genuine belief that working conditions and economic performance would be greatly improved. Signs reflecting the new confidence were erected in all nationalised pits, bearing the words, 'This Colliery is now managed by The National Coal Board on behalf of the people'. The British coal industry was seen to be of immense strategic importance, and was a vital part of efforts to achieve post-war reconstruction. A very significant proportion of the state's massive investment in the coal industry occurred in Scotland, typified by

Figure 1.1: This famous image is of Auchincruive 4, 5 and 6 Colliery, and was taken not long after nationalisation. Also known as 'Glenburn' Colliery, it was situated close to Prestwick in Ayrshire. It was one of many pits earmarked for reconstruction by the NCB, and continued production for 26 years in the NCB era. Harrison Collection. SC446168

the successful new sinking at Killoch in Ayrshire (Figure 1.2). The excitement and optimism was clearly evident when Queen Elizabeth II and Prince Philip opened Rothes Colliery in 1958 (Figures 1.3 and 1.4).

By the 1960s, however, rapid change was about to engulf the British energy market, and the once totally dominant coal industry began to face serious competition from petroleum, natural gas, nuclear, and to a lesser extent, hydro-electric power. It also had to contend with clean-air legislation, which had been introduced in response to rapidly worsening air pollution in the 1950s. Traditional markets began to melt away, and what had once been a problem of demand exceeding supply was evolving into one in which supplies of coal were exceeding demand. The National Coal Board was suddenly faced with a situation in which it had to make an effort to sell its coal. In the domestic sphere, this was exemplified by attempts to promote a new 'clean'

form of coal called 'Nuggets', and to market solid-fuel central heating systems (see Figures 1.5 and 1.6).

From 1960, the opening of Killoch Colliery brought about a new era in which the increased output of the new generation of 'superpits' allowed for the closure of smaller, less efficient mines. In the 1960s alone, three-quarters (112) of Scotland's collieries were closed, but throughout this period, and during the 1970s, their disappearance provoked comparatively little concern or interest outside the industry. There were so many mines that it was inconceivable that the long-term future of the industry as a whole might be under threat. At this time, interest in industrial heritage was only just beginning, and the recording and protection of the historic built environment remained predominantly focused on traditional archaeological and architectural subjects. One significant exception to this pattern was John R Hume, whose extensive photographic recording and documentation of Scotland's industries included many coal mines, such as Cardowan Colliery at Stepps on the east side of Glasgow (see Figures 1.7, 1.8 and 1.9). Sadly, however, the vast majority of collieries that disappeared in this period were never recorded.

As the decline accelerated in the final decades of the 20th century, the scale of the threat to the industry became apparent. A combination of factors threatened to eradicate all traces of mining activity. The surface remains of most collieries were cleared away for environmental and commercial reasons, and records relating to the industry were often lost. In a number of cases, records were rescued, as was the case in Ayrshire where at least two valuable collections of coal-mining images have survived in private hands (see Figures 1.10 and 1.11). As the crisis within the industry worsened in the mid-1980s, RCAHMS began to record the last surviving collieries in Scotland (see Figure 1.12). Perhaps most significant, however, was the formation of the Scottish Mining Museum at Lady Victoria Colliery in Midlothian, whose staff were able to collect records and artefacts from all over the Scottish coalfields.

During the 1990s, the destruction of mining sites and landscapes accelerated due to major afforestation and reclamation schemes in more rural areas, and redevelopment projects in more urban areas, often resulting in the demolition of prominent mining structures, such as the two winding towers at Rothes Colliery in Fife (see Figure 1.13). Although open-cast activity continued, deep coal mining was eventually confined to one colliery, which served the Longannet power station in Fife, near Culross. When in 2002 this mine was forced to close following a major underground flood, deep coal mining in Scotland finally ceased (see Figure 1.14).

Much has been written about the coal industry in the UK during the last 100 years, and some of the more important works are listed in the bibliography. They have included company histories, often written in advance of nationalisation, and a five-volume set of books on the history of the British coal industry commissioned by the NCB. There is also a wealth of literature describing the miners' struggles, the advance of trade unionism, welfare issues and reports on accidents and disasters. In addition there are a number of professional mining journals, and valuable sequences of colliery year books. All of this material can be found in the library of the Scottish Mining Museum which, in addition to inheriting material from the NCB, also incorporates the library of the former Mining Department of Heriot-Watt University, and collections from libraries at the Ministry of Fuel and Power's Mines

Figure 1.2: Killoch Colliery, c.1958. One of the most successful of the new Scottish superpits, it was sunk by the NCB during the 1950s, production commencing in 1960. Its huge concrete towers became a landmark in East Ayrshire. Harrison Collection. SC706224

Figure 1.3: HRH Queen Elizabeth II underground at Rothes Colliery in Fife during the opening ceremony in 1958. Although the project was already in difficulty, coal was still regarded as a fuel of the future, and great faith remained in the new generation of NCB mines. This photograph is one of a huge collection of images held by the Scottish Mining Museum. SMM:1998.603, SC706693

Figure 1.4: HRH Prince Philip, Duke of Edinburgh underground at Rothes Colliery in Fife during the opening ceremony in 1958. Scottish Mining Museum. SMM:1998.603, SC706701

Figure 1.5: In 1947, the National Coal Board took control of an industry which supplied almost all the UK's energy needs. By the 1960s, its dominant position was being challenged by new competing forms of energy, and publicity campaigns attempted to promote the image of coal as a modern and 'progressive' fuel. Scottish Mining Museum. SMM:1998.616, SC706322

Department, Esk Valley College, the Fife Mining School, the National Union of Mineworkers, and the Coal Industry Social Welfare Organisation (formerly the Miners' Welfare Fund).

There are, nevertheless, significant gaps in the coverage of the published literature. In particular, most general historical accounts stop before nationalisation. Furthermore, few if any books provide a systematic account of what the industry comprised, especially in the latter half of the 20th century. Given that in many parts of Scotland it is now impossible to trace the location of most coal mines from physical remains on the ground, and little associated documentary evidence usually survives, there is a very real possibility that the true extent and importance of the industry may fade into obscurity.

The purpose of this book is therefore to provide reliable information on the last extraordinary decades of the Scottish coal industry. At its heart is a county-based gazetteer (Chapter 5) providing information on the nationalised collieries, and also containing lists of the licensed private coal mines that were deemed to be too small to be taken into the state sector. This has been compiled using a variety of sources including Ordnance Survey maps, colliery year books, a colliery questionnaire completed by the managers of all nationalised mines in the late 1940s, and the NCB shaft register for Scotland. With the help of the gazetteer, it is now possible to locate all major coal mines that have operated in Scotland since 1947, and in most cases there are also details of the number of people working at the mines, dates when they were first opened and when they finally closed, and where known, the name of the founding company. The opportunity has also been taken to produce comparative information on Scottish collieries using labour-force statistics published

Figure 1.6: A float promoting coal-fired central heating at a miners' gala in the 1960s. The NCB attempted to staunch the loss of demand for household coal markets to oil by promoting solid-fuel central heating systems. Scottish Mining Museum. SMM:1998.615, SC706318

Figure 1.8: The compilation of the gazetteer in Chapter 5 was greatly assisted by Professor John R Hume, whose extensive research and photographic work in the 1960s and 1970s included many collieries, such as Cardowan at Stepps on the east side of Glasgow. SC587893

Figure 1.7: In 1983, Cardowan Colliery was the last pit to close in Lanarkshire, the county which in 1900 dominated the Scottish coal industry. This view shows one of the two headgear 20 years earlier in May 1966. John R Hume. SC587881

Figure 1.9: The immense infrastructure of the coal industry included extensive railway sidings, and huge fleets of wagons as well as large numbers of steam locomotives. This view shows a Gibb & Hogg of Airdrie tank engine operating outside Cardowan Colliery in 1966. John R Hume. SC587894

Figure 1.10: This photograph of Highhouse Colliery in Ayrshire is entitled 'Bait Time', and is part of an extraordinary collection of images saved from destruction by Terry Harrison of Mauchline, and made available for inclusion in the gazetteer. Harrison Collection. SC446730

in colliery year books. This is also included in tabular form in the appendices, along with a list of the most important private coal companies prior to nationalisation, as well as a closure timeline charting the decline of the industry year-by-year. There is, in addition, an alphabetical list of approximately 550 deep coal mines known to have operated in Scotland since 1947, and a glossary of terms relating to the coal industry.

Inevitably, some of the data contained in the book is incomplete, and any new information relating to the Scottish coal industry is therefore most welcome. All the information and records gathered so far have been deposited in the National Monuments Record of Scotland (NMRS), and can be updated and amended as necessary. Online access to the NMRS catalogues can be gained through the RCAHMS website at www.rcahms.gov.uk.

One of the principal aims of this book has been to revive interest and pride in the Scottish coal industry, and to encourage a flow of new information from which it will be possible to create a fuller record of what is, without doubt, a hugely important part of Scotland's recent history. Above all, it is hoped that it will help to ensure that the huge contribution of the coal industry to modern Scotland in the last 60 years will not be forgotten.

Figure 1.11: This view of one of the two winding engines at Polquhairn Colliery in Ayrshire was taken by the late John McKinnon of Mauchline in the mid-1950s, and is part of a wonderful collection of images which his family have made available for inclusion in this publication. J McKinnon Mining Collection. SC706175

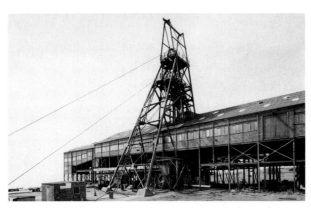

Figure 1.12: In 1991, Frances Colliery in Fife was the last colliery to be recorded by RCAHMS. It had been mothballed but never re-opened, and was later demolished with the exception of the headgear, which survives as a monument to the Fife coal industry. SC369015

Figure 1.13: The demolition of Rothes Colliery in March 1993 in advance of coal privatisation symbolised the end of the NCB era, and the moment when the clearance of coal-mining remains accelerated. Monktonhall Colliery was later demolished (in 1997–8) having failed to survive in the private sector. *The Herald*, Stan Hunter. SC894671

Figure 1.14 left: Longannet Mine and Power Station, Fife. The Longannet Mine was the last colliery to operate in Scotland, its closure in 2002 signalling the cessation of deep coal mining in Scotland. The power station now burns imported low-sulphur coal. Scottish Mining Museum. SC382467

Figure 1.15: Close to the Longannet Power Station complex in Fife is Preston Island, on which some of Scotland's most significant early 19th-century coalmining remains can be found, alongside well-preserved salt pans. Aerial view taken in 1988. SC588821

# Chapter 2: Before Nationalisation

The coal industry played a central part in Scotland's history during the last three centuries, and was the driving force that fuelled the many industries for which Scotland had become famous by the end of the 19th century. As was the case with most of Britain's traditional heavy industries, coal reached its peak in 1913, immediately before World War I. At this time, all industries had one major factor in common – they were totally dependent either directly or indirectly upon coal as their principal source of energy and illumination. Without coal, Scotland's astonishing industrialisation would have been impossible. Stanley Jevons, whose book on the coal trade was published in 1915, observed that coal should be regarded as standing not beside but entirely above all other commodities, noting that, 'The importance of coal as a raw material lies, however, in the fact that it is of service not in one industry but in all; and at the present time, it is the original source of practically all our artificial heat and light and power'.[1]

By 1913, Scottish mines accounted for approximately 15% of annual British coal output.[2] The coals were generally of a high quality, and suitable for a wide range of uses within Scottish industry.[3] Almost all this production was concentrated in the Carboniferous central belt of the country between the Firth of Clyde in the west and the Firth of Forth in the east (see Figure 2.1). The largest concentrations of mines were to be found to the west in Ayrshire, Renfrewshire and northern Dumfriesshire, in Lanarkshire (as far north as Glasgow), in the counties directly to the east of Glasgow including East Dunbartonshire and Stirlingshire,

and on either side of the Firth of Forth in the Lothians, Clackmannanshire and Fife.

The most heavily exploited of these areas was Lanarkshire,[4] but perhaps the oldest and most intensive areas of coal extraction were based initially on monastic industries, especially salt evaporation from sea water.[5] Major concentrations of salt pans were to be found in the Firth of Forth, such as those at Prestonpans, Bo'ness, Kennet Pans, Culross and St Monans,[6] but salt extraction was also carried out elsewhere on the coasts of Scotland, including the Firth of Clyde, and as far north as Brora in the Highlands. Indeed, salt was the main impetus behind one of the most innovative early coal-mining ventures, at Culross Colliery in Fife. The mine had been leased by Sir George Bruce in 1575, and in 1604 he completed the construction of an artificial island on the foreshore, through which a shaft emerged from the undersea workings below, permitting the shipment of coal directly by boat when the tides were high.[7] Although there are few visible remains of the Moat Pit, some of the finest surviving remains of salt pans and associated colliery buildings can be found nearby at Preston Island (see Figure 2.2). These date from the early 19th century, and were established by Sir Robert Preston of Valleyfield. By strange coincidence, this site has been profoundly affected by the activities of Scotland's last and most modern coal mine at neighbouring Longannet. The island has been surrounded by fly-ash produced by coal burned at Longannet power station, and is now effectively part of the Fife mainland.

The presence of a steam-engine house on Preston Island indicates the extent to which, by the beginning

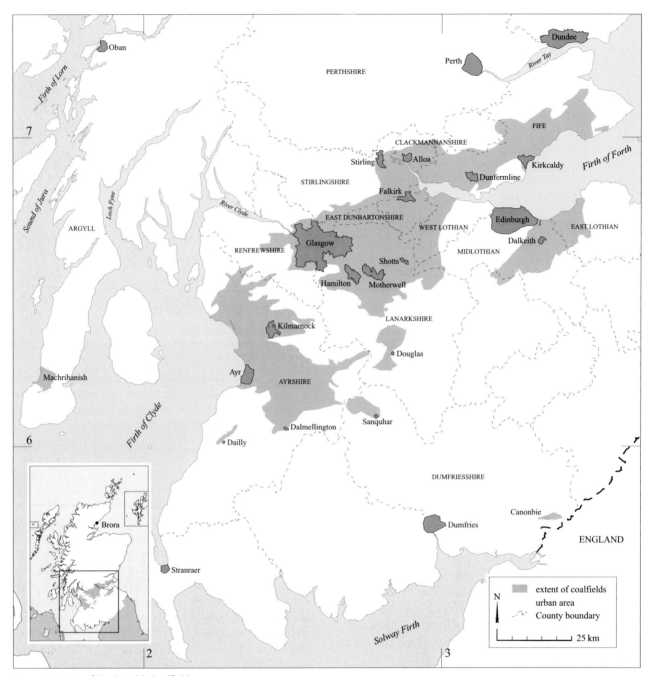

Figure 2.1: Map of the Scottish Coalfields

of the 19th century, steam power had revolutionised coal mining. The depth and extent of mining had hitherto been severely restricted by drainage problems, but with the introduction of Newcomen atmospheric engines from 1713, the scale of potential mining operations was greatly enhanced. By 1720, several engines had already been installed by the Cunninghame's of Saltcoats in the Irvine area,[8] and others were being built elsewhere in the Scottish coalfields. In the 1770s,

Boulton & Watt's improved steam engine greatly enhanced the reliability and potency of steam power, and when applied in Scotland from 1800 onwards, provided not only the means to build even more productive coal mines, but also generated a large and growing market for the coal itself. In the ensuing years, the installation of Trevithick's Cornish engines in the Scottish coalfields was especially significant.

Other factors also greatly assisted the expansion of production in Scotland's coalfields. The smelting of iron with coke in 1709 by Abraham Darby at Coalbrookdale in England allowed the iron industry to break free from its dependence on charcoal. In Scotland, this led to the establishment of the Carron Iron Works in 1760 and the beginning of the Scottish industrial revolution.[9] Carron and other iron companies in Scotland became prodigious consumers of coal, and in many cases, took control of coal production by buying mines or establishing collieries of their own. The expansion of coal mining was also boosted by rapid industrialisation and urbanisation. In addition, the construction of canals and railways were especially important in the late 18th century, and all through the 19th century passenger and mineral railways were built throughout central Scotland, especially in the coalfields.[10]

Figure 2.2: Preston Island, near Culross, Fife, 1988. Aerial view of the early 19th century remains of the salt works and colliery. The island, which was situated in the Firth of Forth, has been surrounded by deposits of fly-ash from neighbouring Longannet Power Station, and is now effectively part of the mainland. SC710852

Figure 2.3: Hamilton Palace Colliery, Bothwellhaugh. Sunk by the Bent Colliery Company in 1884 at a time when the Lanarkshire coal industry was flourishing. This view shows the visit of the Mining Institute of Scotland to the colliery on 10 August 1899. The pit survived nationalisation, eventually closing in 1959. South Lanarkshire Libraries, SC706522

There followed, from 1850 to 1913, the golden era of coal mining in which an extraordinarily complex coal-dependent industrial base developed in Scotland.[11] Driven by the success of Neilson's 'Hot Blast' process, the iron industry flourished in several coalfield areas, notably the Monklands to the east of Glasgow, but also elsewhere in Lanarkshire, in Ayrshire, Stirlingshire, Fife and Clackmannanshire, and in East and West Lothian. Closely associated with coal mining were heavy ceramics industries, which ranged from brick and tile manufacture through to high-quality fireclay ceramics that were vital to the iron, steel and chemical industries.[12] Other big coal users included the town gas industry, the primary use of which was initially illumination, and increasingly from the 1890s, electricity generation. In more rural areas, important consumers included distilleries and lime kilns, the latter producing lime for mortar as well as fertiliser for fields, and later for flux in the steel industry.

Coal therefore fuelled the evolution of Scotland's manufacturing industries, but these industries also transformed coal production by supplying new, more powerful winding engines as well as a wide range of other machines for generating electricity, and hydraulic and pneumatic power; for cutting, conveying, grading and cleaning coal; for providing drainage and ventilation; and for other equipment such as roof supports, locomotives, and rails. At the same time, coal mining was transformed from what had been for some a part-time rural craft into an intensive professional industry, with its own community of often highly trained and experienced engineers, surveyors and colliery managers (see Figure 2.3).

At the same time, a large industrial sector that was dependent upon or linked to coal mining had evolved by 1913. This included engineering companies such as Grant Ritchie and Andrew Barclay, both of Kilmarnock, and Murray & Paterson of Coatbridge, builders of winding engines, and Sir William Arrol & Company of Dalmarnock, one of many structural steel contractors who built head frames and steel-framed pit-head buildings for collieries. Other companies, such as Colvilles, produced a wide range of steel products for use both above and below ground, whilst Hurst Nelson of Motherwell was one of a number of companies who built railway wagons. Bruntons of Musselburgh specialised in the production of wire rope, and the Blantyre Engineering Company's products included coal preparation plants. Outside engineering, one of the most significant mining-related Scottish manufacturers was the Nobel's Explosive Company (later a founding part of ICI), whose huge factory at Ardeer

in Ayrshire produced flame-retardant high explosives especially for use in coal mines.[13]

By 1900, the sophistication of Scotland's industrial economy was reflected in the comparatively advanced nature of many of its collieries, particularly in the Lanarkshire coalfield. Several of Scotland's coal companies had been quick to apply new technology as it became available. This had included the use of wire ropes from the 1840s, compressed air from 1849 at Govan Colliery in Glasgow,[14] and electricity from the 1880s.[15] Indeed, in 1881, Earnock Colliery near Hamilton was the first in Britain to be lit by electric light,[16] and by 1910, portable electric lamps were being introduced in some mines. Meanwhile, the potential for the application of electricity to underground mechanical tasks had become apparent. Scottish mining engineering companies soon grew to prominence. One of the pioneers, Muir and Mavor (founded in 1883), who became Mavor and Coulson in 1890, supplied its first electric coal cutter with a totally enclosed motor, to a pit in Staffordshire. The company eventually merged with its principal rival, Anderson Boyes of Motherwell.[17] Together they produced a wide range of flame-proof coal-mining machinery and equipment, and were particularly famous for their coal cutters (see Figure 2.4). Other Scottish companies producing mining equipment included The Belmos Co Ltd of Bellshill, and The Wallacetown Engineering Co Ltd of Ayrshire.

When it reached its peak in 1913, the Scottish coal industry produced over 42 million tons per annum, and employed 148,000 miners, exporting to Europe and Ireland in particular. At that time, a large colliery employed up to 1,000 miners, ten times the norm for a large pit in the 18th century. Despite the application of mechanisation, coal mining remained a very labour-intensive activity, and the proportion of coal extracted by mechanical means was probably little more than 10%. At this stage, the Lanarkshire coalfield was still dominant, but was already in comparative decline, having produced 56% of Scottish output in 1900, slipping back to 41% by 1913. Ayrshire too appeared to be declining, the principal area of growth being in Fife.[18] Coal remained the dominant fuel for all industry, the only exceptions being road and marine transport, where the petrol-fuelled internal-combustion engine was taking over.[19]

The coal industry emerged from World War I in a dilapidated state, having suffered severe labour shortages and a lack of investment. By 1920, Scottish output had fallen by 25%, and as the economic depression deepened, sequences of wage cuts, strikes, lockouts, and hunger marches ensued.[20] It was therefore estimated that in the 20 years before World War II,

Figure 2.4: The Flemington Electrical Works, Craigneuk Street, Motherwell, c.1910. This view of the Anderson Boyes erecting shop shows the assembly of electrically-powered coal-cutting machines in the early 20th century. The company continued to produce mining machinery until 1998. SC886506

Figure 2.5: Neilsland Colliery, near Hamilton. A fine example of a Lanarkshire Colliery which did not survive the inter-war years, closing in 1932. SC384795

productivity (as measured by output per man shift) in British mines rose on average by only 14% (despite relatively benign mining conditions), as compared with 118% in the Netherlands, 84% in the Ruhr (Germany), and 54% in Poland. It was also calculated that 33% of all time lost to strikes in Britain during the same period occurred within the coal industry.[21] As the slump continued into the 1930s, the decline of the Lanarkshire coalfield accelerated, and showpiece collieries such as Neilsland (see Figure 2.5)[22] were closed and subsequently demolished.

Although the British productivity figures were disappointing compared with other European countries, several of the Scottish coal companies responded to low coal prices by attempting to reduce costs through mechanisation. They seem to have been much more successful than their Welsh and English counterparts because, by 1924, 48% of Scottish coal output was mechanically-cut, as compared with only 19% for the UK as a whole. The same pattern prevailed in 1937, by which time 79% of Scottish coal was machine-cut, the UK total having risen to only 57%. However, despite the mechanisation, Scotland's share of British coal output had shrunk back to 14%, and by the end of World War II, had retreated further to only 12%.[23]

Despite improvements in working conditions, mining remained an extremely hazardous and arduous occupation. Scotland's worst mining disaster had occurred at Blantyre Colliery near Hamilton in 1877 when an explosion killed at least 207 miners. Lax procedures at the pithead resulted in uncertainty over how many miners were underground at the time of the explosion, so the final death toll was uncertain. The industry continued to be plagued by fatal accidents, and 22 years later a fire at Mauricewood Colliery near Penicuik in Midlothian killed 63 men. Although there was enhanced awareness of the dangers of gas and coal dust, many collieries remained 'naked light' pits long into the 20th century, the law eventually being changed after the Kames disaster in 1957. This particular Ayrshire pit had previously been considered safe enough for the continued use of naked light, and the resulting gas explosion killed 17 men. Thereafter, approved portable electric lamps became mandatory.

Meanwhile, improvements in ventilation achieved by the installation of mechanical fans greatly reduced the risk from accumulations of explosive gas. After Mauricewood, the most serious disaster to occur was at Redding Colliery near Falkirk in 1923, but it was a flood rather than an explosion that killed 40 miners. Subsequently, the worst Scottish disaster was that at Valleyfield Colliery in Fife, shortly after the outbreak of World War II in 1939, when 33 miners were killed by an explosion.

In general, welfare facilities at the pithead were very poor. Even in 1948, most pits had little more than a first-aid room or a 'morphia administration scheme', and some merely admitted to relying on the 'nurse in the village' in the event of an emergency.[24] Some collieries had benefited from the activities of the Miners' Welfare Fund (MWF, see Figure 2.8), which had been founded in 1920, and which was steadily helping to fund the construction of pithead baths in collieries throughout Britain. In Scotland, only two coal companies had built baths prior to the establishment of the MWF.[25] After 1920, however, several fine art-deco style pithead baths designed by the architect J A Webster were built, and included those at Arniston Gore Colliery in Midlothian (Figures 2.6 and 2.7), and Castlehill No. 6 near Shotts in Lanarkshire (Figure 2.9). Canteens were also comparatively rare until World War II, during which many were constructed to ensure that extra food rations for the 'Bevin Boys' were not so easily diverted to less strategic stomachs in the family.[26]

At the end of World War II in 1945, it was commented that, 'No other major British industry carried so many unsolved problems into the war, and none brought more out'.[27] These problems were identified as: depletion of resources; low productivity; poor layout and mechanisation; lack of integration; declining coal quality; price and sales instability; poor recruitment; poor working conditions; poor living conditions; and poor labour relations.[28] The fragmentation was exacerbated by the ownership structure. In 1946, it was estimated that there were 1,470 collieries in Britain, of which 481 were small mines (employing less than 30 people underground). There were also over 800 registered coal owners, ranging from individuals to large corporations, but it was also calculated that 14 owners controlled 25% of the industry's workforce.[29]

Prior to nationalisation in Scotland, the industry was dominated by a small group of predominantly iron and steel producers, led by Bairds & Dalmellington, with 23 pits in Ayrshire. The Fife Coal Company had 17 pits, and the Alloa Coal Company, 11 (mostly in Clackmannanshire). The Coltness Iron Company also had 11 pits, spread through Lanarkshire, West Lothian and Fife, whilst the Shotts Iron Company had ten, mostly in Lanarkshire and West Lothian, but with a significant cluster to the south of Edinburgh in Midlothian. Bairds and Scottish Steel owned nine pits, mostly in the anthracite and coking coal areas of Dunbartonshire, with additional pits in Lanarkshire, Stirlingshire and West Lothian. United Collieries operated seven

pits in West Lothian, Midlothian and Lanarkshire, and New Cumnock Collieries Ltd had six in south-east Ayrshire. Amongst the five collieries in Fife operated by the Wemyss Coal Company were two of the largest in Scotland, Michael and Wellesley. Other important companies included A G Moore (five pits in Ayrshire, Lanarkshire and Midlothian), the Lothian Coal Company (four pits in Midlothian, including Lady Victoria Colliery), the Lochgelly Iron and Coal Company (five pits in Fife), James Nimmo and Company (four pits in Lanarkshire and Stirlingshire), William Dixon and Company (four pits in Lanarkshire and West Lothian), the Summerlee Iron Company (three collieries in Lanarkshire and East Lothian, including Prestongrange), and Archibald Russell (the two Polmaise collieries in Stirlingshire).[30]

The diverse fragmented structure of the coal industry meant that, in general, it was unable to take control and advance. The conflicting and competing interests of companies often hampered collaboration and development, preventing the amalgamation of neighbouring mines, and the co-ordination of capital investment. In addition, working conditions were variable but usually poor, and company-owned housing was often equally poor, although many miners had been re-housed in local authority housing between the wars. Against this background, many, including the mining unions, were in favour of destroying the existing ownership structure of the industry. It was with these conditions in mind that the Reid Report of 1945 recommended a united command of the coal industry.[31]

Figure 2.6 (inset) and Figure 2.7: The pithead baths at Arniston Gore Colliery, in Midlothian completed in 1936. Like many of the baths built in Scotland, they were built with the assistance of the Miners' Welfare Fund in the 1930s, and were designed by J A Webster. SC675019. Figure 2.7: The National Archives, Kew London, COAL80/37/1, SC614063

Figure 2.8: The Miners' Welfare Fund (MWF) was founded in 1920. By levying a small fee on every ton of coal produced, it was able to fund a variety of projects at collieries, and in the mining communities, perhaps the most notable of which were pithead baths. It operated until nationalisation in 1947, and was subsequently replaced by the Coal Industry Social and Welfare Organisation (CISWO). SC674985

Figure 2.9: The pithead baths at Castlehill No. 6 Colliery in Lanarkshire, owned by the Shotts Iron Co. The baths were built with the support of the Miners' Welfare Fund, and were designed by J A Webster and completed in 1935. The National Archives, Kew London, COAL80/224/3, SC614093

## Endnotes

1 Jevons, S (1915), pp. 8–9.

2 Statistics compiled by staff at the Scottish Mining Museum.

3 In some areas, such as the Monklands to the east of Glasgow, the coals contained blackband ironstone which was crucial to the development of the iron industry. In addition, after the invention of the 'Hot Blast' process by J B Neilson in 1829, some splint coals were particularly well suited to use in blast furnaces, preventing the need for much more expensive coke.

4 Scottish Home Department, *Scottish Coalfields: Report of the Scottish Coalfields Committee*, cmd. 6575, HMSO (Edinburgh 1944), pp. 54–5, which suggests that Lanarkshire dominated Scottish coal production by the end of the 19th century.

5 For a general account of the Scottish salt industry, see Whatley, C A (1987).

6 For an account of the salt pans at St Monans in Fife, see Yeoman, P *et al* (1999)

7 The Moat pit was one of the first documented examples of the successful exploitation of undersea coals. It is also of interest because of the use of horse-powered machinery to drain the mine using technology similar to that depicted in Agricola's *De Re Metallica*, published only a few decades earlier in 1556. The venture created such interest that it attracted a royal visit from King James VI in 1617, eight years before the pit was destroyed by a fierce storm. For more details, see Bowman, A I (1970).

8 See Whatley, C A (1977) for an account of the introduction of Newcomen engines in Ayrshire.

9 For information on Carron, see the company's own history published c.1959 to commemorate 200 years of production, and Campbell, R H (1961).

10 The RCAHMS publication, *Forts, Farms and Furnaces, Archaeology in the Central Scottish Forest* (1998) provides useful information on the archaeology of industrialisation in the central Scottish coalfields, and especially the expansion of mining and mineral railways in the second half of the 19th century.

11 For general information on Scotland's industrial revolution, see John R Hume's two volumes of *The Industrial Archaeology of Scotland (1976–7)*, and for engineering in particular, see the book Moss, M S and Hume, J R (1977), *Workshop of the British Empire: Engineering and Shipbuilding in the West of Scotland*. For broad coverage of Scotland's industrial monuments, see also *Monuments of Industry* by Hay, G D and Stell, G P (1986).

12 See Douglas, G J and Oglethorpe, M K (1993) for further information on Scotland's heavy ceramics industries.

13 For a brief account of the history of Nobel's Ardeer works, see Dolan, J and Oglethorpe, M K (1996).

14 See the NCB's *A Short History of the Scottish Coal-mining Industry* (1958), p. 81. William Baird & Company introduced the prototype 'Gartsherrie' coal cutter in 1864, which was also powered by compressed air.

15 One of the most useful sources of information on the evolution of mining technology is A Hill's *Coal Mining: a technical chronology 1700–1950*, published by The Northern Mine Research Society in 1991.

16 Loynes, E (1984), p. 7.

17 For a history of Anderson Boyes, see Carvel, J L (1949).

18 Information on coal production in Scotland's coalfields compiled by the Scottish Mining Museum.

19 Hutton, G (2000), p. 19.

20 The unrest was particularly severe in the Fife coalfield, and resulted in the election in 1935 of Willie Gallacher as communist member of the Westminster parliament for West Fife, see MacDougall, I (ed.) (1981).

21 See Ashworth, W (1986).

22 Neilsland Colliery was established in the mid-1890s at the height of the Lanarkshire coal industry, and illustrates the scale and sophistication of the industry in the late 19th and early 20th centuries. The colliery closed in 1932, and was subsequently completely demolished.

23 Coal production statistics compiled by the Scottish Mining Museum, and the National Coal Board's Newbattle Archive.

24 Information from NCB Colliery questionnaire (1948).

25 The Wemyss Coal Co constructed baths at its Denbeath Colliery (later renamed Wellesley Colliery) 1915, and Wilsons and Clyde Coal Co Ltd provided baths at Douglas Castle Colliery in 1919.

26 The 'Bevin Boys' were conscripted miners who, from 1943, were drafted each week on the basis of a number randomly selected by the then Minister for Labour and National Service, Ernest Bevin. Those selected were then sent to work in the mines rather than serve in the armed forces.

27 Court, W H B (1951), *Coal (History of the Second World War: UK Civil Series)*, HMSO: London, p. 391, quoted in Ashworth, W (1986), p. 5.

28 See Haynes, W W (1953).

29 Ashworth, W (1986), p. 6.

30 Information from the NCB Colliery Questionnaire of 1948, and from the NCB shaft register.

31 Ministry of Fuel and Power (1945).

# Chapter 3: Nationalisation and After

On 12th July 1946, 53 years after nationalisation of the coal industry was first proposed to parliament by James Keir Hardie, the newly-elected Labour Government acted on the findings of the Reid Report, and successfully passed the 'Coal Industry Nationalisation Act'. This provided for the creation on 1st January 1947 of the National Coal Board (NCB), in which all the assets of the coal industry would be vested.

Initially, there was an atmosphere of great optimism (see Figures 3.1 and 3.2), particularly in the Government and amongst the Trade Unions, who believed that huge improvements in working and living conditions were now possible. In reality, the creation of the NCB was a huge task, and there was no guidance based on previous experience on how to operate such a massive State-owned organisation. Of the many complex issues, one of the most difficult was the compensation of former owners, a process which was settled through the creation of a Central Valuation Board.

In the UK as a whole, the total assets transferred on vesting day included 1,500 working collieries with their stocks of products and stores, plant, equipment, and land, 30 fuel and briquetting plants, 55 coke ovens and associated chemical by-product plants, 85 brickworks, 1,803 farms, 140,000 houses, 27,000 farm houses and cottages, numerous offices, 275 shops and business premises, several swimming baths, a cinema and a slaughterhouse. Other assets included private mineral railways, wharves, depots, retail milk rounds, a holiday camp, a bicycle track, and 177,000 mainline railway wagons and locomotives.[1] Despite the transfer of these assets, the NCB inherited no working capital. In practice, as a State-owned industry, all funding for future development had to emanate from the Treasury. The extent of this burden became clear when the NCB published its *Plan for Coal* in 1950.[2] Based on the earlier findings of the Reid Report in 1945, it planned to spend £635 million over the following 15 years. A surprisingly large proportion of this investment occurred in the Scottish coalfields.

As well as the actual production of coal, the NCB's remit covered a huge range of activities, one of the most important of which was welfare. The NCB's Medical Service, for example, was charged with a wide range of responsibilities both within and beyond the workplace. In particular, it was involved in advising management on the medical aspects of safety, and the introduction of health education. Also important was the fact that the NCB was obliged by statute to provide first-aid rooms at the surface of all coal mines. In addition, co-ordinated safety programmes were introduced to reduce high rates of accidental death and injury in the mines. Meanwhile, basic surface facilities such as baths and canteens, also widely unavailable before nationalisation, were introduced by the NCB, who took over many of the responsibilities of the Miners' Welfare Fund.[3] Outside the mines, perhaps the largest of the NCB's social responsibilities was for housing, but this was gradually transferred to local government, along with ownership of the housing stock itself.

As was the case in 1913, coal remained the primary fuel for nearly all energy needs, only road transport and shipping relying more heavily on petroleum products. Gas and electricity, which were later to become competitors in the domestic energy market, were entirely dependent upon coal. The NCB was therefore faced with an energy crisis, the cause of which was an inability of its mines to produce enough coal. The challenge was therefore one of increasing coal production to meet domestic demand, and of satisfying the high export targets required to earn foreign exchange to service overseas debt incurred through the war years.

In Scotland, this involved a combination of strategies. Nationalisation had taken 206 collieries into State ownership[4] (see Figure 3.6), leaving at least another 105 small mines (some of which also produced clays and fireclays) in the private sector.[5] For the NCB, there was no option but to close many of its old uneconomic

Figure 3.1: The scene on 'vesting day' at Newcraighall Colliery on the east side of Edinburgh, where miners celebrated the nationalisation of the British coal industry on 1st January 1947. East Lothian Council Museums Service, David Spence Collection SC706648

Figure 3.3: Bothwell Castle 1 and 2 Colliery, Lanarkshire. A William Baird colliery dating from 1875, this was one of many Lanarkshire collieries earmarked for early closure after nationalisation in 1947, production ceasing in 1950. South Lanarkshire Libraries, SC706677

Figure 3.4: Forthbank Colliery, Alloa, Clackmannanshire. Developed by the NCB in 1947, this was an example of a new breed of short-term drift mine designed to exploit relatively shallow deposits of coal. The lifetime of such developments was usually anticipated to be little more than ten years, buying time before production from the big new sinkings commenced. Forthbank closed in 1958. Norval of Dunfermline, SMM:1998.348, SC445863

Figure 3.2: The National Coal Board flag flies from the headframe of Newcraighall Colliery in celebration of the nationalisation of the industry in 1947. East Lothian Council Museums Service, David Spence Collection, SC706616

Figure 3.5: Sorn Colliery, Ayrshire. One of the most successful of the National Coal Board's short-term post-war drift mines, its 31 years of operation being far longer than anticipated, the mine eventually closing in 1983. SMM:1996.2565, SC381806

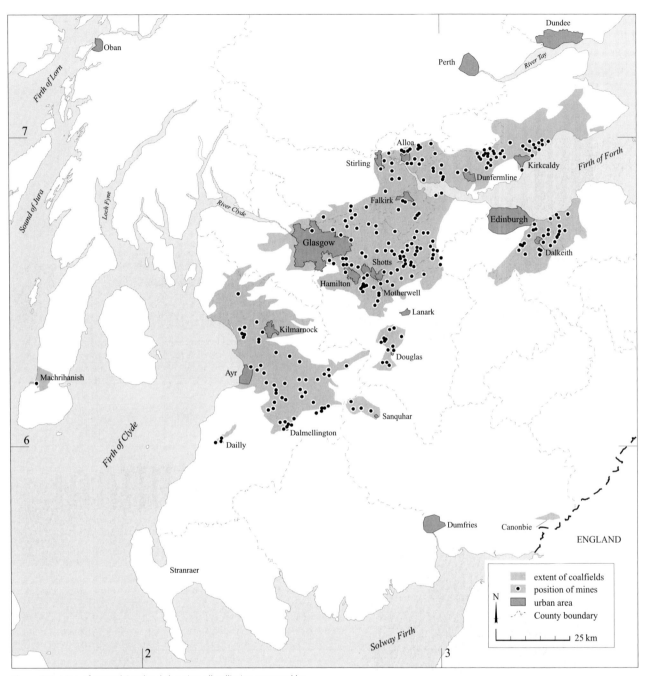

Figure 3.6: Map of central Scotland showing all collieries operated by the National Coal Board between 1947 and 1995.

mines, particularly those that had struggled on in the Lanarkshire coalfields (see Figure 3.3). The lost capacity was initially partly replaced by 40 short-term surface drift-mine projects (see Figures 3.4 and 3.5), so on balance, the number of mines operating during the late 1940s and early 1950s stabilised at around 175 before dramatically dropping in the late 1950s after a spate of closures (Figure 3.7).[6] In contrast, the mining workforce in Scotland grew during the first ten years

(Figure 3.8), reflecting the push to achieve maximum output. However, although coal production for Britain as a whole peaked in 1956, output from Scotland was disappointing (Figure 3.9), declining steadily after 1950 despite the new investment,[7] shrinking further to 10.44% of UK production by 1955.

In the same year, the NCB published *Scotland's Coal Plan*, which responded to the continuing decline in productivity by categorising its mines. Those that

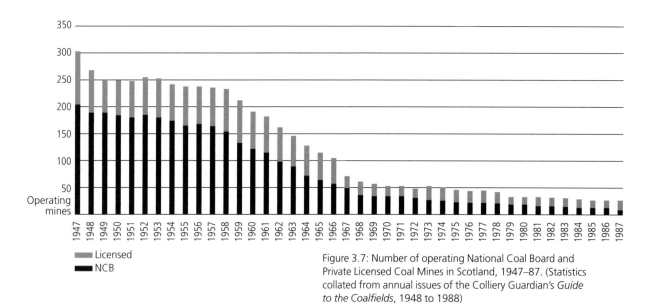

Licensed
NCB

Figure 3.7: Number of operating National Coal Board and Private Licensed Coal Mines in Scotland, 1947–87. (Statistics collated from annual issues of the Colliery Guardian's *Guide to the Coalfields*, 1948 to 1988)

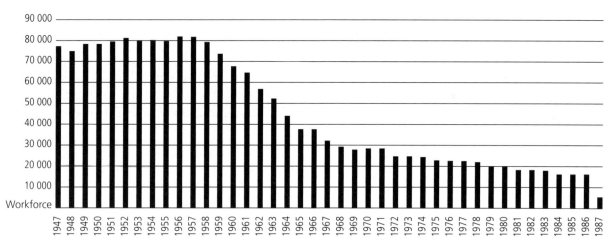

Figure 3.8: Total workforce in National Coal Board mines in Scotland, 1947–87. (Statistics collated from annual issues of the Colliery Guardian's *Guide to the Coalfields*, 1948 to 1988)

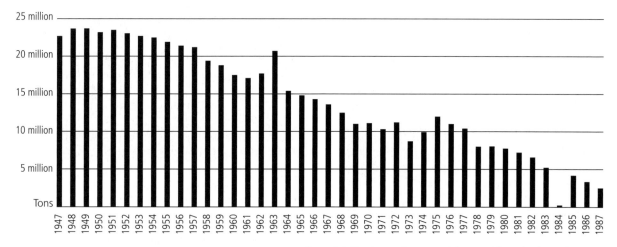

Figure 3.9: Total annual output of the Scottish coal industry from 1947 to 1987. Statistics compiled by staff at the NCB's Newbattle Archive.

were beyond help were to close, whilst others considered to be worth retaining but not worthy of significant further investment were kept open in the short term. A further group of more promising mines, such as Bowhill (Figure 3.10) and Valleyfield in Fife, Kames (Figures 3.11 and 3.12) and Barony in Ayrshire (Figure 3.13), Polmaise 3 and 4 (Figure 3.14) and Manor Powis (Figure 3.15) in Stirlingshire, and Polkemmet and Kinneil (Figure 3.16) in West Lothian were selected for reconstruction, which in some cases involved merging

with neighbouring pits as well as extensive surface and underground development. However, most significant was a commitment to 15 'new sinkings'.[8]

Much of the responsibility for the design of the surface structures of the new collieries, and for some of the more ambitious reconstructions, fell to the newly appointed 'Production Architect' for the NCB's Scottish Division, Egon Riss, who had been recruited from the Miners' Welfare Fund in 1947. Born in Austria in 1901, Riss was educated at the Weiner Technische

Figure 3.10: Bowhill Colliery, Cardenden, Fife. In the early years of nationalisation, many pits considered to have a future were included in a reconstruction programme. In the case of Bowhill, this involved the sinking of a third shaft, and the cutting of the first sod in 1952 which is seen here. Reconstruction continued at Bowhill and other pits throughout the 1950s. SMM:1996.01513, SC378657

Figure 3.12: Kames Colliery, Muirkirk, Ayrshire, 1962. Reconstruction in the 1950s replaced the conventional headgear and winding arrangement (see Figure 3.10) with tower-mounted friction winders. The colliery subsequently survived a fatal explosion in 1957 in which 17 people died, continuing to produce coal until closure in 1968. J McKinnon Mining Collection, SC381680

Figure 3.11: Kames Colliery, Muirkirk, Ayrshire. This view shows the colliery prior to major reconstruction in the mid-1950s. J McKinnon Mining Collection, SC706182

Figure 3.13: Barony Colliery, near Auchinleck, Ayrshire. This view was taken after the completion of the reconstruction programme in the 1950s, but before the collapse of Shafts 1 and 2 in 1962. It shows the new 'A' frame over No. 3 shaft, and the two older shafts beyond. The colliery was originally established in 1910, and became one of Bairds and Dalmellington's most important pits. In 1989, it was also the last deep mine to close in Ayrshire, and only the 'A'-frame now survives. Harrison Collection, SC706234

Figure 3.14: Polmaise 3 and 4 Colliery, near Stirling. One of Scotland's most famous collieries, Polmaise was sunk in 1904 by Archibald Russell Limited, and the quality of its coal, which included anthracite, was such that it merited a reconstruction programme in the 1950s. A focal point in the 1984 miners' strike, Polmaise closed not long afterwards in 1987. SMM:1996.3414, SC381878

Figure 3.15: Manor Powis Colliery, near Stirling. Another important anthracite producer, this view shows the surface buildings during reconstruction work in the 1950s, which included the sinking of a third shaft (temporary headgear visible far right). SMM:1996.0500, SC381866

Figure 3.16: Kinneil Colliery, Bo'ness, West Lothian. An important source of gas and coking coals, this was the scene of one of the larger reconstruction projects overseen by the NCB in the 1950s. SC446477

Hochschule in Vienna, and developed interests in architectural engineering and social projects, including urban housing. His early work included a trade union building in 1923, and projects associated with the Silesian coalfields, but he was best known for the Tuberculosis Sanatorium in the Lainz hospital complex in Vienna, which was completed in 1931.[9] After escaping to the UK in 1938 where he joined many other famous refugees such as Walter Gropius, he was briefly interned (despite being Jewish) before the Royal Institute of British Architects (RIBA) assisted his recruitment to the Auxiliary Military Pioneer Corps. In 1941, he was transferred to the Royal Marines as a sapper, and whilst at Lockerbie, gained his commission at the rank of second lieutenant.

After commencing work for the NCB, his first new project was Rothes Colliery in Fife, to the north of Kirkcaldy, which had been planned initially by the Fife Coal Company (Figure 3.17). This afforded him an opportunity to express his ideas from the 1920s and 1930s, which were strongly influenced by the Modern Movement. His design of the surface buildings and ancillary components of the pithead influenced all subsequent Scottish collieries, and probably projects elsewhere in the UK as well as new sinkings in Germany and the Netherlands. Perhaps the most important factor was the NCB Scottish Division's desire to create a new dynamic and modern image (see Figure 3.18).

This was certainly achieved magnificently at Rothes, which was opened by Her Majesty Queen

Figure 3.17: Rothes Colliery, Glenrothes, Fife. The first of the big new sinkings of the NCB era, this view shows the surface buildings approaching completion in the mid-1950s, production eventually commencing in 1958. SMM:1996 0525, SC378681

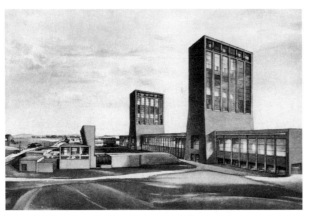

Figure 3.18: Rothes Colliery, Glenrothes, Fife. Rothes was not only the first of Scotland's new superpits, but was also the first entirely new colliery to be completed to a design of the NCB Scottish Division's architect, Egon Riss, whose work until 1958 had been confined to reconstructions such as Kinneil (see Figure 3.16 ). SC706493

Figure 3.19: Rothes Colliery, Glenrothes, Fife. Amid scenes of great optimism, Her Majesty Queen Elizabeth II formally opened the colliery on 30th June 1958. However, serious problems were already being encountered as the sinking progressed, and after a four-year struggle the colliery closed in 1962 amid great controversy. SMM:1998.603, SC706681

Elizabeth II on 30th June 1958 amid great optimism (Figure 3.19). Such was the initial confidence in the project that an entire new town, Glenrothes, had been planned, and many miners and their families were moved from declining areas of the Scottish coalfields in the expectation of a long working life for the colliery. In reality, the project had already begun to encounter drainage difficulties, and it was said that as the sinking of the shafts progressed and pumping problems worsened, so the water levels in neighbouring pits such as Kinglassie receded. In the ensuing four years, desperate attempts were made to solve the drainage problems, but ultimately it proved to be too costly to maintain, and Rothes closed in 1962 amid great political embarrassment.

Meanwhile, to the west in Clackmannanshire, Glenochil Colliery was to be the largest surface mine complex in the UK. Here too, serious geological problems were encountered, and once again, after only a few years of disappointingly low output, the complex was closed prematurely, also in 1962.[10] The problems experienced at both Rothes and Glenochil prior to their closure were sufficient to cause the NCB to reconsider its strategy, and also to face the fact that the market for coal was changing rapidly in the face of growing competition. This prompted the decision in 1959 to terminate another major new project at Airth in Stirlingshire, only a year after sinking had commenced.

These setbacks were, however, not typical of the new projects, as was demonstrated by the success of the other 'superpits' such as Killoch in Ayrshire, which opened in 1960 (Figures 3.20 to 3.25), and Bilston Glen in Midlothian, which commenced production in 1963 (Figure 3.26). All reflected Riss's belief in the synthesis of architecture and engineering, and also the importance of putting miners' welfare at the heart of colliery design. The immensity of the investment was also evident from the scale of underground development, which required large permanent roadways at and near to the pit bottom to handle much larger produc-

tion than was normal at traditional pits. (see Figures 3.24, 3.25 and 3.27).

Amongst many reconstruction projects, Riss was closely involved with that of Barony Colliery in Ayrshire. Sinking of the third shaft had begun soon after the end of World War II (see Figures 3.28 and 3.29), and was later to be taken on by Riss, whose own sketch of the new project also includes the neighbouring power station (2 x 30 megawatts), to which it supplied fuel in the form of slurry (see Figures 3.30 and 3.31). The colliery remained reliant upon the continued op-

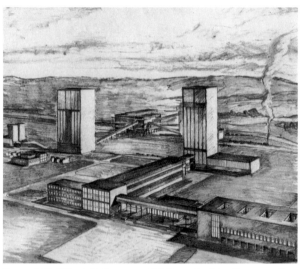

Figure 3.20: Killoch Colliery, near Ochiltree, Ayrshire. The second of Egon Riss's new sinkings, this drawing by the architect himself faithfully depicts the final appearance of the colliery, which commenced production in 1960, closing in 1987. SC706731

Figure 3.22: Killoch Colliery, c.1960. The surface arrangement of the new pit included a large canteen, which sold not only hot and cold drinks, food and snacks, but also sweets and chewing tobacco. The latter was widely available in collieries as smoking tobacco underground was strictly forbidden because of the potential presence of explosive gases. Harrison Collection, SC706245

Figure 3.21: Killoch Colliery, c.1960. Riss's colliery layouts included much improved facilities for the miners themselves. The showers in the pithead baths were designed to accommodate the needs of a workforce which peaked at over 2,300 in 1965. Harrison Collection, SC706240

Figure 3.23: Killoch Colliery, c.1960. The surface complex included extensive office facilities, including that of 'manpower deployment'. Harrison Collection, SC706244

eration of the earlier two shafts throughout the 1950s, but these became inoperable after a collapse in 1962 (see Figure 3.32). However, the complex was thought to be sufficiently valuable to merit investment in the sinking of a fourth shaft for ventilation and emergency access, which was completed in1965 (see Figure 3.33), permitting continued mining operations for a further 24 years.

In keeping with the modernist tradition, Riss's colliery designs appear to suggest a fascination with glazing, usually a prominent feature of the winding towers of his projects, all of which were different (Figures 3.34,

3.35 and 3.36), despite containing modern multi-rope friction winders. Some were more successful than others, and the tall towers at Killoch in Ayrshire (Figure 3.34) were particularly well known for their failure to exclude wind and rain, especially in the winter.[11] They were, nevertheless, potent industrial landmarks which represented a period of optimism within the coal industry itself, and one of economic and political reconstruction both in Britain and in Europe as a whole. Riss died on 20 March 1964, before the final completion of his last two major projects at Seafield in Fife and

Figure 3.24: Killoch Colliery, c.1960. This view of the tippler station demonstrates the extent of the investment in the colliery's underground facilities. Harrison Collection, SC706242

Figure 3.26: Bilston Glen Colliery, Loanhead, Midlothian. A drawing by Egon Riss depicting the layout of the new colliery. Bilston Glen commenced production in 1966, the same year as Seafield Colliery in Fife. It was unusual amongst Riss's new projects for having ground-mounted friction winders, the other new sinkings being equipped with tower-mounted winders. Having been the focus of some of the most bitter scenes in the 1984 miners' strike, Bilston Glen closed five years later in 1989. SC706505

Figure 3.25: Killoch Colliery. c.1960. The scale of the underground main roadways at this junction demonstrate that the anticipated scale of production was much greater than at traditional Scottish pits. Harrison Collection, SC706243

Figure 3.27: Monktonhall Colliery, Midlothian, c.1967. Situated on the east side of Edinburgh, Monktonhall was the last of the new generation of Scottish superpits, production commencing in 1967. This view of one of the main roadways shows the extent and scale of the project underground. Continental Conveyors Ltd. SC706525

Monktonhall in Midlothian (Figures 3.35 and 3.36), which opened in 1966 and 1967, respectively.

An unexpected consequence of the new sinking and reconstruction projects was that British mining engineering companies were unable to meet the massive demand for winding and haulage equipment. As a result, a significant proportion of the new equipment was purchased from abroad. For example, the new mine-car handling system at Barony was imported from Germany, and the new multi-rope friction

winders at Kames and Seafield were manufactured by ASEA of Sweden.

With the assistance of the new equipment, the enhanced productivity of the new 'superpits' and of many of the reconstructions allowed for a more systematic cull of the less productive pits. Figure 3.37 (net closures) shows that following the initial surge immediately after nationalisation in 1947, the heaviest period of pit closures was during 1959 and throughout the 1960s. Whilst many of these had been anticipated for

Figure 3.28: Barony Colliery, near Auchinleck, Ayrshire. Here, the sinking of a new third shaft commenced as early as 1945, and the temporary headframe can be seen just beyond the boilerhouse chimney, the original two shafts being in the foreground. SMM:1997.0412, SC381602

Figure 3.30: Barony Colliery. Although the sinking of the third shaft commenced before nationalisation, the reconstruction of Barony was taken on by Egon Riss, whose drawing also includes the adjacent power station, which was built to consume coal slurry from the colliery. SC706502

Figure 3.29: Barony Colliery. This aerial view from the south was taken at around the time of the completion of the reconstruction in 1950, and shows the new 'A'-frame above the third shaft, and the original pithead centred on Shafts 1 and 2 in the foreground. Also prominent at the north edge of the complex is a brand new coal preparation plant. The two older shafts were irreparably damaged by a collapse in 1962 (see Figure 3.32). Harrison Collection, SC706235

Figure 3.31: Barony Colliery and Power Station, Ayrshire. The power station was opened in 1953, and was designed to burn coal dust in slurry form. As the National Grid expanded to include rural areas, electricity generation became an increasingly important consumer of coal. However, the power station eventually closed in 1982 and was subsequently demolished. SC446042

Figure 3.32: Barony Colliery. This view shows the aftermath of the collapse of No. 2 shaft in 1962. With both the original shafts permanently disabled by the disaster, the colliery was left with only one shaft, and was therefore legally unable to operate until the successful sinking of a new shaft. SC706226

Figure 3.34: Killoch Colliery, Ayrshire, 1988. Each of Egon Riss's new collieries were distinctively different. The towers at Killoch were particularly striking because of the huge areas of glazing. SC376899

Figure 3.35: Seafield Colliery, near Kirkcaldy, Fife. As at Killoch and Rothes, the two majestic reinforced-concrete towers at Seafield incorporated large areas of glazing and housed powerful electrically-powered friction winders. SC446440

Figure 3.33: Barony Colliery. The commencement of the sinking of No. 4 shaft, which provided ventilation and a vital potential means of escape. No. 3 shaft's 'A'-frame and Barony Power Station are visible on the hill beyond. The new shaft was completed in 1965. SMM:1998.0322(a), SC381526

Figure 3.36: Monktonhall Colliery, Midlothian. In 1967, this was the last of the new superpits to come on stream, and became a landmark when approaching Edinburgh from the east. Large glazed areas again featured prominently in the towers, No. 1 shaft (right) being used to wind coal, No. 2 being used for men and materials. With the exception of Rothes, where mine cars were used, all Riss's superpits employed skip winding technology. SC379662

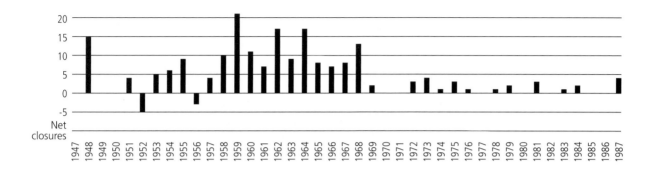

Figure 3.37: Net closures of nationalised coal mines from 1947 to 1987. In all but two years, more mines closed than were opened throughout this period.

some time, the loss of Michael Colliery in September 1967, only months after the planned closure of neighbouring Wellesley, certainly was not. Michael had been the Wemyss Coal Company's showpiece colliery, and even after nationalisation and the big new sinkings, was easily the largest mine in Scotland at the time of nationalisation. At its peak in 1957 it employed over 3,000 people. The end of the colliery was brought about by a disastrous underground fire (Figure 3.38) which killed nine miners and destroyed a development that was supposed to have exploited 12 million tons of new reserves.[12]

In 1967, the number of collieries operated in Scotland by the NCB stood at 65, and shrank by almost two-thirds in the ensuing decade (see map in Figure 3.39). By 1987, the number had fallen further to only 11, and by 1997, 50 years after 206 Scottish collieries had been taken over by the state, only two (Longannet and Monktonhall) were still operational. Inevitably, the number of miners employed in the industry followed a similar pattern (see Figure 3.8), but the downward trend was exacerbated by the mechanisation process, the steepest period of decline being in the 1960s when the production of the new 'superpits' took effect. This is reflected in output-per-man statistics, which show a marked improvement in productivity in the 1960s (see Figure 3.40). Also apparent from the latter is a rise in productivity immediately after the oil crisis in the mid-1970s, followed by a decline in subsequent years, the huge impact of the 1984 miners' strike also being very clear. Another significant pattern can be seen in the average workforce of NCB mines, which rose steadily as the number of mines reduced. In 1947, the average workforce in newly nationalised Scottish collieries was 400, and by the mid-1980s, it had risen to 1,200.

The reasons for the spectacular decline in the Scottish coal industry relate primarily to massive changes in the demand for coal. At the end of World War II, the market for coal was totally secure, and the need for more coal to assist in the reconstruction process was paramount. The key to its demise was the diversification of energy sources, and the internationalisation of the coal trade.[13] Even in the mid-1950s, coal continued to dominate British energy consumption, and demand for coal was so great that Government policy actively prevented alternative male-employing industries from locating in the coalfields and recruiting labour away from the mines. However, before the end of the decade, alternatives to coal such as oil and nuclear power were beginning to make an impact. This is perhaps reflected in the loss of confidence following the failure of the Rothes and Glenochil projects in the late 1950s, and the cancellation of the new sinking at Airth. Indeed, the rapidity of the change in the subsequent years is demonstrated by the fact that 140 Scottish pits were closed in not much more than a decade, and in the UK as a whole during the 1960s alone, 400,000 miners left the industry.

The drop in demand was caused by a combination of factors both at home and abroad. In Britain, these included the decline in heavy industrial users, especially the iron and steel industries who consumed significant quantities of coking coal. Simultaneously, there was a rapid contraction of the railway network in favour of road transport, combined with the rapid conversion away from coal-fired steam locomotives to diesel and electric power in the 1950s and 1960s. Perhaps more significant was intense competition both in industry and in the home from oil, and later from natural gas. Indeed, one of the biggest setbacks occurred

Figure 3.38: Michael Colliery, East Wemyss, Fife. Easily the biggest colliery in Scotland, Michael employed over 3,350 miners at its peak in 1967. In the same year, after extensive redevelopment work had opened up millions of tons of new reserves, spontaneous combustion triggered an underground fire which killed nine men and prematurely ended the life of the colliery. This aerial photograph was taken as the disaster unfolded on 9 September 1967. *Daily Express*, SC378676

with the arrival of natural gas from the Frigg field in the North Sea in 1977, which rapidly removed the town gas industry as one of coal's largest markets, and simultaneously provided a formidable new competitor for the coal industry.

In the aftermath of the introduction of Clean Air legislation in 1956, which followed sequences of notorious smogs earlier in the decade, the domestic and industrial consumption of coal in British cities dropped dramatically, and coal's reputation as a comparatively dirty fuel was established. Households moved swiftly away from coal as a preferred domestic fuel, and in a long battle to retain its markets, the NCB made

strenuous efforts in the 1960s to attract new customers, particularly as the age of 'Central Heating' had arrived. In one particular initiative, attempts were made to portray coal 'Nuggets'[14] and similar smokeless products as clean fuel, and publicity campaigns involved dressing coal delivery men in white uniforms similar to those of milkmen (Figure 3.41). Meanwhile, other experiments included the installation of large coin-operated automatic vending machines dispensing sacks of coal (Figure 3.42).

One of the most important trends in the coal market was the increased dependence upon electricity generators. Demand for electricity grew rapidly after

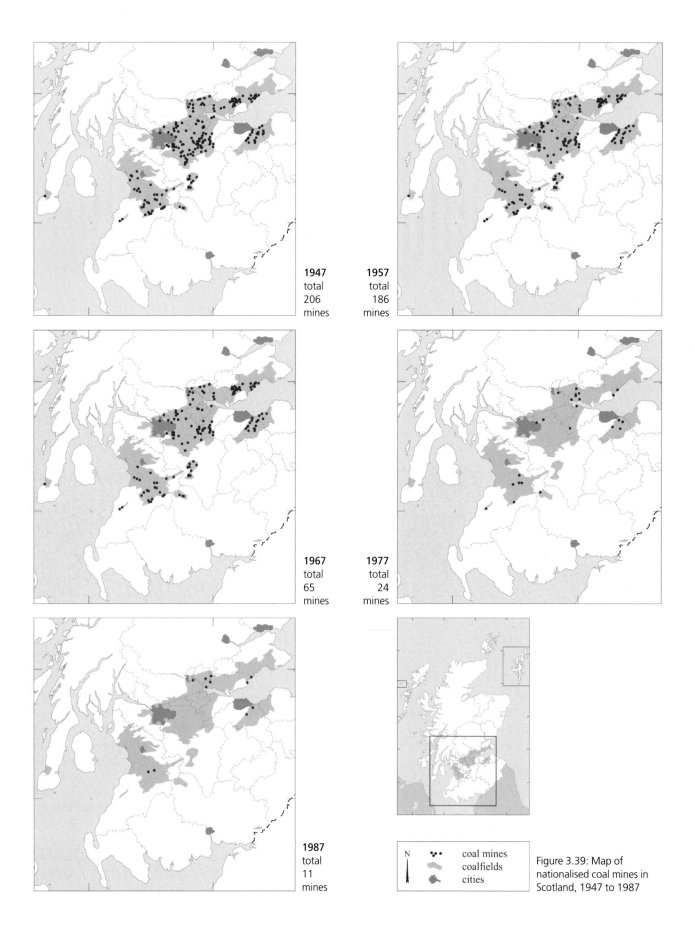

**1947**
total
206
mines

**1957**
total
186
mines

**1967**
total
65
mines

**1977**
total
24
mines

**1987**
total
11
mines

N    coal mines
     coalfields
     cities

Figure 3.39: Map of
nationalised coal mines in
Scotland, 1947 to 1987

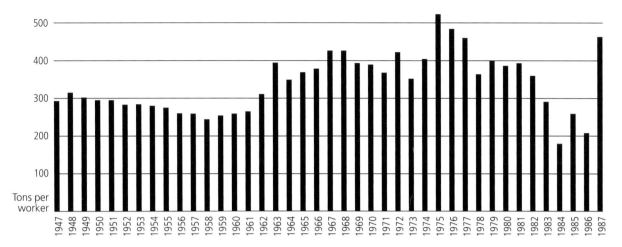

Figure 3.40: Average annual *per capita* coal output in Scotland from 1947 to 1987 (Statistics compiled by staff at the National Coal Board's Newbattle Archive)

Figure 3.41: 'Nuggets at your service' – marketing coal as a 'clean' fuel. In an attempt to compete with oil central heating in the 1960s, the NCB produced new smokeless fuels such as 'Nuggets'. To promote a clean image for the fuel, delivery men were dressed in white overalls similar to those worn by milkmen. SC706320

Figure 3.42: 'The Fuelomatic'. To further promote clean coal 'Nuggets', the NCB also experimented with coin-operated vending machines selling sacks of coal at convenient places such as bus stations. The idea was that customers could pick up a sack of coal on the way home when catching the bus. This view shows Ronald Parker, CBE, Chairman of the NCB's Scottish Division, inaugurating a new prototype machine at the Scottish Omnibus Coach Station in Edinburgh on 10th October 1961. SMM:1998.616, SC706324

1945, driven by electrification programmes (especially in rural areas), new and modernising industry, and from the 1960s, domestic demand was further boosted by the boom in consumer white goods, such as washing machines and refrigerators. In 1947, electricity generation accounted for almost 15% of the coal market, and by 1971 had risen to over 50%.[15] Several of the new power stations were located adjacent or close to reconstructed or new mine complexes, as was the case at Barony Colliery in Ayrshire (Figures 3.30 and 3.31), and at Kincardine in Fife (Figure 1.14). Others, such as Cockenzie in East Lothian and Methil in Fife did not have adjacent mines, but were easily supplied by rail. Those at Methil and Barony were unusual in that they burned coal slurry, but the others used coal supplied directly from the mines. The expansion of the national grid and the growth of coal-burning power stations resulted in the use of the term 'Coal by Wire' to describe the electrification process.

The climax of coal-burning electricity generation was the Longannet project, which emanated from the success of the burning of pulverised fuel at Kincardine Power Station in the 1950s using Hirst coal from Manor Powis, Dollar and Bogside collieries. Longannet was commenced in 1960 as a joint project between the NCB and the South of Scotland Electricity Board (SSEB), and was designed to generate 2,400 megawatts from four 600 megawatt units. It was situated on the north side of the Forth estuary near Culross, a situation that enabled the use of cold sea water to enhance thermal efficiency, and also the disposal of ash in lagoons near to Preston Island. The great advantage of the Upper Hirst seam was that it was of consistent quality, and had a very low sulphur content. It was estimated that the new power station would consume 20,000 tonnes per day.

Longannet required the sinking of several new collieries to meet this demand, and two new mines were sunk at Solsgirth and Castlehill, both commencing production in 1969. These were linked to Longannet itself by a five-mile tunnel within which the coal was conveyed by a cable belt that was said at the time to be the longest in the world. The system was notable because of the new computer technologies applied by the NCB, known as MINOS (Mine Operating System), much of which was supplied by Ferranti. The extent of automation was very advanced for the period, and in order to control the system of conveyors and bunkers, a new control centre was set up in Castlebridge, where work on the sinking of Scotland's last deep shaft commenced in 1978.

Longannet commenced output in 1972, four years after Cockenzie, and became the second-largest coal-fired power station in the UK. Although in the short term it provided a substantial market for Scottish coal, the growing dependence upon electricity generation proved to be problematic, particularly in Scotland where there was already significant hydro-electric capacity, and where the Government chose to site five heavily subsidised nuclear generating stations.[16] The addition of a large oil-fired station at Inverkip might have worsened the situation, had not the OPEC-induced oil crisis of 1974 rendered oil more expensive.

The position of the coal industry worsened in the 1980s for a number of reasons, one of the most important of which was the change in government policy following the return of the Conservative Party in the 1979 general election. The new administration favoured the de-regulation of markets and the privatisation of state-owned industry. It also retained painful memories of the 1974 general election defeat which was brought about by the miners, and with few sitting MPs in mining constituencies, there was little loyalty towards the coal industry. When the opportunity arose, British Steel was permitted to buy coking coal from overseas, which was much cheaper than that offered by the NCB. Worse was to follow as an international market in cheap steam coal flourished, led by big oil producers such as Shell, BP and Exxon, who were anxious to diversify their energy interests following the shock of the oil crises of the 1970s. New coal producing countries such as Australia, South Africa, Colombia, Indonesia and Venezuela increasingly supplied European markets through the ports of Amsterdam, Rotterdam and Antwerp (ARA), and ARA spot prices for coal were much lower than those offered by the NCB. There was therefore increasing pressure from the Central Electricity Generating Board (CEGB) to follow the example of British Steel and buy foreign coal.[17]

The plight of the coal industry was worsened still further by the government's decision to invest further in nuclear energy. This resulted in the approval of the development of Britain's first pressurised water-cooled reactor at Sizewell 'B', and coincided with the establishment of the cross-Channel 'Connector', which allowed the CEGB to buy electricity generated by France's nuclear power stations. The situation was further complicated by the privatisation of the British electricity generating industry, which resulted in the bulk of Scotland's coal production being consumed by one large utility company, Scottish Power. Driven by the interests of shareholders, there was growing pressure to buy cheaper foreign coal.

The mid-1980s proved to be catastrophic for the British coal industry. The NCB, which was renamed the 'British Coal Corporation', lost much of its do-

Figure 3.43: Monktonhall Colliery, Midlothian. The demolition of No. 1 Tower on 8th February 1998. Photograph by Campbell Drysdale. SMM:1998.400.vii, SC445877

Figure 3.44: Monktonhall Colliery, Midlothian. A sequence of photographs taken by Campbell Drysdale of the demolition of No. 2 Tower on 2nd November 1997. SMM:1998.400.iii, iv, v and vi, SC445872, SC445874, SC445875 and SC445876

mestic coking coal market as the steel industry contracted and bought what little it needed from abroad, also using oil-injection in some blast furnaces. In the four years from 1983 to 1987, coal imports jumped from four to 10 million tonnes, and a year-long miners' strike split the mining communities and failed to arrest the flow of pit closures. In the subsequent privatisation and fragmentation of electricity generation, protection for the British coal industry was relaxed still further, whilst nuclear capacity retained its subsidies. Fresh contracts with the new electricity companies had to be negotiated, and pressure on mining costs was increased. New environmental legislation added to the difficulties, forcing English coal-burning power stations to install flue-gas de-sulphurisation units at coal-burning power stations, or to buy low-sulphur foreign coal.[18] Meanwhile, the electricity generators were increasingly issued with licenses allowing them to build new power stations powered by natural gas in what became known as 'the Dash for Gas'.

By 1994, immediately prior to privatisation, only 15 mines had survived in the UK, two of which were in Scotland. At one of these, Monktonhall, miners and management were encouraged to organise a buy-out and the establishment of a miners' co-operative. Although the same model was to succeed in Tower Colliery in Wales, it failed at Monktonhall because the huge scale of the pit and its drainage requirements incurred large maintenance and pumping costs, and these could only be covered if more than one face could be maintained in production at any one time. In the private sector, it proved to be impossible to raise the capital to enhance production, and the project eventually failed, taking with it the personal investment of the miners. The colliery buildings were subsequently cleared away, the two towers being demolished by explosives in November 1997 and February 1998 (see Figures 3.43 and 3.44).

The last surviving Scottish colliery, the Longannet complex, was taken over by Scottish Coal and continued in operation, having successfully negotiated successive contracts with Scottish Power, the private company that took over from the local electricity authority, the South of Scotland Electricity Board (SSEB). However, after attempting to develop reserves to the south of the River Forth in the Airth area, the mine suffered a major flood in 2002, and its subsequent closure signalled the end of deep mining in Scotland.

Since then, all Scottish coal has been produced by open-cast mining, the contribution of which had been negligible until World War II. At that time, attempts were made to enhance war-time production by the opening of open-cast projects throughout the UK,

but the dominance of deep mining remained until the 1980s. In 1947, 2.47% of Scottish output was open-cast, and by 1967 it had grown to 6.04%. Ten years later, it had leapt to 18.61%. Such was the decline of the deep mines that this had increased to 40.87% by 1989.[19]

When considering the rapid decline of the coal industry in Scotland, it would be wrong to exclude the private deep mines, summary details of which have been included at the end of each county section of the gazetteer. At nationalisation in 1947, 105 small mines extracting coal remained within the private sector (under license to the NCB) because their operations were too small to merit transfer to state ownership. Normally employing less than 30 miners, they were often transient in nature, and in some cases, although a license was successfully applied for, the licensee failed to establish a productive mine. By 1957, 92 had retained licences to mine, this figure falling to 50 in 1967, 24 in 1977, and 20 in 1987 (see Figure 3.45). All had ceased to function by the time the Longannet mine closed in 2002. Although significant in number, the output of the licensed mines remained comparatively small, achieving a peak of 2.3% in the UK in 1955, retreating to 1% by 1980.[20]

The nationalised era left behind substantial physical remains throughout central Scotland. Although the surface remains of abandoned mines tended to survive long after closure in the 1940s and 50s, attitudes changed radically after the Aberfan Disaster (in South Wales) on 21st October 1966 when a bing became unstable, the resulting landslide engulfing a local school. Thereafter, the pace of land reclamation and demolition accelerated in the UK. The disappearance of entire mining landscapes had already begun with the expansion of open-cast coal mines. By the 1990s, it was normal for mine surface buildings to be demolished within a year of closure, and even older structures such as the towers at Rothes Colliery were demolished either to allow the realisation of assets prior to privatisation, or to minimise potential liabilities.

In practice, the physical remains of the coal industry have presented huge preservation problems, not least because of punitive long-term maintenance costs, and often hostile environmental conditions. In addition, mining communities have frequently been ambivalent towards preservation, often associating the surviving buildings with bad memories and bitter disappointment. For these reasons, successful attempts at preservation have been rare in Scotland. A few headframes have survived, such as those at Highhouse and Barony in Ayrshire, Mary and Frances in Fife, and Castlebridge in Clackmannanshire (see Figure 3.46),

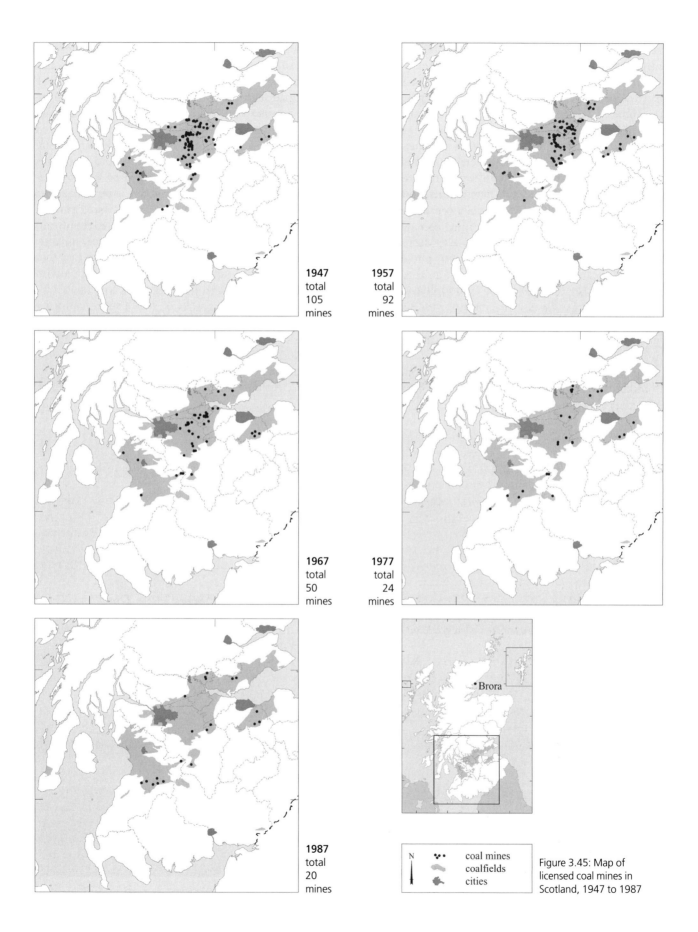

1947
total
105
mines

1957
total
92
mines

1967
total
50
mines

1977
total
24
mines

1987
total
20
mines

Brora

N

• • •
• •  coal mines
    coalfields
    cities

Figure 3.45: Map of
licensed coal mines in
Scotland, 1947 to 1987

but elsewhere scatterings of surface buildings have only survived as a result of random re-use, or in the case of Killoch, conversion to deal with open-cast coal preparation.

For political and practical reasons, it proved impossible to give many mines statutory protection, and in the cases of Barony and Frances, listed building status managed only to save the headframes. However, two significant exceptions to this trend were Prestongrange and Lady Victoria Collieries in East Lothian and Midlothian, respectively. In these cases, pioneering activity by ex-miners and groups of activists in the 1960s and again in the 1980s managed to save many of the colliery buildings, and the two mines have survived to become museums.

Although the mines themselves have disappeared, throughout the coalfields many mining communities have survived, and with them buildings and amenities such as miners' welfare institutions, Gothenburg taverns,[21] sports and leisure facilities, meeting halls, and libraries, colleges and convalescent homes. Many have evolved through time, and their links with the coal industry may no longer be obvious. More poignant are the memorials to the thousands of miners who died as a result of accidents in the mines. Collieries were particularly dangerous places to work, and even after nationalisation disasters occurred, one of the most famous being that at Knockshinnoch Castle in Ayrshire in 1950 (see Figures 3.47 and 3.48). Furthermore, the industry continued to take its toll on significant numbers of the workforce through fatalities, injuries, and respiratory diseases. It is nevertheless the case that nationalisation brought about huge improvements in safety, and that the coal industry pioneered safety culture in the workplace, facets of which spread to many other industries. This is well illustrated by the Mines

Figure 3.46: By 2004, only a small number of colliery headframes or towers survived in Scotland. With the exception of the Scottish Mining Museum at Lady Victoria Colliery (see Chapter 5), these included those at Frances and Mary Collieries in Fife, and Barony and Highhouse in Ayrshire. In all but the latter, a single headgear is all that survives, every other pithead structure having been cleared away.

Frances Colliery, SC750450

Highhouse Colliery, SC750452

Mary Colliery, SC750451

Rescue Service (see Figure 3.49), which, as the coal industry declined, was able to diversify and offer its safety services to other industries.

The rapid and often total disappearance of coal mining remains was one of the principal reasons for the compilation of the coal mine gazetteer which follows. In the process of compiling information for the gazetteer, it has been possible to draw comparisons and assess the relative size and importance of some of the collieries, and of the regions in which they operated. Some of the information can be found in the appendices, where tables of colliery statistics are located.

When measured in terms of average annual workforce, Michael Colliery in Fife was by far the most important colliery in Scotland, employing an average of 2,600 miners in its 20 years of operation after 1947. This is surprising, given the scale of the new superpits, but the largest of these, Seafield (also in Fife), was not far behind with an average of almost 2,200 miners. The only other collieries to achieve an average workforce in their lifetime of over 2,000 miners were Bilston Glen in Midlothian and Killoch in Ayrshire, both of which were NCB superpits.

If longevity is taken into account, the colliery to have provided the highest man-years of employment for miners during the nationalised period is Polkemmet in West Lothian, which produced high-quality coking coal for the steel industry for 40 years after 1947. Thereafter, Bilston Glen in Midlothian was the greatest employer for its 26 years of production, and then Cardowan in Lanarkshire, which mined coal for 37 years after nationalisation. Amongst the other collieries of significance are Comrie (in Fife), which operated for over 40 years, Lady Victoria in Midlothian, which survived for 35 years until closure, and the smaller coking-coal and anthracite-producing pits at

At Highhouse, the winding engine house survives, and at Castlebridge Colliery in Clackmannanshire, much of the surface arrangement adjacent to the winding tower has been retained as offices.

Barony Colliery, SC750454

Castlebridge Colliery, SC750455

Kinneil, Bedlay, Polmaise 3 and 4 and Manor Powis, all of which survived for between 25 and 41 years in the public sector. However, the prize for the longest period of production in the NCB era goes to Barony Colliery in Ayrshire, which maintained production from 1947 through to closure in 1989.

Finally, although regional production figures are sparse, it is clear from those available in 1913 that Fife was already beginning to challenge Lanarkshire's dominance of the Scottish coal industry. At nationalisation, Lanarkshire easily had the largest number of collieries, but the majority were relatively small operations, and in terms of numbers of miners, Fife was already forging ahead. For the remainder of the 20th century, Lanarkshire and Ayrshire continued to fade, whilst Midlothian's stature grew, boosted by the two superpits at Bilston Glen and Monktonhall. Without doubt, however, Fife remained the most important coal-mining area in Scotland, right up until the closure in 2002 of Longannet, Scotland's last deep mine.

Figure 3.47: Knockshinnoch Castle No. 6 Mine, New Cumnock, Ayrshire, 1950. Coal mining is notorious for being a hazardous activity, and although conditions and safety precautions did improve during the nationalised era, accidents and disasters continued to afflict the industry. One of the first major incidents was at Knockshinnoch Castle Colliery on 7th September 1950, when an inrush of water and mud trapped 129 miners underground. This view shows an anxious crowd waiting for news at the surface. Harrison Collection, SC706229

Figure 3.48: Knockshinnoch Castle Colliery, New Cumnock, Ayrshire, 1950. The disaster was caused by the driving of new workings too close to the surface. The resulting collapse of sodden ground left a huge crater at the surface. Rescue teams managed to reach most of the trapped miners through adjacent workings at Bank Colliery, but 13 miners died in the tragedy. SC706232

Figure 3.49: Mauchline Colliery, Ayrshire, c.1960. The modern coal industry nurtured a health and safety culture that permeated into other industries. This was exemplified by the work of Mines Rescue teams such as that shown here at Mauchline Colliery. Harrison Collection. SC446082

## Endnotes

1 Ashworth, The History of the British Coal Industry, p. 23–4.

2 NCB, Plan for Coal, HMSO (London, 1950).

3 After nationalisation of coal, the activities and responsibilities of The Miners' Welfare Fund were temporarily transferred to the Miners' Welfare Commission before falling to a new organisation, the Coal Industry Social Welfare Organisation (CISWO).

4 These figures exclude the oil shale industry, which was not nationalised. Scotland's oil-shale mines were mostly located in West Lothian immediately to the west of Edinburgh.

5 Small mines were generally permitted to remain in the private sector under license, provided that they employed fewer than 30 miners underground.

6 The statistics summarised in Figures 3.7 and 3.8 were collated from data published in the Colliery Guardian's annual publication, *Guide to the Coal Fields*, between 1948 and 1988.

7 These output statistics were compiled by NCB staff, and supplied to the author by the archivist at the NCB's Newbattle Records Centre.

8 NCB, Scotland's Coal Plan (1955).

9 The best source of biographical information on Egon Riss can be found in a dissertation entitled Egon Riss – Architect, by Grant Watson Robertson, which was submitted to the Mackintosh School of Architecture for a B.Arch (Hons) in 1988. Other information can be gleaned from obituaries, including that which appeared in the *RIBA Journal* in May 1964 (p.229), and from an article which appeared in *The Times* on 12th November 1952 entitled, 'Removing Scenes of Squalor from Scottish Pitheads'.

10 Glenochil was expected to be able to exploit the substantial pillars left behind by old stoop-and-room mine workings. These proved to be too costly to extract after compression caused by years of subsidence, and because market conditions for coal had changed so rapidly that the expense of extracting coal in this type of situation had become prohibitive. Information from George Archibald at the Scottish Mining Museum.

11 Information from George Gillespie at the Scottish Mining Museum, who remembers raising this problem with Riss personally.

12 Halliday, R, The Disappearing Scottish Colliery: A Personal View of some aspects of Scotland's Coal Industry since *Nationalisation, Scottish Academic Press (Edinburgh, 1990)*, p. 44.

13 For an analysis of the changing markets of British coal, see Hudson, R (2002), 'The changing geography of the British Coal industry: nationalization, privatisation and the political economy of energy supply, 1947–97.'

14 'Nuggets' were made at the Niddrie Nuggeting Plant on the southern edge of Edinburgh.

15 Ashworth, The History of the British Coal Industry, p. 41.

16 These were Chapelcross, Dounreay, Hunterston A and B, and Torness.

17 Hudson (2002) notes that in 1973, the world trade in steam coal was estimated to be 14 million tonnes, and that by 1989, it had grown to 172 million tonnes.

18 Scotland's largest consumer of coal, Longannet Power Station, escaped this problem because the 'Upper Hirst' coal mined in the Longannet mine complex was almost sulphur-free.

19 Data collated by the Scottish Mining Museum.

20 Data collated by the Scottish Mining Museum.

21 The principal of Gothenburg or 'Goth' taverns dates from the late 19th century when, instead of imposing total temperance to tackle alcohol abuse, the Swedish local authorities established drinking houses, the profits of which were used to fund projects which benefited the community as a whole. The idea spread overseas, and in Scotland many Goth Taverns were built in the coalfield communities. A number have survived, such as the 'Dean Tavern' in Newtongrange, Midlothian, 'The Goth' in Armadale, West Lothian, and the newly restored 'Gothenburg' at Prestonpans in East Lothian.

Figure 3.50: Lady Victoria Colliery, viewed from the south in 1966, 15 years before its closure. In the early 21st century, it has survived to become the Scottish Mining Museum, and is the only coherent surviving colliery complex in Scotland. John Keggie, SC367264

# Chapter 4: Lady Victoria Colliery

Situated on the Lothian coalfield to the south of Edinburgh (see Figure 5.160 in the Gazetteer), Lady Victoria Colliery[1] is an outstanding example of a model colliery dating from the late 19th century when the Scottish coal industry was reaching its peak, both in terms of size and sophistication. It is also remarkable because, in the 110 years since it first began to produce coal, deep coal mining in Scotland has disappeared entirely, and the surface remains of all collieries have either been entirely erased, or have been radically altered to such an extent that they are no longer immediately recognisable as having been collieries. Almost all of Lady Victoria's surface arrangement survived relatively intact following closure in 1981 (see Figure 4.1), and as such the complex provides much information on the content and arrangement of a large colliery site as it evolved through the 20th century.

The creation of the colliery coincided with the birth of the Lothian Coal Company in 1890, which was formed by the amalgamation of the Marquess of Lothian's coal company and that of Archibald Hood, who had acquired Whitehill Colliery at nearby Rosewell in 1860. Hood, who brought with him expertise gained through his extensive work in South Wales,[2] immediately became the managing director of the new company, and the first major project was the development of the new colliery. At Rosewell, Hood had first shown his concern for the living conditions of the workforce, expanding the village with good-quality miners' rows with gardens, and assisting the formation of a local co-operative. These improvements were later repeated at Newtongrange on a substantial scale, where, like Rosewell, the village survives relatively intact.[3]

At Lady Victoria, one of the new project's principal aims was to exploit 'Parrot' and 'Splint' coals, both being much in the demand at the time. These were to be found in the 'Limestone Coal Group', which, in the Lothian coalfield, are often 400 metres below the shallower previously-worked 'Productive Coal Measures'. Contained within the strata there were 24 seams of coal with a thickness of over 32cm, but most were unworkable, only five being genuinely exploitable.

The colliery was named after Lady Victoria Alexandrina Montagu Scott, the wife of the chairman of the company (1890–1900), the 9th Marquess of Lothian. The site at Newtongrange was chosen because of its position in the coalfield basin, and its proximity to the existing pit at Lingerwood, which provided its statutory second shaft. On the surface, the site was immediately adjacent to the Edinburgh to Hawick Railway (opened in 1849), giving access to potential markets both in Edinburgh and the Lothians, but also in the Borders towns and their textile mills. It also permitted the import of materials vital to construction and developments both at the surface and underground. By the time the colliery was fully developed, it had many miles of railway sidings covering 17 hectares of land.

Underground, the site had the advantage of working to the rise in the strata, a situation which in effect involved driving roadways and following the coal seams on an upward gradient away from the pit bottom. The system used gravity to take the hutches full of coal down the slope to the bottom of the shaft (from where they were taken in cages to the surface), whilst also bringing the empty hutches back up at the same time using a common haulage rope looped round a

Figure 4.1: View of Lady Victoria Colliery, shortly after the cessation of mining in 1981. SC894658

Figure 4.2: Lady Victoria Colliery, 1929. A postcard showing the pit bottom, suggesting that pit ponies were still in use at the time. SMM:1997.0909, SC382531

Figure 4.3: Lady Victoria Colliery, 27th March 1981. The last ton of coal to by mined at the colliery, which was said to have yielded over 39 million tons of coal in its lifetime. *Dalkeith Advertiser*, SC384780

pulley wheel with a brake to govern the speed. This saved a great deal of effort and capital when compared with 'working to the dip', which required underground haulage to take the coal to the pit bottom. The other principal saving was the need for only one shaft. By 1890, legislation dictated that collieries should have at least two shafts or mines to ensure means of escape and adequate ventilation.[4] In this case, the plan was to link the new workings with those of Lingerwood, whose No. 1 (Dixon's) shaft became the upcast shaft for Lady Victoria.

The sinking of the shaft commenced in August 1890, continuing until Spring 1894, after which work commenced on the opening out of the pit bottom. It had been planned to have a shaft diameter of 5.5 metres but this was expanded to 6.09 metres, making it bigger than any other contemporary project in Scotland. Because of water continuously draining into the shaft, it was lined with tapered red brick progressively as the deepening process continued. Cast-iron water-catching rings (garlands) were installed at intervals of 54.86 metres to capture water, which was subsequently drained by steam pumps from a main catchment point at a depth of 244 metres. By September 1893, the shaft had reached a depth of 448 metres, the sinking progressing at a rate of between 1.5 and 2.7 metres per week. Sinking was finally completed at a depth of 501 metres, two men having been killed during the project.

The shaft was then equipped with ropes and two double-decked cages, each being designed to accommodate six hutches or 24 men on either deck. A large steam-powered winding engine supplied by steam from six Lancashire boilers was designed to lift a payload of 10 tons, but the actual total weight of a full cage in the pit bottom position was 18 tons, including rope, chains and full hutches. With the attachment of a balance rope to the floors of the cages, this was reduced to an effective payload of 10 tons. A complete wind from bottom to top of the shaft took 48 seconds, and the capacity of the winder was stated to be in excess of 1,200 tons per day. A double decking arrangement at the pithead and pit bottom allowed both the upper and lower decks of the cages to be loaded and unloaded simultaneously.

After the pit bottom became operational, roadways were driven to the Parrot and Splint seams. The Parrot was the first to be developed because of the growing demand for gas-producing cannel-type coal for gas works,[5] and was called 'Parrot' because of the chirping noise it made when put on a fire as the gas escaped and was ignited by the fire. The Splint seams provided excellent steam and house coal which was robust, travelling well and producing little in the way of dust and dross. In later years, the Kailblades, Smithy and Coronation seams were also worked, the colliery mostly producing steam and house coals, a substantial proportion of its output also continuing to go to the town gas industry.

In its second year of operation, 500 tons of parrot coal were being hauled up the shaft daily, and a year later the addition of splint coals raised the total to 600 tons. In the years that followed, the management were quick to modernise, introducing, for example, new designs of face support which helped to reduce the accident rate from rock falls to one fifth the Scottish average. In the 1920s, miners were provided with electric lamps in the dustier working areas of the pit where coal-cutting machinery was in use, but naked lights were still used elsewhere for many years. The original haulage roadways were constructed to be 3.05 metres wide and 2.59 metres high, but were enlarged to 3.66 by 3.05 metres after an increase in the size of the hutches. Underground endless-rope haulage was also introduced, replacing 120 ponies (see Figure 4.2).

Hood had experimented with early types of coal cutter at Whitehill Colliery in 1877, but these had been unpopular, especially because they created a lot of dust. At Lady Victoria, Mavor and Coulson bar-type cutters were successfully introduced in 1905, and were later replaced by chain-type cutters by the 1950s. These in turn were replaced in the 1960s by shearer loaders along with armoured conveyors. In 1947, the first full year of nationalisation, the colliery produced 340,000 tons, which was equivalent to 1,246 tons per day, rising to 2,000 tons per day by 1951. At its peak in 1953, the labour force reached 1,765 people, of whom 1,360 worked underground. By 1969, face mechanisation had led to a reduction in manpower to 1,247 men, and by 1979, to 895, greatly reducing the hard manual labour involved in extracting the coal. Two years later, on 27th March, 1981, the last ton was mined and the colliery closed after 87 years of operation, in which time almost 40 million tons of coal had been extracted (see Figure 4.3).

In subsequent years, although the underground workings were abandoned and the shaft was filled, Lady Victoria's surface buildings were saved from demolition, eventually becoming the Scottish Mining Museum. A key to its survival was the statutory protection provided by the Scheduling of the steam-winding engine, the entire colliery subsequently being Listed Category 'A' by Historic Scotland. The museum has since evolved, but much of the original fabric has survived. The layout of the colliery can be seen in Figure 4.5, an annotated photograph on which the principal elements of the surface arrangement have been identified (in an anti-clockwise direction). An accompanying table (Table 1) provides a brief description of each component of the colliery, and with the assistance of Ordnance Survey maps (see Figure 4.4), it has been possible to date most elements of the surface arrangement. In addition to RCAHMS survey photographs, images from the museum's own collections have been used to illustrate various areas and processes, and are arranged in the same order as that in which they appear in Table 1.

When production first commenced in 1894, apart from the colliery's winding-engine house, boiler house and a small arrangement around the shaft itself, much of the colliery complex had yet to be developed (see Figures 4.4 and 4.33). For several years, therefore, coal was not washed, merely being sorted over screens and then dispatched by rail. By 1907, much of the current pithead had been constructed, the tub circuit and the picking tables being housed within steel-framed arcaded red-brick buildings, in the bottom of which ran railway sidings connected to the Waverley line (see Figures 4.8 to 4.15). Shortly afterwards (by 1914), the 'Old Power Station' (Figure 4.23) and the old 'Washer' (Figure 4.18) had been erected, the latter containing a 'bash-tank' coal preparation plant which operated until 1968, in combination with 'The Dredger' (Figure 4.30), a pitch-pine wooden structure at the north end of the colliery which treated dirty water from the washer, extracting coal dust which was later used to fuel the boilers. In the same period, the dross storage 'Hopper' (Figure 4.18) was built adjacent to the washer.

Between 1914 and 1932, the most important developments included the construction of the New Power Station in 1924 (Figure 4.24), and a of new Boilerhouse (Figure 4.29) which doubled the number of Lancashire boilers to 12 in order to provide extra steam for electricity generation. The colliery workshops were added in the 1950s (Figures 4.27 and 4.28), as were the pithead baths (Figure 4.25), canteen and Scientific Laboratory. These were situated on the opposite side of the road from the pithead, and were connected by an overhead reinforced-concrete walkway or 'Gantry' which allowed miners to cross the road in safety. The last major development to occur at the colliery was the installation of the 'Dense Medium Plant' in 1963-4, a modern coal washery which was situated at the southern end of the pithead (Figures 4.16 and 4.17). This was accompanied by the construction of the 'Fines Treatment Plant' and 'Thickener' at the north end of the complex (Figure 4.32), this facility eventually rendering the old 'Washer' and 'Dredger' redundant.

Since closure, only a small part of the colliery complex has been demolished. This includes the canteen and baths, which were lost in the 1980s, and part of the adjacent Central Workshops in the 1990s. Much of the machinery in the old and new power stations had already been removed by the time museum status was secured, and so these buildings were subsequently converted for use as permanent exhibition space and for the museum's visitor facilities (see Figure 4.31). In 1994, The Scottish Mining Museum Trust purchased the colliery from Lothian Estates with the assistance of a loan from the National Memorial Heritage Fund, the loan being converted to a grant on completion of the restoration of the Old Washer building in 1998. From 1999, a highly successful development sponsored by the Heritage Lottery Fund and Historic Scotland has incorporated central parts of the pithead, including the mine tub circuit, which can now accommodate visitor access and has extensive interpretation facilities. Elsewhere, however, the picking tables, dense-medium plant, boilerhouse and fines-treatment plant all lie unused and await attention. Indeed, restoring, interpreting and maintaining these buildings in the future is one of the greatest challenges facing the museum.

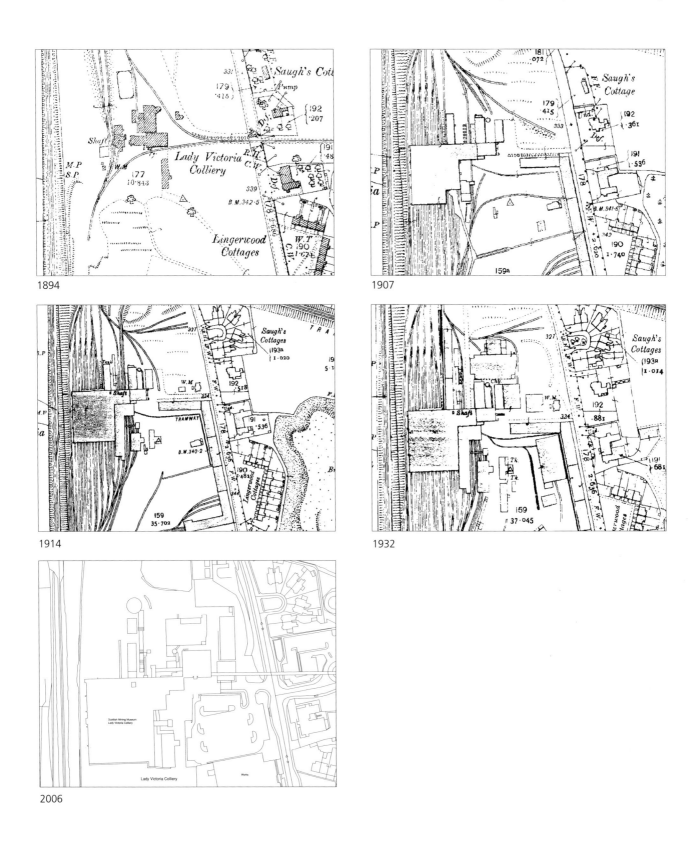

1894

1907

1914

1932

2006

Figure 4.4: The evolution of Lady Victoria Colliery, as depicted on Ordnance Survey maps, (25 inch : 1 mile). Crown Copyright 2003, All rights reserved: RCAHMS GD GD03135G0011

## Table 1:
## The Principal Components of
## Lady Victoria Colliery

(see Figures 4.4 and 4.5)

**1 Headgear:** 26 metres high, and constructed from fabricated mild steel by Sir William Arrol and Company of Glasgow in 1893, with two sheaves (or pulleys), each of 5.79 metres diameter (see Figure 4.6).

**2 Shaft:** sunk and simultaneously lined with brick during a four year period between 1890 and 1894 to a depth of 501 metres, and with a diameter of 6.09 metres (20 feet) which was at the time the largest circular shaft in Scotland. In addition to containing the two cages, this was the downcast shaft, air from the colliery being exhausted through an upcast shaft at neighbouring Lingerwood Colliery. After closure in 1981, the shaft was blocked, and all the mineworkings below are now inaccessible (see Figure 4.7).

**3 Pithead:** a mostly two storeyed complex of steel-framed buildings with red-brick walls (see Figure 4.9), which at ground level accommodate railway sidings (formerly connected to the mainline railway) separated by arcaded brick walls, between which railway wagons were filled with coal from above (see Figures 4.8, 4.9, 4.10). Much of this structure was built between 1894 and 1907. The upper level contains a circuit around which tubs or hutches of coal circulated, having been unloaded from split-level cages (see Figure 4.11). The circuit contains a weighing station, and tipplers which overturned the hutches, their contents being collected below on conveyors (see Figure 4.12). The circuit was designed to operate on a gentle slope to allow gravity to do much of the work, but a creeper (see Figure 4.13) took the hutches back to the shaft (see Figure 4.14). The original arrangement had double-decking at both the pit-head and the pit-bottom to permit simultaneous loading and unloading of both decks of each cage. Despite the reconstruction of the tub circuit in the 1960s and the reduction from double to single decking, the slight slope towards the winding engine house (to the east) can still be detected.

**4 Picking Tables:** accommodated at the upper level of the main pithead building's south-west end, and built partly between 1894 and 1907, and extended southwards by 1913, embracing the Dross Hopper. The most southerly portion was added after 1932. The tables were arrangements of slow moving conveyors from which lumps of stone and other waste were manually removed by teams of coal pickers (see Figure 4.15).

**5 Dense Medium Plant:** a form of 'washer' or wet coal-preparation plant added to the south end of the pithead buildings in 1963–4 (see Figures 4.16 and 4.17), and designed by Simonacco of Carlisle. It incorporated a washing process which used magnetite to raise the specific gravity of the 'waterfluid' in a large bath, permitting the separation of coal and stone. The magnetite was recovered and re-used using a magnetite recovery system.

**6 Hopper:** a tall brick-built building, the walls of which have battered buttresses between which concave reinforcing panels helped to contain the weight of large quantities of coal dross within. Map evidence suggests it was built between 1907 and 1914 (see Figure 4.18).

**7 Old Washer:** a brick-built rectangular-plan building, constructed between 1907 and 1914, containing a felspar washer, to which was added on its south gable (by 1932) a smaller brick building containing a 'rewasher' designed to treat dross from the washer. The machinery of the latter has been removed, but the washer itself was retained after it fell out of use in the mid-1960s, and was renovated by the museum in the late 1990s (see Figure 4.18 and 4.19).

**8 Smithy:** situated on the east side of the pithead, a brick-built gabled structure erected in 1910. It contained a blacksmith's hearth and associated metal-working equipment, but was stripped and converted in the late 1990s to accommodate an interactive-exhibition area (see Figure 4.20).

**9 Winding-Engine House:** a gabled single-storeyed and basement polychrome brick rectangular building built between 1890 and 1894 to accommodate a 2,400 horse-power steam winding engine built by Grant Ritchie and Company Limited of Kilmarnock, said to be the largest ever to be erected in Scotland. A fire in 1902 destroyed the timber lining, and the internal walls were subsequently covered with ceramic tiles, and the floors with chequer-plate steel. After closure and the filling of the shaft, the engine was converted into an exhibit, and is now operated by an electric motor instead of steam (Figure 4.21 and 4.22).

**10 Old Power Station:** from early on, pioneering mining technology, including electrical machinery, was used at the colliery. A gabled brick-built rectangular single storeyed building, the colliery's first power station was built between 1894 and 1907, and generated electricity using two 1,000 kW Curtiss steam turbines. These were later relegated to reserve status after the introduction of a new 5,000 kW generator. In the late 1990s, the building was converted to house visitor and conference facilities for the museum (see Figure 4.23).

**11 New Power Station:** a tall brick-built rectangular-plan gabled building built in 1924 to replace the old power station, and to provide more electricity as mechanisation intensified at the colliery, and in the other two pits in the Newbattle group. After the mine ceased to generate its own electricity and was connected to the National Grid, the building was converted to offices, and in the 1990s, was used to house a range of new permanent exhibitions (see Figures 4.23 and 4.24).

**12 Central Workshops:** built by the National Coal Board in 1957–8 to serve all the collieries in the Lothians region. Several workshop buildings were demolished in the late 1990s to make way for the new museum car park.

**13 Gantry from pithead to the baths and canteen:** a re-inforced concrete overhead walkway built in 1954 to the design of NCB Scottish Division architect Egon Riss. It provided covered access over the A7 trunk road from the baths and canteen, first to the lamproom (in the upper level of the 'New Power Station'), continuing on to the pithead (see Figure 4.24 and 4.25).

**14 Lothian Coal Company Offices (out of picture):** possibly built in 1873 originally as a school, a two-storeyed 'U'-plan rubble-built block, now occupied by the Scottish Mining Museum's administration, and its extensive library and archive (see Figure 4.26).

**15 Baths (out of picture):** a single-storeyed brick-built building constructed in 1953, containing showers and lockers for up to 3,020 miners, and situated at the far end of the overhead gantry. The baths were demolished in 1986 (see Figure 4.25).

**16 Colliery Workshop:** dating from the 1950s, a gabled brick building where the maintenance and repair of the colliery's machinery was carried out. Following the closure of the colliery, an extra floor was inserted and the building was converted to become the NCB's Newbattle Archive. It has since been rented out for document storage (see Figures 4.27 and 4.28).

**17 Boiler House:** built in 1924, the current building retains eight of its original 12 Lancashire boilers and associated economisers, and replaced an earlier boiler house (which contained six Lancashire boilers). They were bought from the Gretna munitions factory, which had ceased to operate at the end of World War I. Steam was

Figure 4.5: Lady Victoria Colliery: aerial view of the surface
arrangement, highlighting the major components of the colliery.

| | | | |
|---|---|---|---|
| 1 | Headgear | 13 | Gantry |
| 2 | Shaft | 14 | Lothian Coal Company offices (out of view) |
| 3 | Pithead | 15 | Baths (out of view) |
| 4 | Picking tables | 16 | Colliery workshop |
| 5 | Dense medium plant | 17 | Boiler house |
| 6 | Hopper | 18 | Boiler-house chimney |
| 7 | Old washer | 19 | Dredger |
| 8 | Smithy | 20 | Thickener |
| 9 | Winding-engine house | 21 | Fines treatment plant |
| 10 | Old power station | 22 | Engine house and workshops |
| 11 | New power station | A | Waverley Line |
| 12 | Central workshops | B | Bing |

produced to power the adjacent steam-winding engine, to generate electricity, and to provide heat and process steam in various parts of the colliery complex. The boilers remained in operation until the closure of the colliery in 1981, and used coal dust retrieved from the 'Dredger' and 'Fines Treatment Plant' (see Figure 4.29).

**18 Boiler-house Chimney:** a circular-section brick stack, 1.6 metre diameter at its base, connected by a long brick-built flue to the boiler house, originally 46 metres high, but now reduced to 33 metres having been shortened on two occasions for safety reasons. This appears to be the same chimney as that built for the original boiler house, and probably therefore dates back to the establishment of the colliery between 1890 and 1894 (see Figure 4.21).

**19 Dredger:** a wooden pitch-pine structure within which dirty water from the 'Old Washer' was treated, bucket elevators extracting coal dust from the bottom of a settling tank. Also dating from between 1907 and 1914, it supplied coal dust for use in the colliery's boilers, but ceased to operate in the mid-1960s not long after the installation of the 'Dense Medium Plant' and the new 'Fines Treatment Plant' (see Figure 4.30).

**20 Thickener:** built in the mid-1960s, a circular concrete settling pond within which coal dust from the adjacent 'Fines Treatment Plant' was collected before being burned in the colliery's boilers (see Figure 4.32).

**21 Fines Treatment Plant:** built in 1963–4 on the north side of the colliery, it comprises a tall rectangular brick-clad structure within which dirty water from the 'Dense Medium Plant' was washed (see Figure 4.32). It operated in conjunction with the Thickener, waste material being taken away by overhead conveyor to the colliery's bing.

**22 Engine Houses and Workshops:** a partially truncated brick-built single-storeyed, gabled range situated on the north side of the pithead, and thought originally to have housed the winding engine used to sink the shaft in the early 1890s. It later housed an electric engine used to drive continuous-rope underground haulage (see Figure 4.6).

**A – Waverley Line:** the Edinburgh to Hawick Railway which was opened by the North British Railway in 1849, and was a vital means by which materials were brought in for the development of the colliery. Once production commenced, the railway provided efficient access to Edinburgh and Leith docks, to the Borders mill towns, and as the electricity market began to dominate, to some of Scotland's power stations. Wagons were taken into sidings passing under the pithead buildings where they were filled with coal from above (see Figure 4.10). The railway was closed during the Beeching cuts in 1969.

**B – Bing:** the spoil heap of the colliery. This has been much reduced because of spontaneous combustion problems which caused noxious fires. Although they provided the museum and village with an authentic odour, they were considered to be an environmental hazard, and the fires were extinguished and much of the spoil removed in the late 1990s.

Figure 4.6: Lady Victoria Colliery, 1988. View from north west showing headgear, part of the north side of the pithead complex, and to the left, the old engine house originally used to house the temporary winding engine during the sinking of the shaft in the early 1890s. SC758215

Figure 4.7: Lady Victoria Colliery, c.1920. View at the pit bottom showing hutches ready to be loaded into the cages. SMM:1997.0876, SC382546

Figure 4.8: Lady Victoria Colliery, 1999. View beneath the picking tables showing the arcades of brick arches separating the railway sidings which passed under the colliery. SC894274

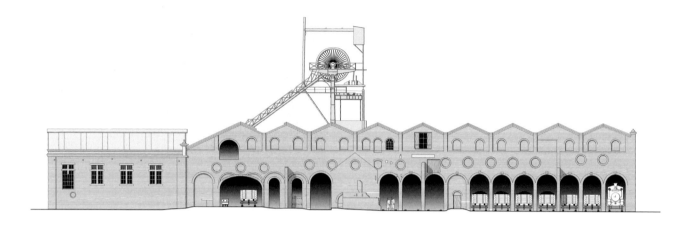

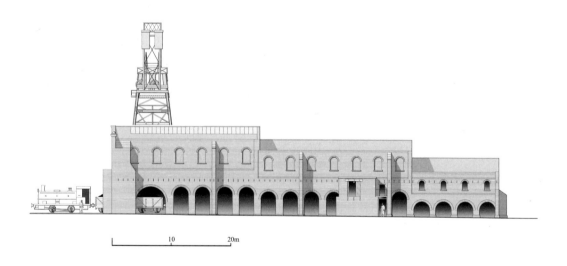

Figure 4.9 Lady Victoria Colliery, 2000: the north (top) and west (bottom) elevations of the main pithead building, including the headframe.

Figure 4.10: Lady Victoria Colliery, 1964. View of coal discharging into railway wagons beneath the picking tables. SMM:1998.0397(xiii), SC384759

Figure 4.12: Lady Victoria Colliery, 1963. Interior view of conveyor from pit-head tippler, which can be seen in the background. SMM:1998.0397(iii), SC384754

Figure 4.11: Lady Victoria Colliery, 1924. View showing the pithead, including the tub circuit and the top of the shaft. SMM:1998.0601.4, SC384784

Figure 4.13: Lady Victoria Colliery. View of the 'creeper', which, after gravity had taken the tubs (or hutches) to bottom of the tub circuit, brought them back up to its summit at the pithead. SMM:1997.0893, SC382550

Figure 4.14: Lady Victoria Colliery, c.1981. View of the tub circuit and banksman's cabin, situated close to the top of the shaft, from where operations at the pithead were controlled.SMM:1997.0548, SC382549

Figure 4.15: Lady Victoria Colliery, 1951. View looking down onto one of the picking tables. Sodium lighting was installed by the General Electric Company Limited in the 1950s because it made the coal glisten, in sharp contrast to the dull surface of the waste and dirt. SMM:1996.0861, SC384753

Figure 4.16: Lady Victoria Colliery, 1998. View of the south end of the colliery, showing the Dense Medium Plant (centre), which was constructed in 1963–4. SC894661

Figure 4.17: Lady Victoria Colliery, 1999. View inside the Dense Medium Plant, showing the bottom of the agitating vessels. SC894660

Figure 4.18: Lady Victoria Colliery, 1993. View of the south-east end of the colliery showing the dross Hopper (left) and the Old Washer (right). SC894265

Figure 4.21: Lady Victoria Colliery, 1991. View of the Winding Engine House (left) and Old Power Station (right), with the boilerhouse chimney beyond. SC894659

Figure 4.19: Lady Victoria Colliery, 1924. View within the Old Washer. SMM:1998.0601.7, SC384786

Figure 4.22: Lady Victoria Colliery. View of the steam-powered winding engine, which was built by Grant Ritchie and Company Limited of Kilmarnock. SMM:1997.918, SC382554

Figure 4.20: Lady Victoria Colliery, 1998. View from south east of the Smithy, prior to the conversion of its upper floor to an interactive exhibition area. SC894271

Figure 4.23: Lady Victoria Colliery, 1991. View showing the Old Power Station (left) and the New Power Station (right), before their conversion to house museum exhibits and visitor facilities. SC894272

Figure 4.24: Lady Victoria Colliery, 1991. The New Power Station, before its conversion to house new museum exhibition areas and visitor facilities. The gantry to the baths can also be seen (top right). SC894669

Figure 4.27: Lady Victoria Colliery, 1998. View of the former Colliery Workshops, converted into an archive centre for the NCB's Scottish Division, and later rented out to external users. SC894667

Figure 4.25: Lady Victoria Colliery, 1955. The pithead baths, which were situated on the opposite side of the main road (the A7 from Edinburgh to Carlisle). They, and the gantry connecting them to the pithead, were completed in 1954. The baths were demolished shortly after the closure of the colliery in the mid–1980s, but the gantry has survived. SMM:1998.0395, SC384761

Figure 4.26: Lady Victoria Colliery, 1998. The former offices of the Lothian Coal Co. Separated from the pithead by the A7 trunk road (from Edinburgh to Carlisle), and used latterly as the offices of the Scottish Mining Museum. The building also houses the museum's excellent library and archive. SC894269

Figure 4.28: Lady Victoria Colliery, 1924. Interior view of the Colliery Workshop, prior to its conversion to an archive centre for the NCB's Scottish Division. SMM:1998.0601.12, SC384790

Figure 4.29: Lady Victoria Colliery, c.1970. Interior view of the Boiler House, which at its peak contained 12 Lancashire boilers producing steam to power the steam-winding engine, the power stations' generators, and for a variety of other uses in the colliery complex. The boilers burned dross retrieved by the 'Dredger' from waste water emanating from the 'Washer', and later from the 'Fines Treatment Plant'. SMM:1997.0888, SC382558

Figure 4.30: Lady Victoria Colliery, 1988. View of the 'Dredger', situated between the boilerhouse chimney (left) and the headgear (right). Its function was to extract by means of a bucket elevator coal dross at the bottom of a settling tank containing water from the old Washer, the dross being used to fuel the boilers in the adjacent Boiler house. SC894668

Figure 4.32: Lady Victoria Colliery, c.1965. View of the Fines Treatment Plant and Thickener, the construction of which in 1963–4 coincided with the installation of the Dense Medium Plant at the south end of the colliery. SMM:1996.0666, SC384760

Figure 4.33: Lady Victoria Colliery, c.1895. This view shows the pithead shortly after the colliery came into production, and before the addition of other buildings around the pithead, including the Old Power Station. SMM:1997.0980, SC382551

Figure 4.31: Lady Victoria Colliery, 1999. View of the pithead in 1999 after the completion of a £5 million renovation and redevelopment programme by the Scottish Mining Museum. SC894268

## Endnotes

1 Much of the information included in this chapter has been collated by staff and volunteers at the Scottish Mining Museum, whose library and archives contain records and information relating to this and many other collieries in Scotland.

2 see Blyth, A (1994), From Rosewell to the Rhondda: *The Story of Archibald Hood, A Great Scots Mining Engineer*, Midlothian District Library Service, Loanhead

3 In addition to Newtongrange and Rosewell, one of the finest surviving model mining villages in Scotland is Coaltown of Wemyss in Fife.

4 Following the aftermath of the Hartley Colliery disaster in Northumberland in 1862 when 220 miners perished underground after being trapped by the collapse of the colliery's only shaft, the Coal Mines Act of 1872 stated that no person should be employed underground in a mine unless there were at least two shafts in communication with each seam being worked, providing separate potential means for both escape and the entry of rescue teams.

5 Gas from the Parrot coal burned with a bright flame, and was therefore particularly well suited for gas lighting.

# Chapter 5: Gazetteer

## Introduction

This gazetteer of Scottish collieries in the nationalised era is arranged alphabetically by former (pre-1975) county, and a short introduction to each county provides a brief historical overview of the collieries in that area. Standard locational data for each colliery includes the name of the parish, the nearest town or village, the current local authority council area, and an eight-figure grid reference. Where possible, the latter has been extracted from the National Coal Board shaft and mine register, which, in addition, has often yielded information on the company that founded the colliery, and dates of sinkings, closure and abandonment of underground workings. Also attached to each entry is a unique site number linking it to the RCAHMS Canmore database (accessible at www.rcahms.gov.uk).

Each entry also contains all or most of a range of standard fields providing information on the type of coal produced, the average workforce during the colliery's working life, the largest number of workers to have been employed at any one time, and the year in which this peak was achieved. From this information, which was derived from annual Colliery Year Books covering the UK, it has been possible to rank the collieries in order of importance, as measured by size of workforce (see Appendices). Unfortunately, comparable information on the output of each colliery is not available.

For many of the older collieries, detailed information was available on their state at the time of nationalisation in 1947. This data was generated by a National Coal Board questionnaire which was sent to all Scottish colliery managers in 1947, and which was completed during the following two years. The detailed responses to the questionnaire provide a fascinating insight into the contemporary state of the industry, and the scale of the task ahead if the industry was to be modernised successfully. For example, despite the work of the Miners' Welfare Fund in the 1920s and 1930s, it seems that the majority of Scottish collieries did not have proper pithead baths, and the lack of medical facilities is also very apparent.

In general, information from secondary sources and the post-nationalisation questionnaire is quoted directly. This has led to some inconsistencies, not least because the information has in many cases come from different colliery managers across the coalfields. Their broad classification of the different categories of coal – House, Manufacturing and Coking – has been followed, and though not mutually exclusive or defined in detail, these categories do give a reasonable picture of the types of mostly bituminous coals extracted from the Scottish coalfields.

One of the aims of the gazetteer has been to provide at least one photograph of each colliery. This has been a considerable challenge, given the nature of the more transient collieries mentioned above and given that RCAHMS has itself recorded no more than a dozen of the collieries prior to their closure. Nevertheless, there are images for most of the sites included in the gazetteer, thanks in large measure to the close collaboration with the archives and library of the Scottish Mining Museum at Lady Victoria Colliery, Newtongrange, Midlothian, whose volunteers have been an inspiring source of information. Other important sources of photographs include The National Archives (formerly the Public Record Office), Kew, London, several Council libraries and archives in Scotland, and a number of individuals. The quality of the images varies, but all provide valuable information.

However, the entries for some collieries, especially small, temporary mines sunk in the early years of nationalisation, are very thin indeed. Many survived only a few years into the 1950s, commencing and disappearing before even the Ordnance Survey had had the chance to register their existence on maps. Finding any detailed information on, or photographs of, many of these mines has proved to be an almost impossible task. Some errors and anomalies in the data are also inevitable, given the ambiguities caused by the merging and de-merging of mines, the existence of many shafts and surface mines under the umbrella title of one colliery, name changes, and changes of ownership prior to 1947. RCAHMS is therefore anxious to fill as many as possible of the obvious gaps, and would appreciate hearing from anyone who is in a position to correct or enhance the data.

Finally, a list of the private (non-nationalised) coal mines is included in the gazetteer. Information on these mines is even more difficult to collate on a national basis, but the Colliery Year Books and the shaft register together have made it possible for indexes of summary information to be included at the end of each county gazetteer section. Although the grid references are only six-figure, they do at least help to provide an overview of the geographical distribution of the licensed sector.

# Abbreviations for Scottish Regions and Districts

Although organised by old counties, this gazetteer has included information on the former administrative Regions and Districts in Scotland for each colliery. The Regions and Districts were themselves replaced with unitary Council areas on 1st April 1996, and are still used by some archives to arrange data. In most cases, the transition to new Council areas was smooth, but in some areas such as East Dunbartonshire and North Lanarkshire, there were boundary changes that have caused considerable confusion. Note that the table below only includes the regions and districts within which collieries are known to have operated after 1947.

| County | Region (abbreviation) | District (abbreviation) | New Council Areas |
|---|---|---|---|
| Argyll | Strathclyde (St) | Argyll and Bute (Ar) | Argyll and Bute |
| Ayrshire | Strathclyde (St) | Cumnock and Doon Valley (CD) | East Ayrshire |
| | Strathclyde (St) | Cunninghame (CU) | North Ayrshire |
| | Strathclyde (St) | Kilmarnock and Loudoun (KL) | East Ayrshire |
| | Strathclyde (St) | Kyle and Carrick (Ky) | South Ayrshire |
| Clackmannanshire | Central (Ce) | Clackmannan (Cl) | Clackmannanshire |
| Dumfriesshire | Dumfries and Galloway (Du) | Nithsdale (Ni) | Dumfries and Galloway |
| Dunbartonshire | Strathclyde (St) | Cumbernauld and Kilsyth (CN) | North Lanarkshire |
| | Strathclyde (St) | Strathkelvin (St) | East Dunbartonshire |
| East Lothian | Lothian (Lo) | East Lothian (Ea) | East Lothian |
| Fife | Fife (Fi) | Dunfermline (Du) | Fife |
| | Fife (Fi) | Kirkcaldy (Ki) | Fife |
| Lanarkshire | Strathclyde (St) | Clydesdale (CY) | South Lanarkshire |
| | Strathclyde (St) | City of Glasgow (Gl) | City of Glasgow |
| | Strathclyde (St) | Hamilton (Ha) | South Lanarkshire |
| | Strathclyde (St) | Monklands (Mo) | North Lanarkshire |
| | Strathclyde (St) | Motherwell (MW) | North Lanarkshire |
| | Strathclyde (St) | Strathkelvin (St) | East Dunbartonshire |
| Midlothian | Lothian (Mi) | City of Edinburgh (Ed) | City of Edinburgh |
| | Lothian (Mi) | East Lothian (Ea) | East Lothian |
| | Lothian (Mi) | Midlothian (Mi) | Midlothian |
| Peeblesshire | Borders (Bo) | Tweeddale (Tw) | Scottish Borders |
| Perthshire | Tayside (Ta) | Perth and Kinross (Pe) | Perth and Kinross |
| Stirlingshire | Central (Ce) | Stirling (St) | Stirling |
| | Central (Ce) | Falkirk (Fa) | Falkirk |
| | Strathclyde (St) | Strathkelvin (St) | East Dunbartonshire |
| Sutherland | Highland (Hi) | Sutherland (Su) | Highland |
| West Lothian | Lothian (Lo) | West Lothian (We) | West Lothian |
| | Lothian (Lo) | Midlothian (Mi) | Midlothian |

# Specimen entry

**Name of Colliery**
NS 5275 2175 (eight-figure grid reference) (NS52SW/34) (NMRS site number)
**Parish:** e.g. Auchinleck
**Region/District:** (see list of abbreviations)
**Council:** current Local Authority
**Location:** Auchinleck (the nearest town or village)
**Previous owners:** Companies or individuals who owned the colliery before nationalisation
**Types of coal:** Includes Anthracite, Coking, Gas, House Manufacturing and Steam, as well as bi-products such as fireclay
**Sinking/Production commenced:** Known dates when the colliery or individual shafts or mines were sunk or driven
**Year closed:** The date production ceased
**Year abandoned:** The date the workings were officially abandoned, often a significant time after closure
**Average workforce:** The mean number of surface and underground workers at a colliery during its lifetime after 1947
**Peak workforce:** The largest number of surface and underground workers at a colliery during its lifetime after 1947
**Peak year:** The year after 1947 when the workforce was at its greatest
**Shaft/Mine details:** Any information, if available, on the shafts or surface mines that made up the colliery
**Details in 1948:** Information specifically relating to the questionnaire entry filled in by the colliery manager in the late 1940s after nationalisation. This information does not exist for collieries sunk after 1947
**Other details:** Any other relevant information

# Gazetteer contents

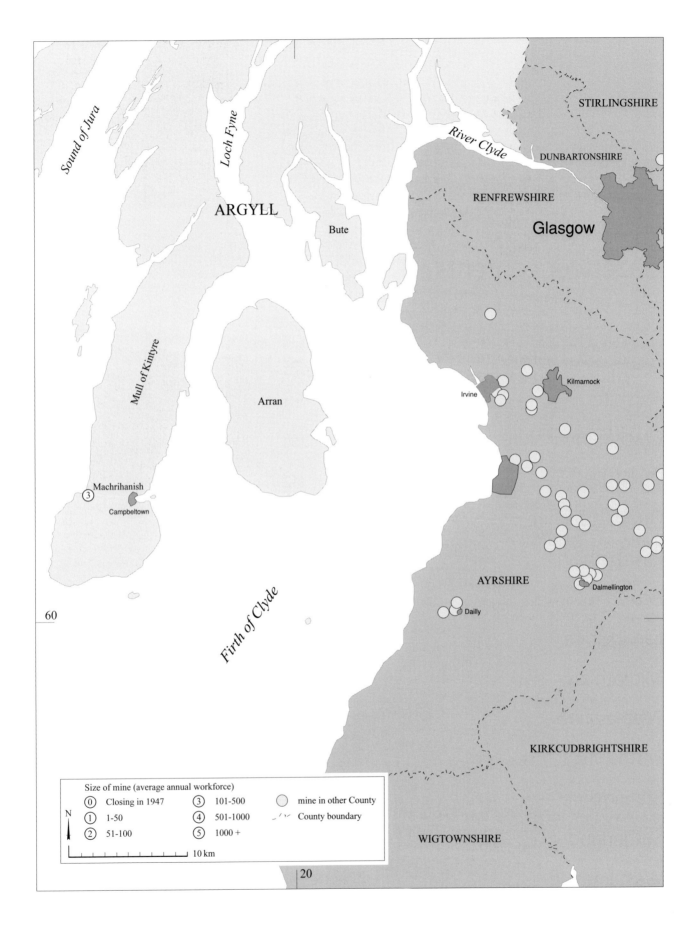

STIRLINGSHIRE

DUNBARTONSHIRE

*River Clyde*

RENFREWSHIRE

Glasgow

*Sound of Jura*

*Loch Fyne*

ARGYLL

Bute

*Mull of Kintyre*

Arran

Kilmarnock

Irvine

Machrihanish

Campbeltown

AYRSHIRE

Dalmellington

Dailly

*Firth of Clyde*

60

KIRKCUDBRIGHTSHIRE

Size of mine (average annual workforce)

| ⓪ | Closing in 1947 | ③ | 101-500 | ◯ mine in other County |
| ① | 1-50 | ④ | 501-1000 | County boundary |
| ② | 51-100 | ⑤ | 1000 + | |

N

10 km

WIGTOWNSHIRE

20

# Argyll

Argyll is the smallest of the Carboniferous coal fields in central Scotland, and is in effect an extension of the Ayrshire coalfield, which re-emerges on the west side of the Firth of Clyde at Machrihanish on the Mull of Kintyre. One colliery comprising two surface drift mines was sunk by the Glasgow Iron and Steel Company in 1946, but coal mining had taken place on the Mull of Kintyre since at least the 18th century.

In the nationalised era, Argyll Colliery employed on average 168 people, peaking briefly at 280 during 1959. It produced steam coal, most of which was exported to Northern Ireland for electricity generation, but shipments also went further afield to customers in Rotterdam and Copenhagen in 1952, for example. However, in addition to domestic local users, the distilleries of Campbeltown (of which there were over 30 at one time) provided a significant market, as did the distilleries on Islay and Jura. Annual output appears to have peaked at almost 100,000 tons in 1957, but was sometimes less than half that due to adverse mining conditions. The mine was prone to spontaneous combustion, which caused serious problems in 1958, and worsening mining conditions, combined with contracting markets for coal, prompted the NCB to close it on 26th March 1967, bringing to an end all deep coal mining in Argyll.

## Argyll (Machrihanish)

NR 6505 2069 (NR62SE/29)

**Parish:** Campbeltown

**Region/District:** St/Ar

**Council:** Argyll and Bute

**Location:** Machrihanish, nr Campbeltown

**Previous owners:** Glasgow Iron and Steel Company Limited

**Types of coal:** Steam

**Sinking/Production commenced:** 1946, production in 1950

**Year closed:** 1967

**Average workforce:** 168

**Peak workforce:** 280

**Peak year:** 1959

**Shaft/Mine details:** 2 surface mines. Main mine struck coal at 365m (1 in 4). Upcast mine at 177m, inclination 1 in 2.5.

**Details in 1949:** Output 60 tons per day, 15,000 tons per annum. 125 employees. Temporary cart screen. No washer. No baths. Small 'packed-lunch' canteen. No medical services. All electricity bought from North of Scotland Hydro-Electric Board. Report dated 15-06-1949.

**Other details:** Pithead baths were opened on 15th January 1953. Supplied power stations in Northern Ireland, production peaking at 500 tons per day. Coal was taken to the harbour by a narrow-gauge railway. Underground fires caused a two-month stoppage on 18th September 1958, and again in 1960.

**Figures 5.2 and 5.3**

Figure 5.2: Argyll Colliery, Machrihanish. View during the development of the new mine showing the setting of the foundations for the yard and surface buildings, c.1947. The new mine commenced production in 1950. SMM:1998.235, SC379682

Figure 5.3: Argyll Colliery, Machrihanish. View of the surface arrangement, with the portal of the main drift mine entrance in the background, c.1947. The mine produced coal from 1950 until its closure in 1967. SMM:1998.235, SC379680

Figure 5.1 left: Map of National Coal Board (NCB) collieries in Argyll

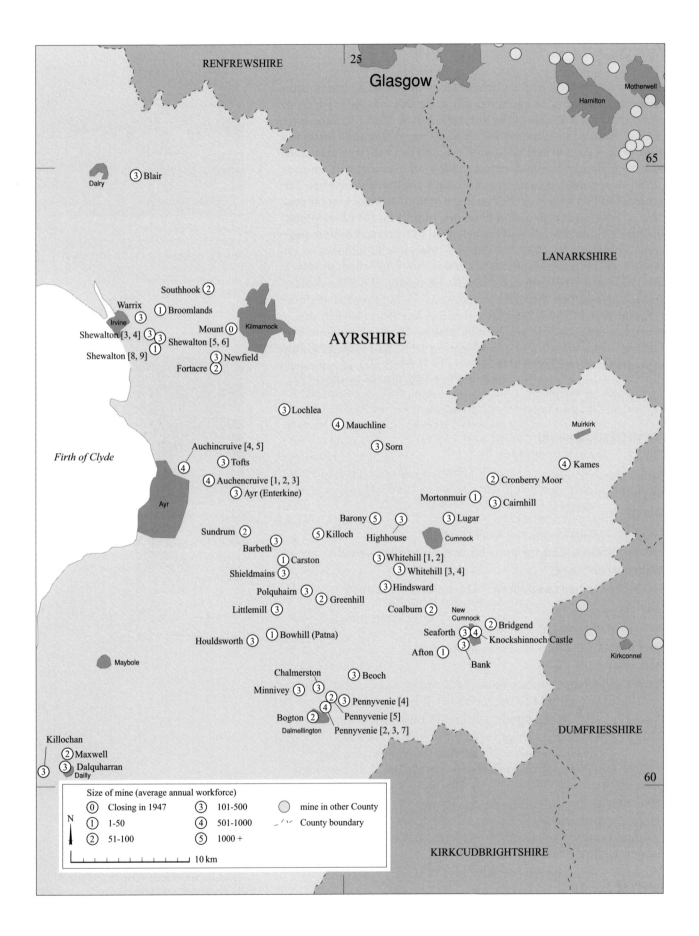

RENFREWSHIRE

25

Glasgow

Hamilton

Motherwell

65

LANARKSHIRE

Dalry

(3) Blair

AYRSHIRE

Southhook (2)

Warrix (1) Broomlands

(3)

Irvine

Mount (0)

Kilmarnock

Shewalton [3, 4] (3)(3)

Shewalton [5, 6]

Shewalton [8, 9] (1)

(3) Newfield

Fortacre (2)

(3) Lochlea

(4) Mauchline

Muirkirk

(3) Sorn

Auchincruive [4, 5]

Firth of Clyde

(3) Tofts

(4) Kames

(4)

(4) Auchencruive [1, 2, 3]

(2) Cronberry Moor

(3) Ayr (Enterkine)

Mortonmuir (1)

(3) Cairnhill

Ayr

Barony (5) (3)

(3) Lugar

Cumnock

Sundrum (2)

(5) Killoch

Barbeth (3)

Highhouse

(1) Carston

(3) Whitehill [1, 2]

Shieldmains (3)

(3) Whitehill [3, 4]

Polquhairn (3)(2) Greenhill

(3) Hindsward

Littlemill (3)

Coalburn (2)

New Cumnock

(2) Bridgend

Houldsworth (3) (1) Bowhill (Patna)

Seaforth (3)(4)

Knockshinnoch Castle

Afton (1)

(3)

Kirkconnel

Maybole

Bank

Chalmerston

(3) Beoch

Minnivey (3) (3)

(2)(3) Pennyvenie [4]

Bogton (2)

(4) Pennyvenie [5]

Dalmellington

Pennyvenie [2, 3, 7]

DUMFRIESSHIRE

Killochan

(2) Maxwell

(3)(3) Dalquharran

Dailly

60

Size of mine (average annual workforce)

(0) Closing in 1947  (3) 101-500  ○ mine in other County

N

(1) 1-50  (4) 501-1000  County boundary

(2) 51-100  (5) 1000 +

10 km

KIRKCUDBRIGHTSHIRE

# Ayrshire

Ayrshire was and remains one of the most important coal-producing counties in Scotland, being exceeded in scale only by Lanarkshire and Fife. Situated to the south-west of Glasgow, it was close to Scotland's industrial heartland, and was also able to supply Northern Ireland's coal-burning power stations. In 1947, 47 mines were taken into State ownership by the vesting process, and the National Coal Board subsequently established 19 new mines of varying size and longevity. The county was also home to two central workshops, one at Lugar and the other at Glenburn (Auchincruive). Much of the production was made up of house, manufacturing and steam coals derived from both the upper and lower groups of Carboniferous coals. Significant quantities of coal suitable for gasworks were also mined in the Dalmellington area.

Three concentrations of mining activity stand out, the largest of which is the Central Ayrshire Coalfield, stretching from Auchincruive near Ayr in the west to Muirkirk in the east, New Cumnock in the south-east, and Dalmellington in the south. This was much the most important area in the post-1947 period, containing all but 12 of Ayrshire's nationalised deep mines. It had, in addition, once been a significant centre of the Scottish iron industry, important 18th- and 19th-century iron works having developed at Muirkirk, Lugar and Dalmellington. A small pocket of mining activity also existed in the south-west of the county around Dailly, where the mines at Dalquharran, Maxwell and Killochan were successfully exploited by the NCB until the late 1970s.

In the north of the county, coal had been mined by the monks of Kilwinning Abbey in the 16th century, and coastal outcrops were successfully exploited for the manufacture of salt, notably in the Saltcoats area. Most famous were the Cunninghames of Auchenharvie, at whose Stevenston Colliery near Irvine the first Newcomen engines in Scotland were built in the 1720s, the ruins of the first engine house surviving in the middle of the Auchenharvie golf course. Later exploitation of coal and black band ores resulted in pockets of mining activity in the Dalry and Kilmarnock areas, fuelling iron and heavy ceramics industries. However, in 1947, the few surviving mines dated from no earlier than the 1920s, and were augmented by the sinking of a number of short-term drifts to assist post-war production. Most of these were relatively small-scale operations, closing in the 1950s, but Blair remained in production until 1969.

Much of the impetus behind the development of the Ayrshire coalfields had been the coincidence of coal and ironstone, and this is reflected in the dominant position of Bairds & Dalmellington prior to nationalisation. The company had been born as a result of the merging of the Dalmellington Iron Company and William Baird & Company in 1931 after the closure of all their iron smelting interests in Ayrshire. It owned over 20 Ayrshire pits in 1947, all of which were situated in the central Ayrshire coalfield. The second largest group of mines was that concentrated in the east of the county, where New Cumnock Collieries operated six pits. Other smaller companies included the Polquhairn Coal Company, A G Moore & Company, and in the north, A Kenneth & Sons, whose pits and brickworks were concentrated between Irvine and Kilmarnock.

Figure 5.4 left: Map of National Coal Board (NCB) collieries in Ayrshire

After nationalisation, Ayrshire benefited from substantial investment from the NCB. As was the case elsewhere in Scotland, this took the form of several new relatively small drift or 'surface' mines designed to satisfy demand in the short-term, whilst unproductive mines were closed. Major investment in colliery reconstructions such as Barony, Kames and Littlemill, and a large new sinking at Killoch, sought to establish much longer-term production. This resulted in a steady decline in the number of NCB mines from 40 in 1947 to 20 by 1964. At the same time, the number of miners employed in these pits rose from 11,000 to a peak of over 12,000 in 1959 before dropping below 10,000 by 1963. This would have been expected as higher output per man was anticipated from the newly reconstructed mines and from the big new superpit at Killoch. However, decline accelerated in the 1970s, with the number of miners falling to under 4,000 and the number of pits to only five. The death of deep mining finally occurred with the closure of Killoch in 1987 and Barony in 1989. However, substantial coal production continues in Ayrshire at a number of open-cast sites, much of the output being washed at the Killoch coal-preparation plant.

Inevitably, perhaps, the largest mine to operate in Ayrshire was Killoch Colliery, which peaked in 1965 with 2,305 employees, and was in the same year the first Scottish pit to produce over one million tons. Killoch was one of the most successful of the NCB's new Scottish sinkings, but began to experience geological difficulties in the 1970s, resulting in a merger with neighbouring Barony Colliery. Barony was easily the second largest mine in Ayrshire and had been the successful outcome of a major reconstruction, despite a shaft collapse in 1962, producing coal for another 27 years. The only other mine to employ more than 1,000 miners was Auchincruive 1, 2 & 3 (Mossblown), which closed in 1960. Other large collieries employing on average more than 500 people included Mauchline, Glenburn (Auchincruive 4 & 5), Pennyvenie, Knockshinnoch Castle, and Kames. Of these, only Pennyvenie and Glenburn survived until the 1970s, alongside smaller but significant mines such as Highhouse and Littlemill. It is also interesting to note that a number of the NCB's surface-mine complexes such as Minnivey and Lochlea survived into the 1970s, as did those in the south-west at Maxwell and Dalquharran. In addition, one of the most successful drift mines was that at Sorn, which operated with an average workforce of 233 from 1953 to 1983.

Outside the public sector, at least 22 small private mines also operated in Ayrshire between 1947 and 2002, some having been worked since 1873. Not all were entirely dedicated to coal, several in the north also being used to extract fireclays and clays for the once thriving sanitary ware, refractory and brick industries. Private or 'licensed' deep mining continued for only two years after the closure of the last nationalised mine, Auchlin, Broomhill, Craigman, Hall of Auchincross, Smithston and Viaduct Mines all ceasing to renew their licences in 1991. Thereafter, with the exception of the coal preparation facility at Killoch, the only surviving monuments to the era of deep coal mining in Ayrshire are the 'A' frame headgear of Barony Colliery's No. 3 shaft, and one of the headframes and its winding engine house at Highhouse Colliery nearby in Auchinleck.

## Afton No.1 Colliery

NS 5835 1081 (NS51SE/20)

**Parish:** New Cumnock

**Region/District:** St/CD

**Council:** East Ayrshire

**Location:** New Cumnock

**Previous owners:** Lanemark Coal Company, then New Cumnock Collieries Limited

**Types of coal:** Manufacturing and Cannel

**Sinking/Production commenced:** 1871–3

**Year closed:** 1948 (production ceased)

**Average workforce:** 23

**Peak workforce:** 26

**Peak year:** 1948

**Shaft/Mine details:** 1 shaft, 168m deep

**Details in 1948:** Output 30 tons per day, 10,249 per annum. 24 employees. No baths or canteen. Some steam power, all electricity from public supply. Report dated 09-08-1948.

**Other details:** Afton No. 2 closed in 1914. Afton No. 1 transferred to NCB in 1948. Site of an experimental oil plant using Cannel coal in the 1930s. Associated brickworks operated until 1978.

**Figure 5.5**

Figure 5.5: Afton Colliery, New Cumnock. Elevated view of surface arrangement, c.1947. Originally dating from the early 1870s, it survived only one year after nationalisation, closing in 1948. SMM:1996.3411, SC381511

## Auchincruive 1, 2 & 3 (Mossblown)

NS 3875 2484 (NS32SE/21)

**Parish:** Monkton and Prestwick

**Region/District:** St/Ky

**Council:** South Ayrshire

**Location:** Annbank Station

**Previous owners:** Geo. Taylor & Company, Wm. Baird & Company from 1916, Bairds & Dalmellington Limited from 1931

**Types of coal:** House and Steam

**Sinking/Production commenced:** 1897 (1 and 2), 1910 (3)

**Year closed:** 1960

**Average workforce:** 875

**Peak workforce:** 1,081

**Peak year:** 1947

**Shaft/Mine details:** 3 shafts, No. 1 165m, Nos. 2 and 3 both 210m

**Details in 1948:** Output 1,115 tons per day, 278,750 per annum, stoop and room and longwall working. 1,077 employees. 4 screens for dry coal, coal breaker and table. Baum-type washer (Blantyre Engineering Company). Baths and canteen available, but no medical services yet. Steam and electricity generated on site, none from public supply. Report dated 08-08-1948.

**Other details:** Shafts 1 & 2 also known as Mossblown. Baths built in 1937 with a capacity for 608 men.

**Figure 5.6**

Figure 5.6: Auchincruive 1, 2 & 3 Colliery, Annbank Station, also known as Mossblown. Oblique aerial view from south, taken in the 1950s. Dating from 1897, it was latterly one of Baird and Dalmellington's collieries, and after Killoch and Barony, the largest employer of miners in Ayrshire, closing in 1967. The pithead baths were built by the Miners' Welfare Fund in 1937. SMM:1996.0599, SC381513

Figure 5.7: Auchincruive 4, 5, 6 & 7 Colliery, Prestwick, also known as Glenburn. Oblique aerial view taken in the 1950s. One of Baird and Dalmellington's many collieries in Ayrshire, the pithead baths were built with the assistance of the Miners' Welfare Fund in 1934. SMM:1996.0600, SC381520

Figure 5.8: Auchincruive 4, 5, 6 & 7 Colliery, Prestwick. General view of surface arrangement, probably taken before nationalisation. Shafts 4 and 5 were sunk originally in 1911–2, and after 1947, it employed on average over 650 men, eventually closing in 1973. SMM:1998.0392, SC382473

## Auchincruive 4, 5, 6 & 7 (Glenburn)

NS 3655 2594 (NS32NE/38)

**Parish:** Monkton and Prestwick

**Region/District:** St/Ky

**Council:** South Ayrshire

**Location:** Prestwick

**Previous owners:** Geo. Taylor & Company, Wm. Baird & Company from 1916, Bairds & Dalmellington Limited from 1931

**Types of coal:** Steam

**Sinking/Production commenced:** 1911–2

**Year closed:** 1973

**Year abandoned:** 1974

**Average workforce:** 659

**Peak workforce:** 958

**Peak year:** 1957

**Shaft/Mine details:** 2 shafts, both 155m

**Details in 1948:** Output 925 tons per day, 231,250 tons per annum, stoop and room (under the urban areas of Prestwick) and longwall working (under the sea). 707 employees. 2 screens for dry coal, coal breaker and table. Baum-type washer (Norton). Canteen and baths (1934), medical services not yet available. Electricity generated on site, none from public supply. Report dated 08-08-1948.

**Other details:** 4 & 5 also known as 'Glenburn'. Area workshops were based here until 1965.

**Figures 5.7 and 5.8**

## Ayr 1 & 2

(see Enterkine 9 & 10)

## Bank 1

NS 6005 1157 (NS61SW/20)

**Parish:** New Cumnock

**Region/District:** St/CD

**Council:** East Ayrshire

**Location:** New Cumnock

**Previous owners:** New Cumnock Collieries Limited

**Types of coal:** House and Steam

**Sinking/Production commenced:** c.1850

**Year closed:** 1969

**Average workforce:** 332 (statistics for Bank 1, 2 & 6 combined)

**Peak workforce:** 400

**Peak year:** 1952

**Shaft/Mine details:** Bank 1 (shaft, 174m in 1948)

**Details in 1948:** Output 250 tons per day, 73,500 tons per annum. 213 employees. 2 dry coal screens, served by Knockshinnoch washer (Baum-type, by Simon Carves). Baths (1931), but no canteen (packed lunches again from Knockshinnoch Castle canteen). Some steam power, and electricity 100% from public supply. Report dated 09-08-1948.

**Other details:** Pithead baths with a capacity of 616 were built in 1931.

**Figure 5.9**

Figure 5.9: Bank colliery, New Cumnock. View of headgear and winding-engine house, c1955. Dating from the 1850s, and one of a group of pits owned by New Cumnock Collieries Limited, it closed in 1969. SMM:1998.639, SC393213

## Bank 2

NS 5961 1176 (NS51SE/23)

**Parish:** New Cumnock

**Region/District:** St/CD

**Council:** East Ayrshire

**Location:** New Cumnock

**Previous owners:** New Cumnock Collieries Limited

**Types of coal:** House and Steam

**Sinking/Production commenced:** 1946

**Year closed:** 1950

**Year abandoned:** 1959

**Statistics:** see Bank 1

**Shaft/Mine details:** surface mine, approximately 37m deep

**Details in 1948:** Output 150 tons per day, 44,100 tons per annum. 70 employees. Screening at Bank No. 1, and washing at Knockshinnoch Castle. Baths available, but packed lunches from Knockshinnoch Castle canteen. Some steam power, but electricity 100% from public supply. Report dated 09-08-1948.

## Bank 6

NS 5954 1150 (NS51SE/21)

**Parish:** New Cumnock

**Region/District:** St/CD

**Council:** East Ayrshire

**Location:** New Cumnock

**Previous owners:** New Cumnock Collieries Limited

**Types of coal:** House and Steam

**Sinking/Production commenced:** 1925

**Year closed:** 1969

**Year abandoned:** 1969

**Statistics:** see Bank 1

**Shaft/Mine details:** surface mine, to approximately 55m

**Details in 1948:** Output 150 tons per day, 44,100 tons per annum. 75 employees. Screening at Bank No. 1 and washing at Knockshinnoch Castle. Packed lunches from Knockshinnoch Castle canteen. Some steam power, 100% electricity from public supply. Report dated 09-08-1948.

**Other details:** Famous because its old disused workings were used as the route through which miners trapped in the Knockshinnoch Castle disaster of 1950 were rescued. Coal and blaes used to fuel associated brickworks until 1960s.

**Barbeth**

NS 4440 2000 (NS42SW/20)

**Parish:** Ochiltree

**Region/District:** St/CD

**Council:** East Ayrshire

**Location:** Drongan

**Previous owners:** A G Moore & Company

**Types of coal:** House

**Sinking commenced:** 1945

**Production commenced:** 1950

**Year closed:** 1955

**Year abandoned:** 1955

**Average workforce:** 137

**Peak workforce:** 149

**Peak year:** 1953

**Shaft/Mine details:** 2 surface mines

**Details in 1948:** Output 25 tons per day, not specified per annum, but said to be building up. 21 employees. Screening and washing at Shieldmains. No baths, and packed lunches only. 100% of electricity from public supply. Report dated 06-08-1948.

**Other details:** Also known as 'Shieldmains Barbeth 2'. Barbeth No. 9 closed in 1928.

**Barony 1, 2, 3 & 4**

NS 5275 2175 (NS52SW/34)

**Parish:** Auchinleck

**Region/District:** St/CD

**Council:** East Ayrshire

**Location:** Auchinleck

**Previous owners:** Bairds & Dalmellington Limited from 1931

**Types of coal:** House and Steam

**Sinking/Production commenced:** 1910 (1 & 2), 1945 (3), 1965 (4)

**Year closed:** 1989

**Average workforce:** 1,078

**Peak workforce:** 1,695

**Peak year:** 1958

**Shaft/Mine details:** 4 shafts, Nos.1 & 2 (NS 5276 2173) both 626m, No. 3 shaft (623m) added in reconstruction (1938–40 & 1945–50), and had 'A'-frame headgear (NS 5267 2187), the only part of the colliery to survive after closure in 1989, despite listed-building status. No. 4 (ventilation and emergency winding, 509m, NS 5264 2134) was sunk in 1965 after the collapse of Nos. 1 & 2 shafts in 1962, during which four men were killed.

**Details in 1948:** Output 1,520 tons per day, 380,000 tons per annum, longwall and stoop and room working. 1,264 employees. 5 screens for dry coal, Jig washer (Campbell Binnie and Reid). Canteen and Baths (built 1931, later replaced and converted into training centre).

Morphia administration scheme. Steam and electricity, all generated at mine in its own power station. Report dated 09-08-1948.

**Other details:** Barony Power Station opened by the SSEB on an adjacent site in 1953, burning slurry from the washery. It closed in 1982 and was subsequently demolished. The Coppee coal preparation plant was the first full-scale dense-medium plant in Scotland, and was opened in 1957. Linked underground to Highhouse. The last working pit in Ayrshire at closure in 1989.

**Figures 5.10 and 5.11**

## Beoch 3

NS 5099 0918 (NS50NW/5.02)

**Parish:** New Cumnock

**Region/District:** St/CD

**Council:** East Ayrshire

**Location:** Dalmellington

**Previous owners:** William Baird, Bairds & Dalmellington Limited from 1931

Figure 5.10: Barony Colliery, near Auchinleck. View taken in 1988 of No. 3 shaft headframe and one of the two winding-engine houses, built in the late 1940s as part of a major reconstruction project. The headframe is the only part of the colliery that survived after its closure in 1989. SC376909

Figure 5.11: Barony Colliery, near Auchinleck. View of No. 4 shaft and fan house, built in 1965 after the collapse of Nos. 1 and 2 shafts in 1962. The 'A'-shaped headframe of No. 3 shaft can be seen in the background. SC376926

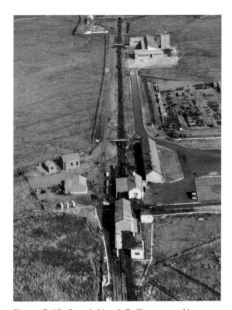

Figure 5.12: Beoch No. 4 Colliery, near New Cumnock. Aerial view, probably taken in the 1950s. A drift mine development by Bairds and Dalmellington and dating from 1937, it closed in 1968. SMM:1996.0596, SC381614

**Types of coal:** Coking, House, Manufacturing and Steam
**Sinking/Production commenced:** 1866
**Year closed:** 1968
**Year abandoned:** 1968
**Average workforce:** 408
**Peak workforce:** 460
**Peak year:** 1960
**Shaft/Mine details:** surface mine
**Details in 1948:** Output 170 tons per day, 42,500 per annum. 112 employees. Screening at Pennyvenie, washing at Dunaskin. No baths (available at Pennyvenie), and pit snacks provided. Electricity on site, but none from public supply. Report dated 07-08-1948.
**Other details:** The highest coal mine in Scotland, consequently suffering from severe winter weather, snow occasionally creating access problems.

## Benbain No. 5
(see Beoch 4)

## Beoch 4 (Benbain No. 5)
NS 5112 0917 (NS50NW/5.01)
**Parish:** New Cumnock
**Region/District:** St/CD
**Council:** East Ayrshire
**Location:** Dalmellington
**Previous owners:** Bairds & Dalmellington Limited
**Types of coal:** Coking, House, Manufacturing and Steam
**Sinking/Production commenced:** 1937
**Year closed:** 1968
**Year abandoned:** 1968
**Statistics:** see Beoch 3
**Shaft/Mine details:** surface mine
**Details in 1948:** Output 330 tons per day, 90,000 tons per annum. 169 employees. Screening at Pennyvenie and washing at Dunaskin. No baths (available at Pennyvenie), and only pit snacks available. Electricity, but none from public supply. Report dated 07-08-1948.
**Figure 5.12**

## Blair 11 & 12
NS 3257 4947 (NS 34NW/20)
**Parish:** Dalry
**Region/District:** St/CU
**Council:** North Ayrshire
**Previous owners:** National Coal Board
**Sinking commenced:** 1953
**Production commenced:** 1954

**Year closed:** 1969
**Average workforce:** 157
**Peak workforce:** 201
**Peak year:** 1960
**Other details:** no other information
**Figure 5.13**

Figure 5.13: Blair Colliery, near Dalry, c.1960. One of only a small number of mines to operate in North Ayrshire in the nationalised era, it was a small drift mine sunk by the NCB in 1954, employing on average 150 miners. It closed in 1969. SMM:1998.657, SC393059

### Bogton Mine

NS 4753 0575 (NS40NE/38)
**Parish:** Dalmellington
**Region/District:** St/CD
**Council:** East Ayrshire
**Location:** Dalmellington
**Previous owners:** Bairds & Dalmellington Limited
**Types of coal:** Coking, House and Manufacturing
**Sinking/Production commenced:** 1931
**Year closed:** 1954
**Year abandoned:** 1954
**Average workforce:** 76
**Peak workforce:** 99
**Peak year:** 1947
**Shaft/Mine details:** surface mine
**Details in 1948:** Output 75 tons per day, 20,000 tons per annum. 92 employees. Screening at Pennyvenie, washing at Dunaskin. No baths, only pit snacks available. Electricity, but no public supply. Report dated 07-08-1948.

### Bowhill (Patna)

NS 4402 1238 (NS41SW/7)
**Parish:** Dalrymple
**Region/District:** St/CD
**Council:** East Ayrshire
**Location:** Patna
**Previous owners:** Dalmellington Iron Company Limited, Bairds & Dalmellington Limited from 1931
**Types of coal:** Anthracite
**Sinking/Production commenced:** unknown
**Year closed:** 1948
**Average workforce:** 37
**Peak workforce:** 37
**Peak year:** 1947
**Other details:** Described as an ironstone pit on 3rd Edition Ordnance Survey maps

Figure 5.14: Bridgend Colliery, New Cumnock, c.1960. Originally licensed to the Nith Coal Company Limited, but then replaced by a new NCB drift mine in 1949, producing coal with an average workforce of 90 miners until closure in 1964. SMM:1998.662, SC393061

**Bridgend**
NS 6234 1323 (NS61SW/21)
**Parish:** New Cumnock
**Region/District:** St/CD
**Council:** East Ayrshire
**Location:** New Cumnock
**Previous owners:** National Coal Board
**Sinking/Production commenced:** 1949
**Year closed:** 1964
**Average workforce:** 89
**Peak workforce:** 146
**Peak year:** 1961
**Other details:** Originally licensed to the Nith Coal Company Limited, but then replaced by an NCB drift mine in 1949
**Figure 5.14**

**Broomlands**
NS 3468 3872 (NS33NW/58)
**Parish:** Irvine
**Region/District:** St/CU
**Council:** North Ayrshire
**Location:** Dreghorn
**Previous owners:** National Coal Board
**Types of coal:** House
**Sinking/Production commenced:** 1950
**Year closed:** 1952
**Year abandoned:** 1953
**Average workforce:** 49
**Peak workforce:** 59
**Peak year:** 1952

**Cairnhill 1 & 2**
NS 6265 2304 (NS62SW/29)
**Parish:** Auchinleck
**Region/District:** St/CD
**Council:** East Ayrshire
**Previous owners:** National Coal Board
**Sinking commenced:** 1952
**Production commenced:** 1957
**Year closed:** 1976
**Year abandoned:** 1977
**Average workforce:** 221
**Peak workforce:** 250
**Peak year:** 1970
**Figure 5.15**

## Carston

NS 4500 1829 (NS41NE/17)

**Parish:** Ochiltree

**Region/District:** St/CD

**Council:** East Ayrshire

**Location:** Drongan

**Previous owners:** National Coal Board

**Types of coal:** House

**Sinking/Production commenced:** 1950

**Year closed:** 1956

**Year abandoned:** 1957

**Average workforce:** 39

**Peak workforce:** 44

**Peak year:** 1952

## Chalmerston 4 & 5

NS 4799 0812 (NS40NE/22.01)

**Parish:** Dalmellington

**Region/District:** St/CD

**Council:** East Ayrshire

**Location:** Dalmellington

**Previous owners:** Dalmellington Iron Company Limited, Bairds & Dalmellington Limited from 1931

**Types of coal:** House and Steam

**Sinking/Production commenced:** 1925

**Year closed:** 1959

**Year abandoned:** 1960

**Average workforce:** 256

**Peak workforce:** 300

**Peak year:** 1951

**Shaft/Mine details:** surface mine

**Details in 1948:** Output 180 tons per day, 49,000 tons per annum. 128 employees. 4 screens for dry coal. Washing at Dunaskin. Baths, and small canteen for tea and 'aerated waters', pit snacks available. Electricity, but no public supply. Report dated 07-08-1948.

**Figure 5.16**

## Chalmerston 7

NS 4808 0815 (NS40NE/22.02)

**Parish:** Dalmellington

**Region/District:** St/CD

**Council:** East Ayrshire

**Location:** Dalmellington

**Previous owners:** Bairds & Dalmellington Limited

**Types of coal:** House and Steam

Figure 5.15: Cairnhill Colliery, High Gaswater near Auchinleck. View of the surface in 1956, prior to the commencement of production in 1957. The mine was worked by an average workforce of 220 miners until closure in 1976. SC381623

Figure 5.16: Chalmerston Colliery, Dalmellington. A Dalmellington Iron Company colliery, sunk in 1925 and closed in the late 1950s. This view shows the new pithead baths, probably in the late 1930s. The National Archives, COAL80/ 234/4, SC614089

**Sinking/Production commenced:** 1934
**Year closed:** 1952
**Statistics:** see Chalmerston 4 and 5
**Shaft/Mine details:** surface mine
**Details in 1948:** Output 160 tons per day, 44,000 tons per annum. 135 employees. Screening at Chalmerston 4/5, washing at Dunaskin. Baths, and small canteen for tea and 'aerated water', with pit snacks available. Electricity, but no public supply. Report dated 07-08-1948.

### Coalburn

NS 5740 1440 (NS51SE/22)
**Parish:** New Cumnock
**Region/District:** St/CD
**Council:** East Ayrshire
**Location:** New Cumnock
**Previous owners:** William Nicol Limited
**Types of coal:** House
**Sinking/Production commenced:** 1924
**Year closed:** 1962
**Average workforce:** 72
**Peak workforce:** 109
**Peak year:** 1952
**Shaft/Mine details:** surface mine, to approximately 37m
**Details in 1948:** Output 52 tons per week, 16,288 tons per annum. 35 employees. One screen for dry coal, no coal washing. No baths, no canteen. Electricity 100% public supply. Report dated 09-08-1948.

### Cronberry Moor

NS 6247 2499 (NS62SW/30)
**Parish:** Auchinleck
**Region/District:** St/CD
**Council:** East Ayrshire
**Location:** Cronberry Moor
**Previous owners:** Bairds & Dalmellington Limited
**Types of coal:** Manufacturing and Steam
**Sinking/Production commenced:** *c.*1920
**Year closed:** 1957
**Year abandoned:** 1957
**Average workforce:** 163
**Peak workforce:** 222
**Peak year:** 1947
**Shaft/Mine details:** 2 shafts, each 82m deep
**Details in 1948:** Output 240 tons per day approximately, 60,000 tons per annum, longwall and stoop and room working. 219 employees. 2 screens for dry coal. Baum-type washer (Blantyre Engineering Company). No baths or canteen, but morphia administration

scheme. Steam and electricity, none from public supply. Report dated 09-08-1948.

## Dalquharran 1 & 2

NS 2665 0182 (NS20SE/32)

**Parish:** Dailly

**Region/District:** St/Ky

**Council:** South Ayrshire

**Location:** Dalquharran, in the Girvan Valley

**Previous owners:** National Coal Board

**Types of coal:** Steam

**Sinking/Production commenced:** 1951

**Year closed:** 1977

**Year abandoned:** 1977

**Average workforce:** 144

**Peak workforce:** 237

**Peak year:** 1958

**Other details:** Coal washed at Killochan. After closure and the cessation of pumping in 1977, water drained from the entire basin into Dalquharran, causing highly polluted water to overflow from the mine into the River Girvan. The problem was solved by the NCB in a unique award-winning scheme involving the tapping of purer water in the strata, reducing the water draining into the mine, and greatly reducing the flow of polluted water into the river.

## Drongan

(see Shieldmains 6, 7 & 14)

## Dunaskin Central Washer

NS 4471 0808 (NS40NW/22/01)

**Parish:** Dalmellington

**Region/District:** St/CD

**Council:** East Ayrshire

**Location:** Waterside

**Previous owners:** Bairds & Dalmellington Limited

**Opened:** 1941

**Year closed:** 1988

**Other details:** A coal preparation plant, complete with slurry defloculation plant, built by the Coppee Company (Great Britain) Limited, described in *Iron and Coal Trades Review*, 25th April 1941. An associated brickworks was built nearby in the disused buildings of the Iron Works, operating until 1976, and subsequently preserved to form part of the Dunaskin Heritage Centre.

## Enterkine 9 & 10 (Ayr 1 & 2)

NS 4100 2391 (NS42SW/21)

Figure 5.17: Enterkine 9 & 10 Colliery, Annbank, also known as Ayr 1& 2. Dating originally from 1878, latterly one of Baird and Dalmellington's pits, it had a workforce of over 400 miners, and operated until 1959. SMM:1997.452, SC445861

**Parish:** Tarbolton
**Region/District:** St/Ky
**Council:** South Ayrshire
**Location:** Annbank
**Previous owners:** Geo. Taylor & Company, Wm. Baird & Company from 1916, Bairds & Dalmellington Limited from 1931
**Types of coal:** Manufacturing and fireclay
**Sinking/Production commenced:** 1878 (9 and10), 1952
**Year closed:** 1959
**Year abandoned:** 1960
**Average workforce:** 434
**Peak workforce:** 665
**Peak year:** 1948
**Shaft/Mine details:** 2 shafts, Nos. 9 & 10, both 219m deep. Drumley Pit used for ventilation upcast and escape.
**Details in 1948:** Output 695 tons per day, 172,750 tons per annum, stoop and room and longwall working. 677 employees. 3 screens for dry coal, Baum-type washer (Blantyre Engineering Company). Baths, canteen (pieces only), no medical services. Steam and electricity, none from public supply. Report dated 08-08-1948.
**Other details:** Also known as AYR 1 & 2. Supplied Annbank Brickworks (located nearby) with fireclay.
**Figure 5.17**

### Fortacre

NS 3925 3405 (NS33SE/33)
**Parish:** Dundonald
**Region/District:** St/Ky
**Council:** South Ayrshire
**Location:** Symington
**Previous owners:** Robert Semple & Company
**Types of coal:** Steam and House
**Sinking/Production commenced:** 1924
**Year closed:** 1957
**Year abandoned:** 1957
**Average workforce:** 88
**Peak workforce:** 109
**Peak year:** 1952
**Shaft/Mine details:** 2 shafts, 43m deep
**Details in 1949:** Output 65 tons per day, 17,250 tons per annum, stoop and room working. 43 employees. Bar screens. No washer, canteen or baths. Electricity from public supply. Report dated 22-02-1949.
**Other details:** Transferred to NCB in 1949.

## Glenburn

(see Auchincruive 4 & 5)

## Greenhill

NS 4822 1530 (NS41NE/09)

**Parish:** Ochiltree

**Region/District:** St/CD

**Council:** East Ayrshire

**Location:** Ochiltree

**Previous owners:** Polquhairn Coal Company Limited

**Types of coal:** House

**Sinking/Production commenced:** 1936

**Year closed:** 1958

**Average workforce:** 70

**Peak workforce:** 74

**Peak year:** 1954

**Shaft/Mine details:** 2 surface mines

**Details in 1948:** Output 80 tons per day, 21,000 tons per annum. Longwall advancing working. 69 employees. Coal sent to Polquhairn by aerial ropeway for screening. No washing. No baths. Canteen facilities available at Polquhairn. 100% electricity from public supply. Report dated 06-08-1948.

**Other details:** Amalgamated with Polquhairn *c.*1952

**Figure 5.18**

Figure 5.18: Greenhill Colliery, near Ochiltree. View of mouth of main mine, 1957. Sunk by the Polquhairn Coal company in 1936, its two surface mines were worked by an average of 70 miners until 1958. SC381629

## Highhouse

NS 5495 2173 (NS52SW/29)

**Parish:** Auchinleck

**Region/District:** St/CD

**Council:** East Ayrshire

**Location:** Cumnock

**Previous owners:** Bairds & Dalmellington Limited from 1931

**Types of coal:** House and Steam

**Sinking/Production commenced:** 1894

**Year closed:** 1983

**Average workforce:** 369

**Peak workforce:** 467

**Peak year:** 1947

**Shaft/Mine details:** 2 shafts, No. 1 177m, No. 2 174m deep. Latterly, linked underground with Barony.

**Details in 1948:** Output 420 tons per day, 185,000 tons per annum, longwall and stoop and room working. 473 employees. 2 screens for dry coal, washing done at Barony. Baths, canteen and morphia administration scheme. Steam and electricity, none from public supply. Report dated 09-08-1948.

Figure 5.19: Highhouse Colliery, Auchinleck. Dating from 1894, and latterly a Bairds and Dalmellington colliery, it was worked by an average of 370 miners until closure in 1983. One of its two headgear and its associated winding engine house have been preserved, and now form part of a small industrial estate. SMM:1996.2549, SC 381636

**Other details:** In *c.*1890, a beam engine built in Bridgeton, Glasgow for Dalry Collieries, was moved to Highhouse. It was subsequently transferred to Heriot Watt University in Edinburgh in the 1950s, and now can be seen at the Scottish Mining Museum at Lady Victoria Colliery, Newtongrange. One surviving winding engine (Grant, Ritchie & Company of Kilmarnock) and engine house dating from *c.*1896, and its associated headframe (dating from the 1960s when it replaced an earlier wooden structure) survive *in situ* within an industrial estate, and have been listed by Historic Scotland.
**Figure 5.19**

### Hindsward 3 & 4
NS 5366 1633 (NS51NW/08)
**Parish:** Old Cumnock
**Region/District:** St/CD
**Council:** East Ayrshire
**Location:** Skares
**Previous owners:** National Coal Board
**Sinking commenced:** 1954
**Production commenced:** 1956
**Year closed:** 1959
**Year abandoned:** 1959
**Average workforce:** 78
**Peak workforce:** 83
**Peak year:** 1957
**Other details:** Earlier colliery, Hindsward 1 & 2 operated by William Baird & Company Limited, closed in 1925. Re-opened as private mine in 1969 by Hamilton & Cairn, operating until 1988.

### Houldsworth
NS 4243 1189 (NS41SW/06)
**Parish:** Dalrymple
**Region/District:** St/CD
**Council:** East Ayrshire
**Location:** Patna/Polnessan
**Previous owners:** Dalmellington Iron Company Limited, Bairds & Dalmellington Limited from 1931
**Types of coal:** Manufacturing and Steam
**Sinking/Production commenced:** 1901–5
**Year closed:** 1965
**Average workforce:** 276
**Peak workforce:** 346
**Peak year:** 1954
**Shaft/Mine details:** 2 shafts, 373m and 379m deep
**Details in 1948:** Output 300 tons per day, 80,000 tons per annum. Longwall advancing working. 268 employees. 2 screens for dry coal, washing at Dunaskin. Baths, and canteen for packed meals. First-aid

room. Steam and electricity, none from public supply. Report dated 05-08-1948.

**Other details:** Named after the Houldsworth family from Manchester, the driving force behind the Doon Valley iron industry.

**Figure 5.20**

Figure 5.20: Houldsworth Colliery, Dalmellington. A Dalmellington Iron Company pit, sunk 1901–5, with an average workforce of 276 miners in the nationalised era, closing in 1965. Harrison Collection, SC446190

## Kames

NS 6848 2622 (NS62NE/42)

**Parish:** Muirkirk

**Region/District:** St/CD

**Council:** East Ayrshire

**Location:** Muirkirk

**Previous owners:** Bairds & Dalmellington Limited from 1931

**Types of coal:** Gas and Steam

**Sinking/Production commenced:** *c.*1870

**Year closed:** 1968

**Year abandoned:** 1968

**Average workforce:** 569

**Peak workforce:** 634

**Peak year:** 1957

**Shaft/Mine details:** 2 shafts, No. 1 249m, No. 2 251m

**Details in 1948:** Output 660 tons per day, 165,000 tons per annum, longwall and stoop and room working. 606 employees. 3 screens for dry coal, Baum-type washer (Blantyre Engineering Company). Baths (1933, for 540 men, with 40 shower cubicles), canteen, morphia administration scheme. Steam and electricity, all generated on site. Report dated 09-08-1948.

Figure 5.21: Kames Colliery, Muirkirk. View from the entrance road prior to reconstruction in the 1950s. The pit was originally thought to be relatively safe and gas-free until an underground explosion killed 17 men in 1957. J McKinnon Mining Collection, SC706207

**Other details:** Thought to be a gas-free pit until a gas/dust explosion killed 17 men in 1957. Reconstructed in 1950s, headframes and engine houses being replaced by tower-mounted Koepe winders, and major developments underground. No. 1 shaft was equipped with automatic skip winding, with a single 3.5 ton skip and counter balance. No. 2 shaft had two double-deck cages, accommodating 5 men or 1 hutch per deck. The electric winding equipment was supplied by the Swedish company, ASEA. Clean air was provided by 'forced ventilation', an unusual arrangement compared with most pits, where exhaust ventilation was the norm.

**Figure 5.21**

## Killoch

NS 4795 2055 (NS42SE/15)

**Parish:** Ochiltree

**Region/District:** St/CD

**Council:** East Ayrshire

**Location:** Ochiltree

**Previous owners:** National Coal Board

**Sinking commenced:** 1953

Figure 5.22: Killoch Colliery, near Ochiltree, 1962. One of the most successful new NCB superpits in Scotland, production commenced in 1960. Killoch was the first pit in Scotland to produce more than one million tons of coal in a year, and on average employed over 2,000 miners in its 27 years of operation. SC381683

**Production commenced:** 1960

**Year closed:** 1987

**Average workforce:** 2,014

**Peak workforce:** 2,305

**Peak year:** 1965

**Shaft/Mine details:** Both shafts were equipped with multi-rope friction electric winders, housed in the top of tall concrete towers. No. 1 shaft 7.4m diameter, 757m deep, upcast, coal winding in 2 x 30-ton skips using two four-rope friction winders. No. 2 shaft, downcast (linked to fan house) 6.15m diameter and 736m deep, winding men, materials and stone with one winding engine.

**Other details:** In 1965, Killoch became the first Scottish pit to produce 1 million tons of coal in a year, a large part of the output being exported to fuel Northern Ireland's power stations. In 1968, reserves were estimated at 100 million tons, but subsequent geological problems reduced productivity to half the Scottish average per man shift, prompting plans in 1972 to drive a link to neighbouring Barony. This was eventually achieved, with the bulk of the workforce being concentrated at Barony. However, closure followed in 1987, and the towers were subsequently demolished, although most of the office buildings were retained. The coal preparation plant was still operating in 2002, treating coal from neighbouring open-cast mines.

Killoch was one of NCB architect, Egon Riss's great schemes. It provided excellent surface facilities for its workforce, and was dominated by the two winding towers. Although impressive, design flaws in the large glazed areas of the towers rendered them vulnerable to wind and rain, causing serious maintenance problems and hastening their demise after closure.

**Figure 5.22**

### Killochan

NS 2474 0143 (NS20SW/34)

**Parish:** Dailly

**Region/District:** St/Ky

**Council:** South Ayrshire

**Location:** Dailly, in the Girvan Valley

**Previous owners:** Killochan Coal Company, South Ayrshire Collieries, and South Ayrshire Collieries (1928) Limited

**Types of coal:** Steam

**Sinking/Production commenced:** 1905

**Year closed:** 1967

**Year abandoned:** 1967

**Average workforce:** 247

**Peak workforce:** 289

**Peak year:** 1952

**Shaft/Mine details:** 2 shafts, both 205m deep

**Details in 1948:** Output 265 tons per day, 66,250 tons per annum, stoop and room and longwall working. 199 employees. 2 screens for dry coal, Luhrig-type washer (Dickson and Mann). No baths, canteen or medical facilities. Steam, and electricity 100% public supply. Report dated 08-08-1948.

**Other details:** The furthest south of the Girvan Valley pits, it received coal from Maxwell and Dalquharran pits to be washed. Pithead baths were built in 1955.

**Figure 5.23**

Figure 5.23: Killochan Colliery, Dailly. One of a small group of collieries in the Girvan Valley in South Ayrshire, it was opened in 1905, and in the nationalised era, operated with an average workforce of 250 miners until closure in 1967. SMM:1996.02351(a), SC381686

### Knockshinnoch Castle

NS 6097 1250 (NS61SW/17)

**Parish:** New Cumnock

**Region/District:** St/CD

**Council:** East Ayrshire

**Location:** New Cumnock

**Previous owners:** New Cumnock Collieries Limited

**Types of coal:** House and Steam

**Sinking/Production commenced:** 1940–44

**Year closed:** 1968

**Year abandoned:** 1969

**Average workforce:** 578

**Peak workforce:** 755

**Peak year:** 1956

**Shaft/Mine details:** 2 shafts, 187m and 128m deep

**Details in 1948:** Output 900 tons per day, 264,600 tons per annum, stoop and room working. 580 employees. 3 screens for dry coal. Baum (Simon Carves) type washer. No baths, but canteen available. Steam, electricity 100% from public supply. Report dated 09-08-1948. Pithead baths were built subsequently in 1949, and also served neighbouring pits.

Figure 5.24: Knockshinnoch Castle Colliery, New Cumnock, c.1950. Sunk by New Cumnock Collieries Limited in the early 1940s, it is famous for the disaster of 7th September 1950 when an inrush of water and mud trapped over 100 miners underground. All but 13 were rescued, the disaster being later dramatised in the film, 'The Brave Don't Cry'. The colliery continued to work for another 18 years, with an average workforce of over 570 men. SC446072

**Other details:** Best known for the disaster that occurred on 7th September 1950 when workings driven too close to the surface allowed a peat basin at the surface to burst into the mine, the ensuing inrush trapping 129 miners underground, of whom 13 subsequently died. Most of the trapped men were heroically rescued through workings connected to the neighbouring Bank No. 6 Colliery. The disaster was subsequently dramatised in 1952 with the release of the British film, 'The Brave Don't Cry', starring John Gregson, Alex Keir and Fulton Mackay.
**Figure 5.24**

### Littlemill 2,3 & 5

NS 4435 1428 (NS41SW/18)
**Parish:** Coylton
**Region/District:** St/CD
**Council:** East Ayrshire
**Location:** Rankiston
**Previous owners:** Originally the Coylton Coal Company, Bairds & Dalmellington Limited from 1937
**Types of coal:** House
**Sinking/Production commenced:** from 1860
**Year closed:** 1974
**Average workforce:** 495
**Peak workforce:** 810
**Peak year:** 1960
**Shaft/Mine details:** 3 shafts. No. 5 begun during re-organisation in 1952, 4.88m diameter, 325m deep, concrete-lined, new tandem-type steel headframe 23m high, used standard NCB 2.5-ton mine cars, lifted by new 805 hp electric winder with dynamic braking.
**Details in 1948:** Output 150 tons per day, 40,000 tons per annum. Longwall advancing working. 158 employees. 2 screens for dry coal, no washing. Baths, medical health centre and canteen under construction. Electricity, but no public supply. Report dated 06-08-1948.

Figure 5.25: Littlemill Colliery, Rankiston, 1958. Taken over from Bairds and Dalmellington and reconstructed by the NCB in the 1950s, it operated with an average workforce of almost 500 miners until closure in 1974. SC446187

**Other details:** Baths extended in the 1950s, with full-meal canteen added. The reconstruction of the 1950s was designed to increase production from 180 to 1,000 tons per day. Improvements included locomotive haulage underground, and a new coal preparation plant designed also to serve neighbouring Polquhairn. Compressed air was supplied by five air compressors at the surface, and two underground. From the 1950s, electricity was provided from public supply by the South of Scotland Electricity Board.

**Figure 5.25**

## Lochlea 1 & 2

NS 4512 3056 (NS43SE/17)
**Parish:** Tarbolton
**Region/District:** St/Ky
**Council:** South Ayrshire
**Previous owners:** National Coal Board
**Sinking/Production commenced:** 1949
**Year closed:** 1973
**Year abandoned:** 1974
**Average workforce:** 164
**Peak workforce:** 239
**Peak year:** 1957

## Lugar Mine

NS 5892 2163 (NS52SE/16.03)
**Parish:** Auchinleck
**Region/District:** St/CD
**Council:** East Ayrshire
**Location:** Lugar
**Previous owners:** Bairds & Dalmellington Limited
**Types of coal:** House and Steam
**Sinking/Production commenced:** 1942
**Year closed:** 1953
**Year abandoned:** 1954
**Average workforce:** 116
**Peak workforce:** 130
**Peak year:** 1949
**Shaft/Mine details:** 2 drifts, Main Mine 1 in 5 (NS 5892 2163), Return Mine 1 in 3 (NS 5883 2161).
**Details in 1948:** Output 80 tons per day, 20,000 tons per annum, longwall working. 101 employees. 1 screen for dry coal, washing at Cronberry Moor. No baths or canteen. Electricity, but none from public supply. Report dated 09-08-1948.

## Mauchline 1,2 & 4

NS 4969 2938 (NS42NE/25)

Figure 5.26: Mauchline Colliery. Aerial view of the surface arrangement and bing, probably taken in the 1950s. Sunk by Caprington and Auchlochan Collieries Limited in 1925, it was operated by Bairds and Dalmellington from 1935. In the NCB era, it operated with an average workforce of over 780 miners until closure in 1966. Harrison Collection, SC446158

**Parish:** Mauchline
**Region/District:** St/Ky
**Council:** East Ayrshire
**Location:** Mauchline
**Previous owners:** Caprington and Auchlochan Collieries Limited, Bairds & Dalmellington Limited from 1935
**Types of coal:** House
**Sinking/Production commenced:** 1925
**Year closed:** 1966
**Year abandoned:** 1969
**Average workforce:** 783
**Peak workforce:** 876
**Peak year:** 1958
**Shaft/Mine details:** 2 shafts, both 274m deep
**Details in 1948:** Output 1,000 tons per day, 250,000 tons per annum, longwall and stoop and room working. 820 employees. 5 screens for dry coal, Baum-type (Norton) washer. Baths (1939, for 756 men), packed-meal canteen. Steam and electricity, none from public supply. Report dated 09-08-1948.
**Figure 5.26**

Figure 5.27: Maxwell Colliery, Dailly. Sunk originally in 1905 by the Killochan Coal Co, and subsequently operated by South Ayrshire Collieries. It was a relatively small complex centred on surface mines employing on average 95 miners until closure in 1973. SMM:1996.01132, SC381796

## Maxwell 2

NS 2685 0285 (NS20SE/33)
**Parish:** Dailly
**Region/District:** St/Ky
**Council:** South Ayrshire
**Location:** Dailly, in the Girvan Valley
**Previous owners:** Killochan Coal Co, South Ayrshire Collieries, and South Ayrshire Collieries (1928) Limited
**Types of coal:** Steam
**Sinking/Production commenced:** 1903
**Year closed:** 1973
**Year abandoned:** 1977
**Average workforce:** 95
**Peak workforce:** 133
**Peak year:** 1958
**Shaft/Mine details:** surface mine
**Details in 1948:** Output 115 tons per day, 28,750 tons per annum, stoop and room working. 98 employees. 1 screen for dry coal, washing at Killochan. No baths, canteen or medical facilities. Electricity all from public supply. Report dated 08-08-1948.
**Other details:** Closure in 1973 was caused by heating problems in the strata.
**Figure 5.27**

**Maxwell 4**

NS 2734 0290 (NS20SE/34)

**Parish:** Dailly

**Region/District:** St/Ky

**Council:** South Ayrshire

**Location:** Dailly, in the Girvan Valley

**Previous owners:** National Coal Board

**Types of coal:** Steam

**Sinking/Production commenced:** 1950

**Year closed:** 1973

**Year abandoned:** 1977

**Statistics:** see Maxwell 2

**Other details:** Coal sent to Killochan to be screened.


**Minnivey 4 & 5**

NS 4635 0795 (NS40NE/28)

**Parish:** Dalmellington

**Region/District:** St/CD

**Council:** East Ayrshire

**Location:** Dalmellington, near Burnton

**Previous owners:** National Coal Board

**Sinking/Production commenced:** 1955

**Year closed:** 1975

**Average workforce:** 378

**Peak workforce:** 470

**Peak year:** 1962

**Other details:** Modern NCB complex comprising 2 surface mines, and was expected to work for 40 years, replacing nearby Chalmerston. The drifts were 1 in 2.5, extending down to 213m. The colliery became the site of the Scottish Industrial Railway Centre after closure.

**Figure 5.28**

Figure 5.28: Minnivey Colliery, near Dalmellington, c.1955. A modern complex of surface mines developed by the NCB in the mid-1950s, and worked by an average workforce of 378 miners up until closure in 1975. SC446173


**Montgomeryfield**

NS 3476 3755 (NS33NW/256)

**Details:** A central washery serving the adjacent collieries of A Kenneth & Sons (Newfield, Shewalton and Warrix), with an associated brickworks at NS 3485 3760 NS33NW/69, which closed c.1975.


**Mortonmuir 8 & 9**

NS 6104 2349 (NS62SW/16)

**Parish:** Auchinleck

**Region/District:** St/CD

**Council:** East Ayrshire

**Previous owners:** National Coal Board

**Types of coal:** no data

Figure 5.29: Mortonmuir Colliery, near Cronberry. View of drift mine under construction, c.1950. This NCB development never appears to have produced coal, and was abandoned in 1954. SC393074

**Sinking commenced:** 1947
**Production commenced:** 1951
**Year closed:** 1953
**Year abandoned:** 1954
**Average workforce:** 37
**Peak workforce:** 38
**Peak year:** 1952
**Shaft/Mine details:** 2 surface mines, currently being driven (1948)
**Details in 1948:** Output: being developed. 16 employees. 2 screens being constructed. No facilities. Electricity, none from public supply. Report dated 09-08-1948.
**Other details:** Appears never to have produced coal. Suffered from water problems.
**Figure 5.29**

## Mossblown

(see Auchincruive 1 & 2)

## Mount

NS 4063 3704 (NS43NW/57)
**Parish:** Kilmarnock
**Region/District:** St/KL
**Council:** East Ayrshire
**Location:** Kilmarnock
**Previous owners:** National Coal Board
**Types of coal:** no data
**Sinking commenced:** 1948
**Production commenced:** never produced
**Year closed:** 1950
**Year abandoned:** 1950
**Shaft/Mine details:** 2 surface mines currently being driven (1949) in vicinity of earlier colliery, the remains of which are visible on RAF vertical aerial photographs taken in November 1945.
**Details in 1949:** Output: being developed. 12 employees. Shaker screen and picking table. No washer, baths, canteen or medical facilities. Electricity from public supply. Report dated 22-02-1949.

## Newfield

NS 3930 3450 (NS33SE/34)
**Parish:** Dundonald
**Region/District:** St/Ky
**Council:** South Ayrshire
**Location:** Symington, by Kilmarnock
**Previous owners:** A Kenneth and Sons Limited
**Types of coal:** House
**Sinking/Production commenced:** 1940

**Year closed:** 1956

**Average workforce:** 126

**Peak workforce:** 144

**Peak year:** 1952

**Shaft/Mine details:** re-opened 1948, surface mine

**Details in 1948:** Output 175 tons per day, 43,750 tons per annum, stoop and room working. 135 employees. 1 screen for dry coal, washed at Montgomeryfield. No baths, canteen or medical services. Electricity 100% from public supply. Report dated 08-08-1948.

## Pennyvenie 2, 3 & 7

NS 4874 0690 (NS40NE/26)

**Parish:** Dalmellington

**Region/District:** St/CD

**Council:** East Ayrshire

**Location:** Dalmellington

**Previous owners:** Dalmellington Iron Company Limited, Bairds & Dalmellington Limited from 1931

**Types of coal:** House, Gas, Manufacturing and Steam

**Sinking/Production commenced:** 1872, No. 7 in 1945

**Year closed:** 1978

**Average workforce:** 581

**Peak workforce:** 725

**Peak year:** 1961

**Shaft/Mine details:** 3 shafts, 161m (No. 2) and 70m (No. 3), no details for No. 7.

**Details in 1948:** Output 450 tons per day, 124,000 tons per annum, longwall advancing working. 310 employees. 1 screen for dry coal, washing at Dunaskin. Baths (1934, for 600 men, including 54 shower cubicles), but no canteen, pit snacks being supplied as required from outside. Steam and electricity, none from public supply. Report dated 07-08-1948.

**Figure 5.30**

Figure 5.30: Pennyvenie 2, 3 & 7 Colliery, near Dalmellington. Sunk originally by the Dalmellington Iron Company in 1872, the baths were constructed in 1936, and No. 7 shaft added in 1945. In the NCB era, the colliery operated with an average workforce of 580 until closure in 1978. SMM:1996.0597, SC381800

## Pennyvenie 4

NS 5016 0717 (NS50NW/08)

**Parish:** Dalmellington

**Region/District:** St/CD

**Council:** East Ayrshire

**Location:** Dalmellington

**Previous owners:** Dalmellington Iron Company Limited, Bairds & Dalmellington Limited from 1931

**Types of coal:** House, Gas, Manufacturing, and Steam

**Sinking/Production commenced:** 1911

**Year closed:** 1961

**Average workforce:** 278

**Peak workforce:** 335

Figure 5.31: Pennyvenie No. 4 Colliery, near Dalmellington. A surface mine begun by the Dalmellington Iron Company in 1911, it was operated by the NCB with an average workforce of 280 miners until closure in 1961. SMM:1998.640, SC393115

**Peak year:** 1957

**Shaft/Mine details:** surface mine

**Details in 1948:** Output 300 tons per day, 80,000 tons per annum, longwall advancing. 235 employees. 4 screens for dry coal, washed at Dunaskin. No baths, but baths at Pennyvenie 2/3 available. No canteen, but pit snacks supplied as required. Electricity, but no public supply. Report dated 07-08-1948.

**Figure 5.31**

## Pennyvenie 5

NS 4883 0708 (NS40NE/107)

**Parish:** Dalmellington

**Region/District:** St/CD

**Council:** East Ayrshire

**Location:** Dalmellington

**Previous owners:** Dalmellington Iron Company Limited, Bairds & Dalmellington Limited from 1931

**Types of coal:** House, Gas, Manufacturing, and Steam

**Sinking/Production commenced:** 1911

**Year closed:** 1953

**Year abandoned:** 1955

**Average workforce:** 66

**Peak workforce:** 70

**Peak year:** 1947

**Shaft/Mine details:** surface mine, associated with Pennyvenie 2, 3 & 7

**Details in 1948:** Output 90 tons per day, 23,750 tons per annum, stoop and room working. 72 employees. Screening at Pennyvenie 2/3, washing at Dunaskin. Baths, no canteen, snacks provided when required. Steam and electricity, none from public supply. Report dated 07-08-1948.

## Polquhairn 1 & 4

NS 4699 1595 (NS41NE/11.00)

**Parish:** Ochiltree

**Region/District:** St/CD

**Council:** East Ayrshire

**Location:** Ochiltree

**Previous owners:** Polquhairn Coal Company Limited

**Types of coal:** House

**Sinking/Production commenced:** 1895

**Year closed:** 1962

**Average workforce:** 289

**Peak workforce:** 460

**Peak year:** 1960

**Shaft/Mine details:** 2 shafts, 77m and 40m deep

**Details in 1948:** Output 180 tons per day, 46,500 tons per annum, stoop and room working. 201 employees. 2 screens for dry coal, no washing. Baths and medical health centre under construction. Canteen for packed meals. Steam, and electricity 100% from public supply. Report dated 06-08-1948.

**Other details:** Handled and processed coal from Greenhill. Benefitted from reconstruction in the 1950s.

**Figure 5.32**

## Polquhairn 5 & 6

NS 4695 1595 (NS41NE/11.01)

**Parish:** Ochiltree

**Region/District:** St/CD

**Council:** East Ayrshire

**Location:** Ochiltree

**Previous owners:** National Coal Board

**Types of coal:** House

**Sinking/Production commenced:** 1955

**Year closed:** 1962

**Workforce data:** combined with Polquhairn 1 & 4

## Powharnal

NS 6453 2492 (NS62SW/31)

**Parish:** Auchinleck

**Region/District:** St/CD

**Council:** East Ayrshire

**Previous owners:** National Coal Board

**Sinking/Production commenced:** 1954

**Year abandoned:** 1959

Figure 5.32: Polquhairn 1 & 4 Colliery, near Ochiltree, c.1955. Sunk by the Polquhairn Coal Company in 1911, it was reconstructed by the NCB in the 1950s, the baths having been completed in 1949. After nationalisation, it operated with an average workforce of 289 miners until closure in 1962. J McKinnon Mining Collection. SC706171

### Seaforth 1, 2 & 3

NS 6024 1205 (NS61SW/22)
**Parish:** New Cumnock
**Region/District:** St/CD
**Council:** East Ayrshire
**Location:** New Cumnock
**Previous owners:** New Cumnock Collieries Limited
**Types of coal:** House and Manufacturing
**Sinking/Production commenced:** 1940
**Year closed:** 1953
**Year abandoned:** 1955
**Average workforce:** 163
**Peak workforce:** 241
**Peak year:** 1947
**Shaft/Mine details:** 2 surface mines (1 in 3 and 1 in 5)
**Details in 1948:** Output 200 tons per day, 58,000 tons per annum, longwall and stoop and room working. 125 employees. 1 screen for dry coal, washing at Knockshinnoch Castle. No baths or canteen. 100% electricity from public supply. Report dated 09-08-1948.

### Shewalton 3 & 4

NS 3378 3675 (NS33NW/54)
**Parish:** Dundonald
**Region/District:** St/CU
**Council:** North Ayrshire
**Location:** Drybridge, by Irvine
**Previous owners:** A Kenneth and Sons
**Types of coal:** House
**Sinking/Production commenced:** 1924
**Year closed:** 1955
**Year abandoned:** 1955
**Average workforce:** 179
**Peak workforce:** 266
**Peak year:** 1947
**Shaft/Mine details:** 2 shafts, both 132m deep
**Details in 1948:** Output 225 tons per day, 56,250 tons per annum, longwall working. 276 employees. Surface haulage to Montgomeryfield screening plant, washer also at Montgomeryfield. No baths, canteen or medical facilities. Steam, electricity all from public supply. Report dated 08-08-1948.

### Shewalton 5 & 6

NS 3459 3627 (NS33NW/55)
**Parish:** Dundonald
**Region/District:** St/CU
**Council:** North Ayrshire

**Location:** Drybridge, by Irvine
**Previous owners:** A Kenneth and Sons
**Types of coal:** House
**Sinking/Production commenced:** 1933
**Year closed:** 1950
**Year abandoned:** 1950
**Average workforce:** 153
**Peak workforce:** 205
**Peak year:** 1949
**Shaft/Mine details:** surface mines
**Details in 1948:** Output 165 tons per day, 40,250 tons per annum, longwall working. 116 employees. Surface haulage to Montgomeryfield screening plant and washer. No baths, canteen or medical facilities. Electricity all from public supply. Report dated 08-08-1948.

## Shewalton 8 & 9

NS 3459 3629 (NS33NW/55)
**Parish:** Dundonald
**Region/District:** St/CU
**Council:** North Ayrshire
**Location:** Drybridge, by Irvine
**Previous owners:** A Kenneth & Sons
**Types of coal:** House
**Sinking/Production commenced:** no data
**Year closed:** 1948
**Year abandoned:** 1948
**Average workforce:** 50
**Peak workforce:** 50
**Peak year:** 1947
**Shaft/Mine details:** surface mines

## Shieldmains 2

(see Barbeth)

## Shieldmains 6, 7 & 14 (Drongan)

NS 4505 1739 (NS41NE/13)
**Parish:** Ochiltree
**Region/District:** St/CD
**Council:** East Ayrshire
**Location:** Drongan
**Previous owners:** A G Moore & Company Limited
**Types of coal:** House
**Sinking/Production commenced:** 1927 (No. 14 in 1943)
**Year closed:** 1950

Figure 5.33: Sorn Colliery, one of the NCB's most successful surface mines. Production commenced in 1953, and continued with an average workforce of over 230 miners for 30 years, closing in 1983. SC446780

**Year abandoned:** 1950

**Average workforce:** 62

**Peak workforce:** 101

**Peak year:** 1947

**Shaft/Mine details:** all surface mines

**Details in 1948:** Output 200 tons per day, 50,000 tons per annum, stoop and room working. 127 employees. 1 screen for dry coal, Baum (Blantyre Engineering Company) washer. No baths, but central canteen for packed meals, and first-aid rooms. 100% electricity from public supply. Report dated 06-08-1948.

**Other details:** Mines 1 to 5 and 10 to 13 closed in 1928.

## Sorn 1 & 2

NS 5297 2765 (NS52NW/34)

**Parish:** Sorn

**Region/District:** St/CD

**Council:** East Ayrshire

**Previous owners:** National Coal Board

**Sinking commenced:** 1952

**Production commenced:** 1953

**Year closed:** 1983

**Year abandoned:** 1983

**Average workforce:** 233

**Peak workforce:** 294

**Peak year:** 1970

**Shaft/Mine details:** No. 1 Mine 662m long, 1 in 4, with direct rope haulage. No. 2 Mine 657m long

**Other details:** One of the NCB's most successful drift mines, lasting longer than its expected life of 25 years. Coal was dispatched by road and washed at Killoch.

**Figure 5.33**

## Southhook

NS 3860 3987 (NS33NW/257)

**Parish:** Dreghorn

**Region/District:** St/CU

**Council:** North Ayrshire

**Location:** near Dreghorn

**Previous owners:** Southhook Potteries Limited

**Types of coal:** House and Manufacturing, and fireclay

**Sinking/Production commenced:** pre-1947

**Details:** Statistics cease in1948

**Average workforce:** 91

**Other details:** Southhook was noted for its fireclay sanitary ware, the mine being associated with a large pottery nearby. The works operated until the 1960s, and it is likely that the mine continued to extract fireclay during this period. The brief appearance of the mine in this gazetteer therefore probably reflects a period when coal output prompted a possible takeover by the NCB. Immediately afterwards (1949 onwards), the mine reverted to 'Licensed Mine' status within the colliery year books.

Figure 5.34: Sundrum Colliery, c.1960. A good example of small short-term drift mine opened by the NCB in the 1950s, it closed in 1961 after eight years of production. SMM:1998.664, SC393117

## Sundrum

NS 4178 2078 (NS42SW/22)

**Parish:** Coylton

**Region/District:** St/Ky

**Council:** South Ayrshire

**Previous owners:** National Coal Board

**Sinking commenced:** 1952

**Production commenced:** 1953

**Year closed:** 1961

**Average workforce:** 65

**Peak workforce:** 77

**Peak year:** 1958

**Other details:** Earlier mine at Sundrum operated by the Dalmellington Iron Company (after 1931 Bairds & Dalmellington Limited), but closed in 1933

**Figure 5.34**

## Tofts 1 & 2

NS 4002 2640 (NS42NW/64)

**Parish:** Tarbolton

**Region/District:** St/Ky

**Council:** South Ayrshire

**Location:** Annbank, by Ayr

**Previous owners:** Bairds & Dalmellington Limited

**Types of coal:** House and Steam

**Sinking/Production commenced:** 1914

**Year closed:** 1948

**Year abandoned:** 1952

**Average workforce:** 152

**Peak workforce:** 289

**Peak year:** 1947

**Shaft/Mine details:** 2 shafts, both 341m

**Details in 1948:** Output: none. Life depends on results obtained by proving mine. 86 employees. 3 screens for dry coal, Jig-type washer (Campbell, Binnie and Reid). No baths, canteen or medical facilities. Steam and electricity, none from public supply. Report dated 08-08-1948.

**Warrix 1 & 2**
NS 3301 3808 (NS33NW/57)
**Parish:** Dundonald
**Region/District:** St/CU
**Council:** North Ayrshire
**Location:** Irvine
**Previous owners:** A Kenneth and Sons Limited
**Types of coal:** House
**Sinking/Production commenced:** 1944
**Year closed:** 1950
**Year abandoned:** 1950
**Average workforce:** 111
**Peak workforce:** 130
**Peak year:** 1948
**Shaft/Mine details:** 2 shafts, both 59m deep
**Details in 1948:** Output 130 tons per day, 32,500 tons per annum, longwall working. 110 employees. 1 shaker for dry coal, some coal sent to Montgomeryfield washer. No baths, canteen or medical facilities. All electricity from public supply. Report dated 08-08-1948.
**Other details:** Associated brickworks at NS 329 380, which operated from *c.*1907 probably until the 1950s

**Whitehill 1 & 2**
NS 5315 1822 (NS51NW/06)
**Parish:** Old Cumnock
**Region/District:** St/CD
**Council:** East Ayrshire
**Location:** Skares
**Previous owners:** Bairds & Dalmellington Limited

Figure 5.35: Whitehill 1 & 2 Colliery, Skares. A Bairds and Dalmellington pit sunk originally in the 1890s, and with fine pithead baths built with assistance from the Miners' Welfare Fund in 1934. After 1947, the colliery operated until 1965 with an average workforce of almost 400 miners. Harrison Collection, SC706227

**Types of coal:** House and Steam

**Sinking/Production commenced:** *c.*1893

**Year closed:** 1965

**Average workforce:** 392

**Peak workforce:** 435

**Peak year:** 1949

**Shaft/Mine details:** 2 shafts, both 245m deep

**Details in 1948:** Output 415 tons per day, 103,750 tons per annum, longwall and stoop and room working. 398 employees. 2 screens for dry coal, jig-type washer (Campbell Binnie and Reid). Baths (1934, for 462 men, with 43 shower cubicles), canteen, and morphia administration scheme. Steam and electricity, none from public supply. Report dated 09-08-1948.

**Figure 5.35**

## Whitehill 3 & 4

NS 5469 1770 (NS51NW/09)

**Parish:** Old Cumnock

**Region/District:** St/CD

**Council:** East Ayrshire

**Location:** Skares

**Previous owners:** Bairds & Dalmellington Limited

**Types of coal:** House Steam

**Sinking/Production commenced:** 1946

**Year closed:** 1965

**Average workforce:** 109

**Peak workforce:** 140

**Peak year:** 1957

**Shaft/Mine details:** 2 surface mines, main mine dipping 1 in 6, and return mine 1 in 3

**Details in 1948:** Output 100 tons per day, 25,000 per annum, longwall working. 55 employees. 1 bar screen, washing at Whitehill 1/2. Canteen and baths also at Whitehill 1/2. Electricity, 100% from public supply. Report dated 09-08-1948.

**Other details:** Skares Brickworks was built adjacent to the mine, and utilised blaes from its bing.

## Licensed Mines in Ayrshire 1947–97

| Name | Licensee | Grid Ref | Opened | Closed |
|---|---|---|---|---|
| Afton 1 (New Cumnock) | New Cumnock Collieries Limited | NS 583 108 | 1873 | 1948 |
| Alton (Galston) | James Bell | NS 499 387 | 1954 | 1956 |
| Auchlin (Skares) | J R McLellan | NS 828 377 | 1977 | 1991 |
| Bridgend (New Cumnock) | Nith Valley Coal Company Limited | NS 622 133 | pre-1947 | then NCB |
| Broomhill (Rankinston) | T Cook | NS 453 140 | pre-1988 | 1991 |
| Busbiehead 2b (Kilmarnock) | J & R Howie Limited | NS 386 404 | pre-1947 | 1950 |
| Busbiehead 3 (Kilmarnock) | J & R Howie Limited | NS 386 404 | 1947 | 1951 |
| Craigman (New Cumnock) | Coleston Mining Limited | NS 542 124 | 1986 | 1991 |
| Dowhail (Dalquharran) | T Cook | NS 274 026 | 1978 | 1979 |
| Fardalehill/Newtonhead (Kilmarnock) | Balgray Bauxite Company | NS 407 388 | pre-1947 | 1950 |
| Fortacres (Symington) | R Semple & Company | NS 393 339 | 1924 | then NCB |
| Garallan (Cumnock) | Garallan Brick & Tile Company | NS 547 183 | pre-1947 | 1961 |
| Grasshill (Glenbuck) | W Fisher & Sons | NS 745 295 | 1961 | 1982 |
| Hall of Auchincross (New Cumnock) | BMC Mining Company Limited | NS 587 139 | 1987 | 1991 |
| Hindsward (Skares) | Hamilton & Cain - previously NCB | NS 536 163 | 1969 | 1988 |
| Lochwood 1 (Kilwinning) | Lochwood Coal Company | NS 269 455 | pre-1947 | c. 1948 |
| Lochwood 2 (Kilwinning) | Lochwood Coal Company | NS 269 455 | 1942 | c. 1969 |
| Mayfield (Saltcoats) | Adams Pict Firebrick Company Limited | NS 254 423 | 1956 | 1958 |
| Muirend (Muirkirk) | James S Burns | NS 693 279 | 1954 | c. 1968 |
| Muirhouse (Dalry) | T M Thomson | NS 321 505 | pre-1947 | 1949 |
| Muirside (Kilmarnock) | J & R Howie | NS 375 393 | pre-1947 | unknown |
| Smithston (Patna) | T Love | NS 412 128 | 1966 | 1991 |
| Southhook (Kilmarnock) | Southhook Potteries Limited | NS 386 399 | pre-1947 | c. 1971 |
| Viaduct (Glenbuck) | Graham & Anderson | NS 729 296 | 1959 | 1991 |

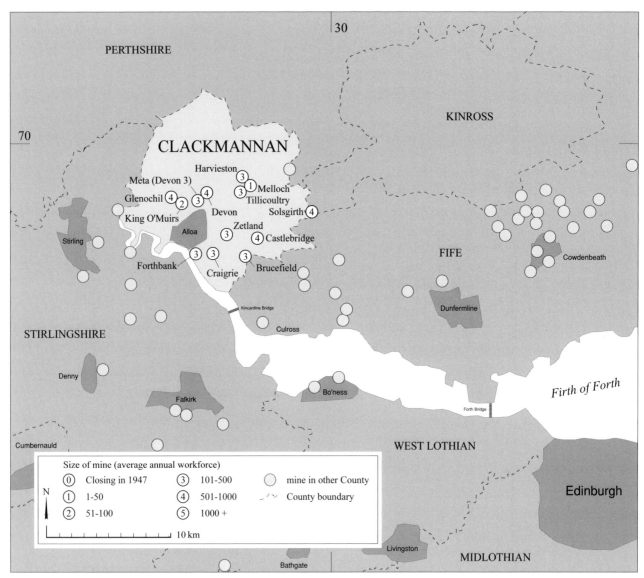

Figure 5.36: Map of National Coal Board (NCB)
collieries in Clackmannanshire.

# Clackmannanshire

Clackmannanshire is situated in the heart of central Scotland, with coal measures running from Stirlingshire to the west and south, through to Fife in the east, and overflowing into neighbouring Perthshire in the north-east near Dollar. The county town of Alloa was where much of the industry was concentrated, and from which the dominant coal company took its name. It was also the home of one of the NCB's central workshop complexes.

Although small in absolute terms, Clackmannanshire's coal industry was hugely important, and was an essential part of the 'wee county's' diverse and highly developed industrial base. Significant local coal consumers included salt pans on the Firth of Forth, ironworks, a large woollen industry reliant upon steam power, grain distilleries, breweries, paper mills, brickworks, sanitary-ware works, engineering works, foundries, and a large glass works at Alloa.

Coal extraction is documented as having occurred in the early 17th century, and expanded rapidly during the 19th century in particular, the largest colliery, Devon, being established by the Alloa Coal Company in 1854 and subsequently re-sunk in 1879. At the time of nationalisation in 1947, nine collieries were taken into public ownership. Of these, the two at Craigrie and Forthbank had recently been revived and redeveloped in the early 1940s, with developments at Dollar, Meta (Sauchie) and Tillicoultry following in 1943, 1946 and 1947, respectively. The older mines such as those at Brucefield and Melloch were retained in the immediate post-war years. Meanwhile, the NCB opened two new mines in the 1950s, the largest being the disastrous Glenochil mine, which operated only briefly between 1956 and 1962. With the closure of Devon Colliery in 1963, deep mining ceased for five years before resuming again with the opening of Solsgirth Colliery in 1969. Solsgirth was part of a large scheme in neighbouring Fife, and was developed to extract Upper Hirst coal for the power station at Longannet. Another new pit, Castlebridge, the last shaft to be sunk in Scotland, commenced production in 1984, closing in 1999.

In terms of manpower, the Clackmannanshire coalfield peaked in 1956 with the opening of Glenochil, almost 3,000 miners being employed by the industry in that year. The mines produced mostly house, steam and manufacturing coal, with small quantities of coal also suitable for town gas works. Devon Colliery had, in addition, produced some ironstone in its early days. By-products used in the heavy ceramics industries, such as by sanitary-ware makers, Charles Buick of Alloa, included fireclays, and blaes was routinely used for brickworks.

The largest mines to operate were the two newest sinkings at Solsgirth and Castlebridge, each employing over 1,000 miners at their peak in 1975 and 1990, respectively. Devon was the most important of the older collieries with an average workforce of over 600 employees, a figure which Glenochil briefly achieved in its short period of operation. There were only four private mines extracting coal after 1947, and these were restricted by the nationalisation legislation to employing no more than 30 miners. The last of the licensed mines had ceased operations by 1991.

## Brucefield

NS 9282 9136 (NS99SW/30)

**Parish:** Clackmannan

**Region/District:** Ce/Cl

**Council:** Clackmannanshire

**Location:** Kennet Village

**Previous owners:** Fordell Mains Colliery Company

**Types of coal:** House and Steam

**Sinking/Production commenced:** 1905

**Year closed:** 1961

**Year abandoned:** 1962

**Average workforce:** 216

**Peak workforce:** 265

**Peak year:** 1957

**Shaft/Mine details:** surface mines, 411m long, dipping 1 in 5.2. No. 1 mine NS 9282 9136, and No. 2 mine NS 9278 9136. Older shafts: No. 1 NS 9301 9130, and No. 2 NS 9303 9129

**Details in 1948:** Output 300 tons per day, 75,000 tons per annum. 199 employees. Jig (bash tank) type washer. Canteen (non-cooked meals), no baths, first-aid room. All electricity AC, bought from Scottish Central Supply Company. Report dated 25-08-1948.

**Other details:** Associated brickworks operated until the 1960s.

## Castlebridge

NS 9403 9267 (NS99SW/41)

**Parish:** Clackmannan

**Region/District:** Ce/Cl

**Council:** Clackmannanshire

**Location:** Gartlove, near Alloa

**Previous owners:** National Coal Board

**Types of coal:** Upper Hirst

**Sinking commenced:** 1978

**Production commenced:** 1984

**Year closed:** 1999

**Average workforce:** 700

**Peak workforce:** 1,200

**Peak year:** 1990

**Shaft/Mine details:** Shaft was 6.1m diameter, concrete-lined, 426m deep, with a 1,050hp multi-rope electrically-powered tower-mounted friction winder. It was sunk at a cost of £57 million as a satellite of the Longannet complex to reduce travelling times, improve ventilation, enhance material handling, and replace the three older drifts in the complex. The last deep-coal mine shaft to be sunk in Scotland in the NCB era, it was also at the time the first new sinking for 20 years. Closed and filled in 2001, production retreating to the Longannet access, which later closed after a catastrophic flood in 2002.

**Other details:** All production was low-sulphur coal dedicated to

supplying Longannet Power Station, to which it was taken by several kilometers of underground conveyor. The low sulphur content of the coal prevented the need for a flue-gas desulphurisation plant at Longannet, but its high ash content led to the creation of fly-ash lagoons in the Forth estuary around Culross, and nearby Preston Island. Castlebridge was one of the most productive collieries in Europe, breaking all productivity records in Scotland.
**Figure 5.37**

Figure 5.37: Castlebridge Colliery, Gartlove. This view shows the mine under construction in the early 1980s when it was developed as an extension to the Longannet complex . It was the last coal-mine shaft to be sunk in Scotland, and commenced production in 1984, operating with an average workforce of 700 men until closure in 1999. SC382458

## Craigrie
NS 9043 9155 (NS99SW/31)

**Parish:** Clackmannan

**Region/District:** Ce/Cl

**Council:** Clackmannanshire

**Location:** Clackmannan

**Previous owners:** Alloa Coal Company

**Types of coal:** House and Steam

**Sinking/Production commenced:** 1942 (re-opened)

**Year closed:** 1952

**Year abandoned:** 1957

**Average workforce:** 134

**Peak workforce:** 136

**Peak year:** 1949

**Shaft/Mine details:** 1 shaft, 56m deep. 1 surface mine connection.

**Details in 1948:** Output 200 tons per day, 50,000 per annum. 157 employees. No washer, no baths, canteen (no cooked meals), first-aid room. AC electricity bought from Scottish Central Supply Company, Report dated 25-08-1948.

## Devon 1 & 2
NS 8975 9587 (NS89NE/40.01)

**Parish:** Alloa

**Region/District:** Ce/Cl

**Council:** Clackmannan

**Location:** Sauchie, by Alloa

**Previous owners:** Alloa Coal Company

**Types of coal:** House and Steam, and originally some ironstone

**Sinking/Production commenced:** Earlier 18th century mine closed 1854, but mine subsequently re-established and in production by 1879.

**Year closed:** 1960

**Year abandoned:** 1963

**Average workforce:** 615

**Peak workforce:** 813

**Peak year:** 1954

**Shaft/Mine details:** The deepest mine in Clackmannanshire. 2 shafts, 190m and 185m deep. No. 1 NS 8975 9587; No. 2 NS 8977 9584.

Figure 5.38: Devon Colliery, Sauchie. The Alloa Coal Co's largest and deepest pit, it was operated by an average workforce of 615 miners in the NCB era. After its closure in 1960, all the surface buildings were demolished, except for a pumping beam-engine house dating from 1864. SMM:1997.439, SC379594

Wet nature of pit required powerful pump, leading to construction of the still-surviving beam-engine house in 1864 by Neilson & Company of Glasgow. It was superseded by electric pumping in 1932.

**Details in 1948:** Output 950 tons per day, 237,500 tons per annum. 488 employees. Baum-type washer. Baths (1931), canteen (1942), medical centre. AC electricity mostly own-generated, but partly bought from Scottish Central Supply Company for two weeks at holidays. Report dated 25-08-1948.

**Figure 5.38**

## Devon 3

(see Meta)

## Dollar

NS 9709 9797 (NS99NE/31)

see Perthshire.

**Note:** Although the surface arrangement is itself a few hundred metres inside Perthshire, the underground workings extend under Clackmannanshire.

## Forthbank 1 & 2

NS 8903 9147 (NS89SE/93)

**Parish:** Alloa

**Region/District:** Ce/Cl

**Council:** Clackmannanshire

**Location:** Forthbank, Alloa

**Previous owners:** Alloa Coal Company (actual drivage NCB, 1947)

**Types of coal:** House, and blaes for local brickworks

**New sinking commenced:** 1947

**Production commenced:** 1949

**Year closed:** 1958

**Year abandoned:** 1958

**Average workforce:** 353

**Peak workforce:** 992

**Peak year:** 1956

**Shaft/Mine details:** 2 surface mines, 311m long, dipping 1 in 4. No. 1 NS 8895 9151; No. 2 NS 8903 9147.

**Details in 1949:** Output 100 tons per day, 62,500 tons per annum. 61 employees. Washed at Devon colliery. Colliery still under construction, baths, canteen and medical services to be provided. Electricity public supply. Report dated 15-02-1949. Surface arrangement included a gantry and picking house with screening plant.

**Other details:** This was a new mine with an estimated life of 24 years, intended to replace Craigrie Colliery, and should not be confused with the earlier Forthbank Colliery, which operated between 1893 and

1902. The earlier Forthbank shafts were at: 1. NS 8868 9180; 2. NS 8867 9177.

**Figures 5.39 and 5.40**

Figure 5.39: Forthbank Colliery, Alloa. Taking on the name of an older pit owned by the Alloa Coal Co, this was a new colliery developed by the NCB from 1947, comprising two surface mines, the portal of one of which is shown here. SMM:1998.349, SC445864

## Glenochil 1 & 2

NS 8727 9600 (NS89NE/80)

**Parish:** Alloa

**Region/District:** Ce/Cl

**Council:** Clackmannanshire

**Previous owners:** National Coal Board

**Types of coal:** House, and fireclay for NCB brickworks

**Sinking commenced:** 1952

**Production commenced:** 1956

**Year closed:** 1962

**Year abandoned:** 1964

**Average workforce:** 587

**Peak workforce:** 908

**Peak year:** 1960

**Shaft/Mine details:** Two parallel mines 2,414m in length, 30m apart, at a 1:5 gradient (No. 1 NS 8727 9600 and No. 2 NS 8726 9605), also utilising adjacent King O'Muirs as an air intake.

**Other details:** Planned to be the largest drift mine complex in the UK at the time, this was one of the NCB's more serious failures in Scotland. The strategy had included mining unworked 76cm seams, and the removal of pillars from previously worked stoop-and-room workings. Years of subsidence and compression resulted in these coals being both difficult and expensive to extract, and in the years that ensued between the planning and the execution of the project, market conditions for coal had changed radically. In that time, new houses were built in anticipation of many years of operation. The neighbouring village of Tullibody was transformed, and miners were brought in from exhausted pits in the Lanarkshire coalfields. The impressive surface complex included substantial coal preparation facilities, but the investment proved to be wasted when the complex underground conditions and falling demand for coal resulted in premature closure. Production had lasted for only six years, and the mine was subsequently incorporated within the site of Glenochil Prison.

**Figure 5.41**

Figure 5.40: Forthbank Colliery, Alloa. The new colliery commenced production in 1949, but worked for only nine years with an average workforce of 353 men, closing in 1958. SMM:998.348, SC445862

## Harvieston

NS 9283 9743 (NS99NW/76)

**Parish:** Tillicoultry

**Region/District:** Ce/Cl

**Council:** Clackmannanshire

**Previous owners:** National Coal Board

**Sinking commenced:** 1955

**Production commenced:** 1957

Figure 5.41: Glenochil Colliery, near Tullibody, 1957. Production commenced in 1956, but it proved to be one of the most spectacular failures of the NCB era, staying in production for only six years, with an average workforce of 587 miners. The site is now occupied by a prison. SMM:1996.02440, SC379618

**Year closed:** 1961
**Year abandoned:** 1962
**Average workforce:** 230
**Peak workforce:** 237
**Peak year:** 1960
**Shaft/Mine details:** No. 1 at NS 9283 9743; No. 2 at NS 9288 9744.
**Other details:** Private mine opened in 1970s.

### King O'Muirs 1
NS 8794 9534 (NS89NE/81)
**Parish:** Alloa
**Region/District:** Ce/Cl
**Council:** Clackmannanshire
**Location:** Tullibody
**Previous owners:** Alloa Coal Company
**Types of coal:** House and Manufacturing
**Sinking/Production commenced:** 1938
**Year closed:** 1956
**Year abandoned:** 1957
**Average workforce:** 82
**Peak workforce:** 153
**Peak year:** 1955
**Shaft/Mine details:** 1 surface mine, 1 air shaft
**Details in 1948:** Output 100 tons per day, 25,000 tons per annum. 55 employees. No washer, no baths, canteen or medical facilities. AC electricity all supplied from Devon Colliery. Report dated 25-08-1948.

### King O'Muirs 2 & 3
NS 8729 9597 (NS89NE/84)
**Parish:** Alloa
**Region/District:** Ce/Cl
**Council:** Clackmannanshire
**Location:** Tullibody
**Previous owners:** National Coal Board
**Types of coal:** House and Manufacturing
**Sinking/Production commenced:** 1950
**Year closed:** 1957
**Year abandoned:** 1964
**Statistics:** see King O'Muirs 1
**Shaft/Mine details:** Third mine at NS 8732 9596
**Other details:** The NCB shaft register states that the main mine and dook was to be connected to Glenochil to serve as a return airway (never completed).

## Melloch

NS 9327 9645 (NS99NW/78)

**Parish:** Tillicoultry

**Region/District:** Ce/Cl

**Council:** Clackmannanshire

**Location:** Tillicoultry

**Previous owners:** Alloa Coal Company

**Types of coal:** House

**Sinking/Production commenced:** *c.*1850

**Year closed:** 1948

**Year abandoned:** 1948

**Average workforce:** 19

**Peak workforce:** 29

**Peak year:** 1947

**Shaft/Mine details:** 1 surface mine, re-opened 1941

**Details in 1948:** Output 40 tons per day, 15,000 tons per annum. 21 employees. Coal washed and screened at Tillicoultry mine, miners using baths and canteen also at Tillicoultry. AC electricity from Devon Colliery. Report dated 25-08-1948.

## Meta (Devon 3)

NS 8940 9577 (NS89NE/82)

**Parish:** Alloa

**Region/District:** Ce/Cl

**Council:** Clackmannanshire

**Location:** Sauchie, Alloa

**Previous owners:** Alloa Coal Company

**Types of coal:** House, dross (unwashed for briquettes)

**Sinking/Production commenced:** 1946 (originally inaugurated on 4th July 1923)

**Year closed:** 1959

**Year abandoned:** 1959

**Average workforce:** 187

**Peak workforce:** 233

**Peak year:** 1956

**Shaft/Mine details:** 1 shaft 115m deep, and 1 surface mine (NS 8902 9587)

**Details in 1948:** Output 200 tons per day, 50,000 tons per annum. 94 employees. No washer; bath and canteen at Devon Colliery 450m away. Medical facilities. AC electricity all from Devon Colliery. Record dated 25-08-1948.

**Other details:** Brickworks established in the 1930s, closed in 1981. A fuel briquetting plant was added in 1943. Meta drift mine at NS 8724 9577 (NS89NE/83).

**Figure 5.42**

Figure 5.42: Meta Colliery, Sauchie, also known as Devon 3. Established by the Alloa Coal Company in 1946, but originally inaugurated in 1923. It was a relatively small colliery, with an average workforce in the NCB era of only 187 miners. This view also shows the adjacent brickworks, which was established in the 1930s, and closed in 1981, 22 years after the colliery. SMM:1997.0433, SC379628

Figure 5.43: Solsgirth Colliery, south-east of Dollar, c.1970. An NCB development, and part of the Longannet complex in neighbouring Fife. Development began in 1964, production commencing in 1969. Thereafter, it operated with an average workforce of 917 miners until closure in 1990. SC446772

Figure 5.44: Tillicoultry Mine. Sunk originally by the Alloa Coal Company in 1876, it was redeveloped by the NCB immediately after nationalisation. It operated until 1957 with an average workforce of 170 miners. This underground view shows a locomotive hauling full wagons of coal in 1952. SMM:2001.0073G, SC706675

## Solsgirth

NS 9839 9469 (NS99SE/06)

**Parish:** Clackmannan

**Region/District:** Ce/Cl

**Council:** Clackmannanshire

**Previous owners:** National Coal Board

**Sinking commenced:** 1965

**Production commenced:** 1969

**Year closed:** 1990

**Average workforce:** 917

**Peak workforce:** 1,007

**Peak year:** 1975

**Shaft/Mine details:** 2 surface mines, each 4.87m by 3.66m, dipping at 1 in 4, and approximately 1,600m in length

**Other details:** All production dedicated to supplying Longannet Power Station via a 8.85km-long tunnel (said to contain the longest conveyor belt in the world), opened in 1970. Previously renowned for high output per face, winning British and European productivity awards (1975). Plans to link underground with Dollar to the north-west never realised. Closed in 1990 but retained as ventilation for Castlebridge (also part of the Longannet Complex).

**Figure 5.43**

## Tillicoultry 1 & 2

NS 9293 9645 (NS99NW/77.00)

**Parish:** Tillicoultry

**Region/District:** Ce/Cl

**Council:** Clackmannanshire

**Location:** Tillicoultry

**Previous owners:** Alloa Coal Company

**Types of coal:** House, Steam and Gas

**Sinking/Production commenced:** 1876 (1) 1947 (2)

**Year closed:** 1957

**Year abandoned:** 1959

**Average workforce:** 170

**Peak workforce:** 346

**Peak year:** 1950

**Shaft/Mine details:** 2 surface mines

**Details in 1948:** Output 220 tons per day, 55,000 tons per annum. 98 employees. Baum-type washer. Baths (1938, for 70 men), canteen. AC electricity from Devon Colliery (NCB). Report dated 25-08-1948.

**Other details:** Well known for high productivity, often twice the national man-shift average. Famous for the first underground live TV broadcast from a coal mine on 25th November 1952.

**Figures 5.44 and Figure 5.45**

## Zetland

NS 9148 9264 (NS99SW/32)

**Parish:** Clackmannan

**Region/District:** Ce/Cl

**Council:** Clackmannanshire

**Location:** Clackmannan

**Previous owners:** Alloa Coal Company

**Types of coal:** House and Steam

**Sinking/Production commenced:** 1935

**Year closed:** 1960

**Year abandoned:** 1961

**Average workforce:** 192

**Peak workforce:** 323

**Peak year:** 1957

**Shaft/Mine details:** 2 surface mines, No. 1 at NS 9153 9295, No. 2 at NS 9159 9307

**Details in 1948:** Output 160 tons per day, 40,000 tons per annum. 89 employees. No washer. No baths, but washing facilities provided. All AC electricity supplied from Devon Colliery. Report dated 25-08-1948.

Figure 5.45: Tillicoultry Mine. Underground view of a BBC camera crew preparing for the first ever live TV broadcast from underground in a coal mine, on 25th November 1952. The mine was considered to be safe enough to permit the use of normal unprotected electrical equipment, and in other photographs in this sequence, the technicians are smoking cigarettes. SMM:2000.0073S, SC706659

## Licensed Mines in Clackmannanshire 1947–97

| Name | Licensee | Grid Ref | Opened | Closed |
|---|---|---|---|---|
| Devon (Grassmainston) | Hillfarm Coal Company Limited | NS 927 933 | 1983 | 1989 |
| Gartinkeir (Coalsnaughton) | J & W Miller | NS 924 948 | 1967 | 1968 |
| Gartmorn (Coalsnaughton) | J Dawson | NS 924 947 | 1967 | 1984 |
| Harviestoun (Dollar) | G Drysdale | NS 937 977 | 1968 | 1991 |

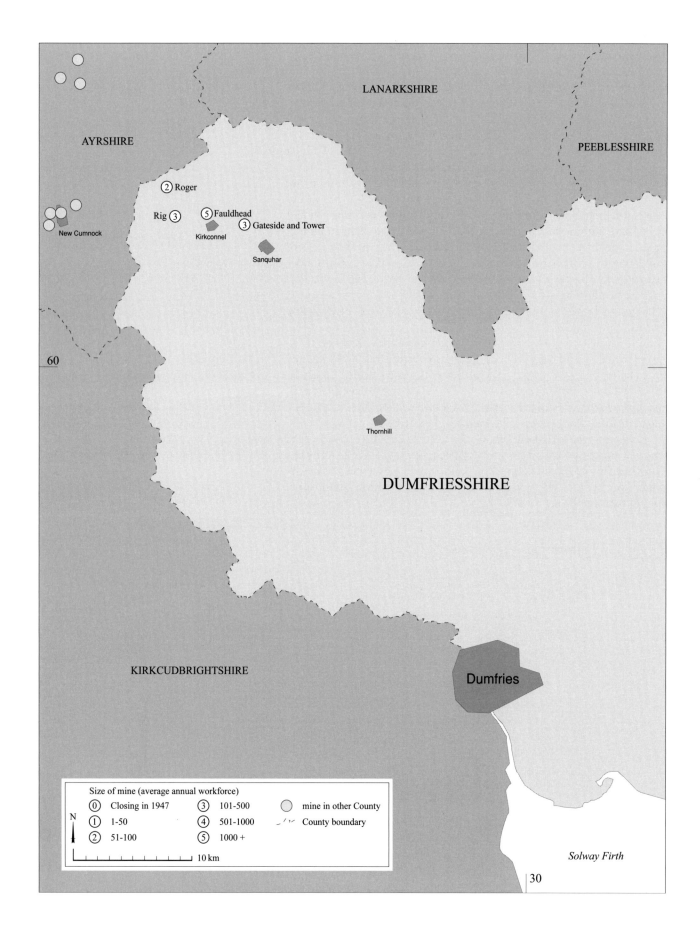

Size of mine (average annual workforce)

| | | | | |
|---|---|---|---|---|
| ⓪ | Closing in 1947 | ③ | 101-500 | ⬤ mine in other County |
| ① | 1-50 | ④ | 501-1000 | County boundary |
| ② | 51-100 | ⑤ | 1000 + | |

10 km

N

LANARKSHIRE

AYRSHIRE

PEEBLESSHIRE

② Roger

Rig ③   ⑤ Fauldhead
         ③ Gateside and Tower
Kirkconnel
         Sanquhar

New Cumnock

60

Thornhill

DUMFRIESSHIRE

KIRKCUDBRIGHTSHIRE

Dumfries

Solway Firth

30

# Dumfriesshire

In geographical and practical terms, coal mining in Dumfriesshire has been an offshoot of the neighbouring New Cumnock coalfield in Ayrshire, and produced a range of house, manufacturing, steam, gas and coking coals. Centred in the north of the county around Sanquhar and Kirkconnel, mining was concentrated in six deep mines, three of which were taken into public ownership by nationalisation in 1947. Of these, Fauldhead in Kirkconnel was by far the most important, employing on average over 1,000 miners, the combined workforce for the county never exceeding 2,000 miners at any one time. It was particularly noted for its high-sulphur coal, which, after the introduction of clean-air legislation, was used predominantly for steam locomotives. The colliery was therefore doomed when British Railways phased out steam locomotives in the 1960s. Other smaller pits were Gateside and Tower collieries in neighbouring Sanquhar. All three were operated by Sanquhar & Kirkconnel Collieries Limited prior to being taken over by Bairds & Dalmellington by 1931.

The NCB sank three new surface drift mines near Kirkconnel in the 1950s, and of these, Roger Colliery operated until 1980. All the others failed to survive a spate of closures in the early 1960s, culminating in the loss of Fauldhead in 1968. There has been extensive opencast mining around Kirkconnel since then. One private mine also operated near Sanquhar between 1969 and the late 1970s.

A second coalfield also exists in the south-east of the county deep beneath Canonbie, and there was coal-ming activity there in the early 20th century. The deep reserves have, however, never been fully exploited, but there is the possibility of future exploitation should suitable market conditions prevail.

Figure 5.46: Map of National Coal Board (NCB) collieries in Dumfriesshire.

## Fauldhead 1 & 3

NS 7323 1237 (NS71SW/27)

**Parish:** Kirkconnel

**Region/District:** Du/Ni

**Council:** Dumfries and Galloway

**Location:** Kirkconnel

**Previous owners:** J I McConnell Esq, Sanquhar & Kirkconnel Collieries Limited from 1903, Bairds & Dalmellington Limited from 1931.

**Types of coal:** House, Manufacturing, Coking and Steam

**Sinking/Production commenced:** 1896 and 1911

**Year closed:** 1968

**Year abandoned:** 1968

**Average workforce:** 1,018

**Peak workforce:** 1,155

**Peak year:** 1949

**Shaft/Mine details:** Previously 4 shafts, two at 55m and two at 137m, but latterly only Nos. 1 & 3, and a drift mine.

**Details in 1948:** Output 989 tons per day, 267,850 tons per annum. Stoop and room and longwall working. 1,028 employees. 6 screens for dry coal, jig (Luhrig) and jig (Dickson and Mann) washers. Baths (1933) with 960 lockers. Canteen (packed meals). Ambulance station. Steam and electricity (0.4% public supply). Report dated 09-08-1948.

**Other details:** Operated by Sanquhar & Kirkconnel Coal Company from 1903, and William Baird from 1925 (later Bairds & Dalmellington Limited). Associated brickworks operated from 1912 until the 1970s.

**Figures 5.47 and Figure 5.48**

Figure 5.47: Fauldhead Colliery, Kirkconnel. Established by J I McConnel Esq in 1896, this became the flagship colliery of the Sanquhar and Kirkconnel Coal Company before being taken over by Bairds and Dalmellington. Kirkconnel Parish Heritage Society, SC706327

## Gateside 4 & 5

NS 7663 1144 (NS71SE/38)

**Parish:** Kirkconnel

**Region/District:** Du/Ni

**Council:** Dumfries and Galloway

**Location:** Sanquhar

**Previous owners:** J I McConnell Esq, Sanquhar & Kirkconnel Collieries Limited from 1903, William Baird from 1925, and Bairds & Dalmellington Limited from 1931

**Types of coal:** Gas, Manufacturing and Steam

**Sinking/Production commenced:** 1891

**Year closed:** 1964

**Average workforce:** 344

**Peak workforce:** 410

**Peak year:** 1952

**Shaft/Mine details:** 2 shafts, 102m deep

**Details in 1948:** Output 240 tons per day, 60,819 per annum. Stoop and room working. 160 employees. 3 screens for dry coal. Jig (Dickson and Mann) washer. No baths. Canteen (snacks only). Steam and electricity, none from public supply. Report dated 09-08-1948.

## Rig

NS 7063 1215 (NS71SW/28)

**Parish:** Kirkconnel

**Region/District:** Du/Ni

**Council:** Dumfries and Galloway

**Previous owners:** National Coal Board

**Sinking commenced:** 1948

**Production commenced:** 1949

**Year closed:** 1966

**Average workforce:** 122

**Peak workforce:** 93

**Peak year:** 1957

**Mine Details:** Two surface mines

## Roger 1 & 2

NS 6985 1453 (NS61SE/15)

**Parish:** Kirkconnel

**Region/District:** Du/Ni

**Council:** Dumfries and Galloway

**Previous owners:** National Coal Board

**Sinking/Production commenced:** 1952

**Year closed:** 1980

**Year abandoned:** 1980

**Average workforce:** 88

**Peak workforce:** 121

Figure 5.48: Fauldhead Colliery, Kirkconnel. By far the largest colliery in Dumfriesshire, it operated in the nationalised era with an average workforce of over 1,000 miners until its closure in 1969. Kirkconnel Parish Heritage Society, SC706330

**Peak year:** 1975
**Figure 5.49**

## Roger 3 & 4

NS 7019 1498 (NS71SW/29)
**Parish:** Kirkconnel
**Region/District:** Du/Ni
**Council:** Dumfries and Galloway
**Previous owners:** National Coal Board
**Sinking/Production commenced:** 1956
**Year closed:** 1968
**Year abandoned:** 1968
**Statistics:** see Roger 1 and 2
**Other details:** none

## Tower Mine

NS 7579 1155 (NS71SE/39)
**Parish:** Kirkconnel
**Region/District:** Du/Ni
**Council:** Dumfries and Galloway
**Location:** Sanquhar
**Previous owners:** Sanquhar & Kirkconnel Collieries Limited, Bairds & Dalmellington Limited from 1931
**Types of coal:** House and Manufacturing
**Sinking/Production commenced:** 1916
**Year closed:** 1964
**Statistics:** see Gateside
**Shaft/Mine details:** 2 surface mines, dipping 213m at 1 in 6
**Details in 1948:** Output 260 tons per day, 61,730 tons per annum, stoop and room working. 174 employees. Screening and washing at Gateside. No baths, but canteen for snacks. Electricity, but not from public supply. Report dated 09-08-1948.

Figure 5.49: Roger 1 & 2 Colliery, west of Kirkconnel. Although humble in appearance, a successful surface mine complex sunk by the NCB in 1952, and operated with an average workforce of 88 miners until its closure in 1980. SMM:1996.02567, SC381048

## Licensed Mines in Dumfriesshire 1947–97

| Name | Licensee | Grid Ref | Opened | Closed |
|------|----------|----------|--------|--------|
| Lady Ann (Sanquhar) | D Thomson | NS 775 127 | 1969 | c. 1977 |

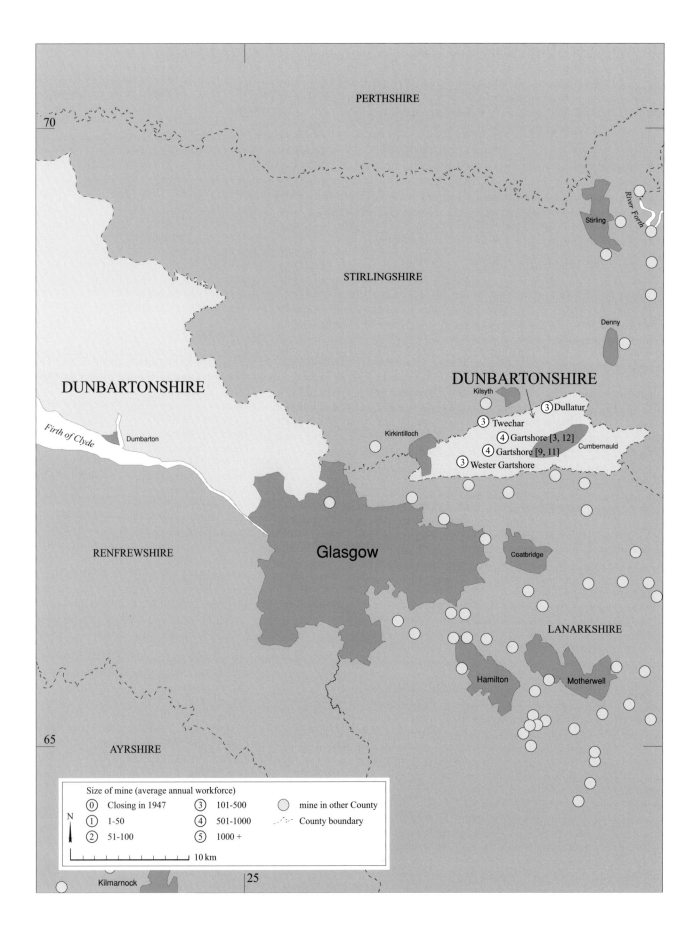

PERTHSHIRE

70

*River Forth*

Stirling

STIRLINGSHIRE

Denny

DUNBARTONSHIRE

Kilsyth

DUNBARTONSHIRE

③ Dullatur

③ Twechar

④ Gartshore [3, 12]

Kirkintilloch

④ Gartshore [9, 11]

Cumbernauld

③ Wester Gartshore

*Firth of Clyde*

Dumbarton

Glasgow

RENFREWSHIRE

Coatbridge

LANARKSHIRE

Hamilton

Motherwell

65

AYRSHIRE

Size of mine (average annual workforce)

| ⓪ | Closing in 1947 | ③ | 101-500 | ◯ | mine in other County |
| ① | 1-50 | ④ | 501-1000 | -··- | County boundary |
| ② | 51-100 | ⑤ | 1000 + | | |

N

10 km

25

Kilmarnock

# Dunbartonshire

Although there was some coal mining activity in western Dunbartonshire in the Clydebank area in the 19th century, more recent mining activities have been concentrated in the eastern detached portion of the county. This is an area of ever-changing local authorities, the confusing political boundaries on the surface being matched by fractured geology below. However, although creating problems of faulting and distortion of strata, the presence of volcanic intrusions also positively affected the coal measures, helping to create significant pockets of anthracite and coking coal. For this reason, although mining was often difficult, expensive and financially risky, iron companies in particular were willing to risk capital and develop collieries.

At the same time, the often limited extent of the coal resulted in the mines being comparatively modest in size compared with other parts of the Scottish coalfields, and together they never employed more than 2,000 people. The largest of the collieries, Gartshore 9 and 11, employed on average 616 miners during its 21 years of operation in the nationalised era. In 1968, it was the last of the Dunbartonshire pits to close, the others having perished during the 1950s and early 1960s.

All five of the county's pits had been taken over as going concerns by the NCB in 1947, the three Gartshore pits near Kilsyth having previously been operated by William Baird, later Bairds & Scottish Steel Limited. The other two, at Dullatur and Wester Gartshore, were run by the Cadzow Coal Company. The Gartshores all dated from the mid-19th century, only the surface mines at Dullatur being a 20th-century development. All produced anthracite, but also yielded coking, house, manufacturing and steam coals. However, it was the cessation of iron smelting at Gartsherrie which triggered the closure of the Gartshore pits in the 1960s. The NCB chose not to embark on any new sinkings in the area, but there were four small licensed mines in the county, three of which were new projects begun after 1947. Of these, the most recent ceased to renew its licence in 1991.

Figure 5.50: Map of National Coal Board (NCB) collieries in Dunbartonshire.

**Dullatur**

NS 7507 7731 (NS77NE/81)

**Parish:** Cumbernauld

**Region/District:** St/CN

**Council:** North Lanarkshire

**Location:** Dullatur

**Previous owners:** Cadzow Coal Company

**Types of coal:** Anthracite

**Sinking/Production commenced:** 1935

**Year closed:** 1964

**Average workforce:** 192

**Peak workforce:** 281

**Peak year:** 1963

**Shaft/Mine details:** 2 surface mines each 220m long, No. 1 (downcast NS 7507 7731) 1 in 5, and No. 2 (upcast NS 7509 7731) 1 in 4

**Details in 1948:** Output 60 tons per day, 16,200 tons per annum. 97 employees. No washer. Baths, canteen, first-aid room. Electricity from Scottish Central Supply. Report dated 19-08-1948.

**Figure 5.51**

**Gartshore 1, 3 & 12**

NS 7154 7492 (NS77SW/23)

**Parish:** Kirkintilloch

**Region/District:** St/St

**Council:** East Dunbartonshire

**Location:** Twechar

**Previous owners:** Bairds & Scottish Steel Limited

**Types of coal:** Anthracite and Coking

**Sinking/Production commenced:** 1865

**Year closed:** 1959

Figure 5.51: Dullatur Colliery, near Cumbernauld. A surface mine sunk by the Cadzow Coal Company in 1935, it operated in the nationalised era with an average workforce of 192 miners until closure in 1964. The National Archives, COAL80/370/4, SC614101

**Year abandoned:** 1968

**Average workforce:** 547

**Peak workforce:** 611

**Peak year:** 1948

**Shaft/Mine details:** 3 shafts, No. 1 106m, No. 3 (NS 7154 7492) 293m, and No. 12 (NS 7154 7493) 305m deep. Mosswater mine 439m long, 1 in 2. No. 3 shaft said to have originally been an ironstone pit, then used for pumping and coal extraction.

**Details in 1948:** Output 480 tons per day, 129,600 tons per annum. 564 employees. No washer, no baths. Canteen at Gartshore No. 3 and Mosswater Mine. First-aid rooms. 90% electricity supplied by Clyde Valley. Report dated 19-08-1948.

**Other details:** Scene of an explosion caused by gas in 1923 which killed eight men. Brick-lined ventilation furnaces seen by George Gillespie (former manager) in 1950 at the foot of one of the shafts. Baths opened in 1952.

**Figure 5.52**

Figure 5.52: Gartshore 3 & 12 Colliery, near Twechar. Dating from 1865, and owned latterly by Bairds and Scottish Steel, this was an important source of anthracite and coking coal. In the nationalised era, it operated with an average workforce of 547 miners until 1959. The National Archives, COAL80/432/5, SC614087

## Gartshore 9, 11, & Grayshill

NS 7034 7389 (NS77SW/24)

**Parish:** Kirkintilloch

**Region/District:** St/St

**Council:** East Dunbartonshire

**Location:** Twechar

**Previous owners:** Bairds & Scottish Steel Limited

**Types of coal:** Anthracite and Coking

**Sinking/Production commenced:** 1875, 1893, Grayshill added in 1930s

**Year closed:** 1968

**Year abandoned:** 1968

**Average workforce:** 616

**Peak workforce:** 740

**Peak year:** 1952

**Shaft/Mine details:** 3 shafts, No. 9 299m (NS 7034 7389), No. 11 302m (NS 7036 7391), Grayshill (man-winding only in 1948) 404m deep (NS 7092 7299). New drift added in 1950s, 1,200m long at 1:4.

**Details in 1948:** Output 750 tons per day, 202,500 tons per annum. 706 employees. 'Blantyre'-type washer. Baths, canteens (at both No. 9 and No. 11), first-aid rooms. 90% electricity from Clyde Valley. Report dated 19-08-1948.

**Other details:** A very gassy pit, but an important source of coking coal.

**Figures 5.53 and Figure 5.54**

Figure 5.53: Gartshore 9 & 11 and Grayshill Colliery, near Kilsyth. Originally established in 1875, but with new shafts added in 1893 and the 1930s, & prior to nationalisation, owned by Bairds and Scottish Steel. Like the other pits in the area, it was an important source of anthracite and coking coal. The National Archives, COAL80/466/2, SC614097

## Twechar No. 1 & Gartshore 10

NS 7006 7615 (NS77NW/57)

**Parish:** Kirkintilloch

**Region/District:** St/St

**Council:** East Dunbartonshire

Figure 5.54: Gartshore 9 & 11 and Grayshill Colliery, near Kilsyth, 1966. The largest pit in Dunbartonshire, it employed on average 616 miners during the NCB era, eventually closing in 1968. J R Hume, SC677850

Figure 5.55: Twechar Colliery, near Kilsyth. Sunk in 1865 and operated by William Baird, later Baird & Scottish Steel, it was an important producer of coking coal. In the NCB era, it operated with an average workforce of 355 men until closure in 1964. The National Archives, COAL80/983/1, SC614095

Figure 5.56: Wester Gartshore Colliery, near Twechar, c.1935. Sunk in 1872 and operated latterly by the Cadzow Coal Co. It produced anthracite and coking coal, employing an average of 299 men in the nationalised era until its closure in 1950. The National Archives, COAL80/1033/3, SC614017

**Location:** Twechar, nr. Kilsyth

**Previous owners:** William Baird, later Bairds & Scottish Steel Limited

**Types of coal:** Steam, Manufacturing, House and Coking

**Sinking/Production commenced:** 1865

**Year closed:** 1964

**Average workforce:** 355

**Peak workforce:** 413

**Peak year:** 1954

**Shaft/Mine details:** 2 shafts, Twechar No. 1 335m, Gartshore No. 10 326m deep

**Details in 1948:** Output 300 tons per day, 71,000 tons per annum. 360 employees. No baths, but canteen and first-aid room. 90% electricity from Clyde Valley. Report dated 19-08-1948.

**Other details:** Initially an ironstone pit supplying Baird's iron and steel operations. Linked to Dumbreck Colliery. Had its own coke ovens.

**Figure 5.55**

## Wester Gartshore

NS 6836 7295 (NS67SE/40.00)

**Parish:** Kirkintilloch

**Region/District:** St/St

**Council:** East Dunbartonshire

**Location:** Waterside, nr. Kirkintilloch

**Previous owners:** Wallace family of Solsgirth House, and from 1914 the Cadzow Coal Company

**Types of coal:** Anthracite, Coking, House and Steam

**Sinking/Production commenced:** 1872

**Year closed:** 1950

**Year abandoned:** 1950

**Average workforce:** 299

**Peak workforce:** 318

**Peak year:** 1947

**Shaft/Mine details:** 2 shafts, both 274m deep, No. 1 downcast (NS 6836 7295), No. 2 upcast (NS 6827 7289)

**Details in 1948:** Output 203 tons per day, 54,810 tons per annum. 292 employees. 2 Sherwood Hunter Baum-type washers. Baths (c.1939), canteen, first-aid room. Electricity generated at colliery. Report dated 19-08-1948.

**Figure 5.56**

## Licensed Mines in Dunbartonshire 1947–97

| Name | Licensee | Grid Ref | Opened | Closed |
|------|----------|----------|--------|--------|
| Kelvin View (Twechar) | J Graham | NS 695 675 | 1963 | 1968 |
| Klondyke (Torfyne, Kilsyth) | Strathkelvin Mining Company Limited | NS 760 798 | 1988 | 1991 |
| Saddler's Brae (Kirkintilloch) | Union Coal Company | NS 684 744 | 1939 | 1964 |
| Tannoch (Cumbernauld) | Rumford Coal Company Limited | NS 783 727 | 1952 | 1952 |

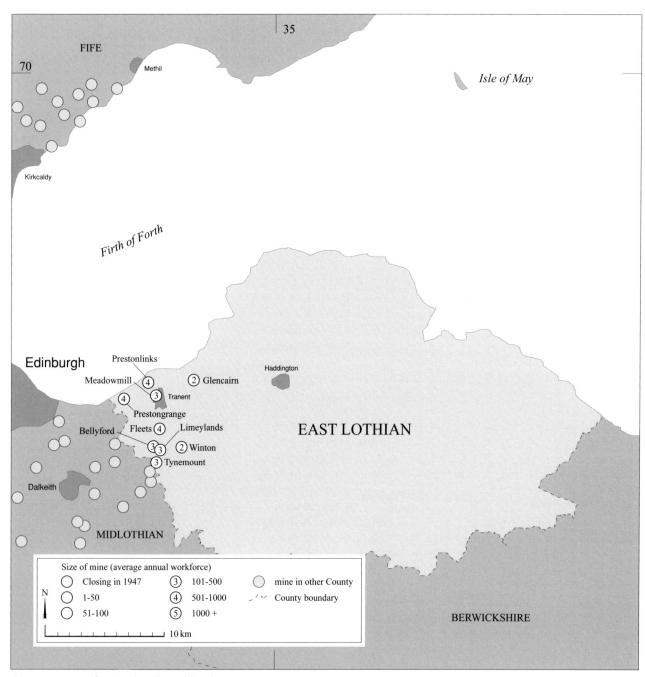

Figure 5.57: Map of National Coal Board (NCB) collieries in East Lothian.

# East Lothian

Forming the eastern edge of the Lothian coalfield, and squeezed into the western corner of the county, East Lothian's coal mines were relatively modest both in number and scale compared with those of neighbouring Midlothian and Fife, but also extended under the Firth of Forth. Many centuries of mining activity had fuelled some important local industries, including salt pans, glass, and ceramics works, but only six collieries were taken on by the NCB in 1947. Of these, the two near Prestonpans at Prestonlinks and Prestongrange were easily the largest, the former having an average workforce of almost 800 miners in the ensuing years. Only one other colliery, Fleets (near Tranent), employed more than 500 men annually. The industry peaked in East Lothian in 1948, employing a total of just over 3,000 people.

The mines, which produced a range of house, steam and gas coals, were operated by several different companies prior to nationalisation, including Edinburgh Collieries Company and the Ormiston Coal Company. One of the more significant operators was the Summerlee Iron Company of Coatbridge, who acquired Prestongrange Colliery in 1895. In the late 1940s and early 1950s, the NCB commenced three new drift-mine projects at Bellyford, Meadowmill and Winton, all three lasting little more than a decade. Indeed, with the exception of Limeylands and Tynemount, both of which closed in the early 1950s, the East Lothian coal industry was eradicated in a five-year period between 1958 and 1963.

Prestongrange closed in 1962, but large parts of its surface arrangement were saved from destruction, becoming Scotland's first coal mining museum with the support of East Lothian County Council. It was later to form part of the Scottish Mining Museum before being inherited by the East Lothian Museums Service and operated as an industrial museum. The site has retained its Cornish beam-engine house, and is also notable because of the rare survival of a Hoffmann continuous kiln at its adjacent brickworks. Meanwhile, Prestonlinks closed a year later in 1963, and was demolished shortly afterwards to make way for the new coal-burning power station at Cockenzie.

Four small private mines operated in East Lothian in this period, the earliest dating from 1936. The last of these mines had ceased to operate by 1987. However, substantial open-cast projects, particularly near Tranent, ensured that the county continued to produce significant quantities of coal.

Figure 5.58: Bellyford Mine, near Tranent. Begun by the NCB in 1949, pithead baths were completed in 1951, production commencing in 1954. The mine operated with an average workforce of 334 miners for only seven years, closing in 1961. SC675073

## Bellyford
NT 4029 6947 (NT46NW/79)
**Parish:** Tranent
**Region/District:** Lo/Ea
**Council:** East Lothian
**Previous owners:** National Coal Board
**Sinking commenced:** 1949
**Production commenced:** 1954
**Year closed:** 1961
**Year abandoned:** 1962
**Average workforce:** 334
**Peak workforce:** 335
**Peak year:** 1954
**Other details:** Pithead baths opened in 1951
**Figure 5.58**

## Fleets
NT 4037 7121 (NT47SW/55)
**Parish:** Tranent
**Region/District:** Lo/Ea
**Council:** East Lothian
**Location:** Tranent
**Previous owners:** Edinburgh Collieries Company
**Types of coal:** House, Steam and Gas
**Sinking/Production commenced:** 1866, re-sunk 1880s
**Year closed:** 1959
**Year abandoned:** 1961

Figure 5.59: Fleets Colliery, Tranent. Dating from the 1860s, and owned latterly by Edinburgh Collieries Limited, pithead baths were completed with the aid of the Miners' Welfare Fund in 1936. After nationalisation, it operated with an average workforce of 570 miners until its closure in 1959. The National Archives COAL80/412/10, SC614111

**Average workforce:** 570

**Peak workforce:** 854

**Peak year:** 1947

**Shaft/Mine details:** 2 shafts, 73m and 154m deep

**Details in 1948:** Output 766 tons per day, 191,500 tons per annum. 816 employees. Campbell Reid bash tank washer. Baths (1936), canteen, first-aid and ambulance room. Electricity supplied from SE Scotland Electricity Board. Report dated 15-08-1948.

**Other details:** Cornish beam-engine built originally in 1847 at Perran, transferred from Dolphinston in 1885.

**Figure 5.59**


## Glencairn

NT 4318 7514 (NT47NW/55)

**Parish:** Gladsmuir

**Region/District:** Lo/Ea

**Council:** East Lothian

**Location:** Longniddry

**Previous owners:** Glencairn Coal Company

**Types of coal:** House

**Sinking/Production commenced:** 1936

**Year closed:** 1962

**Year abandoned:** 1962

**Average workforce:** 72

**Peak workforce:** 190

**Peak year:** 1948

**Shaft/Mine details:** 2 surface mines

**Details in 1948:** Output 110 tons per day, 27,500 tons per annum. 57 employees. Canteen, first-aid room, ambulance. Electricity supplied by SE Scotland Electricity Board. Report dated 15-07-1948.

**Other details:** Transferred to NCB in 1948. Baths opened in 1951.


## The Grange

(see Prestongrange)


## Limeylands

NT 4062 6947 (NT46NW/78)

**Parish:** Ormiston

**Region/District:** Lo/Ea

**Council:** East Lothian

**Location:** Ormiston

**Previous owners:** Ormiston Coal Company

**Types of coal:** House and Steam

**Sinking/Production commenced:** 1895

**Year closed:** 1954

Figure 5.60: Limeylands Colliery, near Ormiston. Sunk by the Ormiston Coal Company in 1895, it operated for only seven years after nationalisation, with an average workforce of 269 miners. SC675074

**Average workforce:** 269
**Peak workforce:** 335
**Peak year:** 1952
**Shaft/Mine details:** 2 shafts, 46m deep
**Details in 1948:** Output 290 tons per day, 72,500 tons per annum. 303 employees. Coal washed at Tynemount Colliery. Canteen, first-aid room, ambulance. Electricity supplied by SE Scotland Electricity Board. Report dated 15-07-1948.
**Figure 5.60**

## The Links
(see Prestonlinks)

## Meadowmill Mine
NT 4016 7383 (NT47SW/56)
**Parish:** Tranent, near Prestonpans
**Region/District:** Lo/Ea
**Council:** East Lothian
**Previous owners:** National Coal Board
**Sinking commenced:** 1952
**Production commenced:** 1954
**Year closed:** 1960
**Year abandoned:** 1962
**Average workforce:** 120
**Peak workforce:** 120
**Peak year:** 1954
**Shaft/Mine details:** 2 surface mines sunk by the NCB on site of former Meadowmill coal washer, immediately to the east of Bankton Colliery. Deepest workings approximately 60m below the surface. Average output about 250 tons per day. Coal sent to Dalkeith central preparation plant.
**Other details:** Cleared away to make way for playing fields and artificial ski slope.

## Prestongrange (The Grange)
NT 3728 7365 (NT37SE/78)
**Parish:** Prestonpans
**Region/District:** Lo/Ea
**Council:** East Lothian
**Location:** Prestonpans
**Previous owners:** Originally the Grant Suttie family, and from 1874 the Prestongrange Coal and Iron Company. Taken over in 1895 by the Summerlee and Mossend Iron Company, which became the Summerlee Iron Company in 1898.
**Types of coal:** Steam and House
**Sinking/Production commenced:** from *c.*1820

**Year closed:** 1962

**Year abandoned:** 1963

**Average workforce:** 686

**Peak workforce:** 700

**Peak year:** 1952

**Shaft/Mine details:** 3 shafts, 115m, 166m, and 225m deep

**Details in 1948:** Output 670 tons per day, 167,000 tons per annum. 654 employees. Norton washer. Canteen (sandwich), ambulance room with ambulance. Power generated at colliery, but to be replaced by SE Scotland Electricity Board supply in September 1948. Report dated 15-07-1948.

**Other details:** The beam engine for No. 1 shaft (which was built in Plymouth and survives as part of the museum) was bought from Harvey & Company of Hayle in Cornwall, and pumped water from the mineworkings from 1874 until 1954.

The pithead baths were completed in 1952 and were the 100th to be installed at Scottish collieries.

The colliery was greatly enlarged by the addition of a brick, tile and fireclay works in the 1890s, which operated until the 1970s. Following the mine's closure in the 1960s, it became a pioneering force behind the creation of the Scottish Mining Museum, of which it formed a major part along with Lady Victoria Colliery in neighbouring Midlothian. Following local authority re-organisation in the 1990s and the abolition of Lothian Region, Prestongrange became part of the East Lothian Museums Service, and remains open to the public as 'Prestongrange Industrial Heritage Museum'.

**Figures 5.61 and 5.62**

## Prestonlinks (The Links)

NT 3940 7502 (NT37NE/07)

**Parish:** Prestonpans

**Region/District:** Lo/Ea

**Council:** East Lothian

**Location:** Prestonpans

**Previous owners:** Nimmo & Company of Slamannan, and from 1912, Edinburgh Collieries Company

**Types of coal:** Gas, House and Steam

**Sinking/Production commenced:** 1899

**Year closed:** 1964

**Year abandoned:** 1964

**Average workforce:** 794

**Peak workforce:** 820

**Peak year:** 1950

**Shaft/Mine details:** 2 shafts, 121m and 115m deep

**Details in 1948:** Output 900 tons per day, 225,000 tons per annum. 801 employees. Lurig-type washer, and small coal washer. Baths (1933), canteen, first-aid room, ambulance. Electricity supplied by SE Scotland Electricity Board. Report dated 15-07-1948.

Figure 5.61: Prestongrange Colliery, near Prestonpans. Originating in the early 19th century and established by the Grant Suttie family, the pit was owned by the Summerlee Iron Company at the time of nationalisation in 1947. It was an exceptionally interesting colliery both because of the survival of a Cornish pumping engine, and the adjacent brickworks. SMM:1996.0694, SC381284

Figure 5.62: Prestongrange Colliery, near Prestonpans. After 1947, the pit operated with an average workforce of 686 miners until closure in 1962. Parts of the surface arrangement, including the Cornish pumping engine, were preserved, the site eventually being allied to Lady Victoria Colliery in Midlothian as part of the Scottish Mining Museum. It has since been incorporated into East Lothian Council's museum service, operating as an industrial museum. SMM:1996.01147, SC381285

Figure 5.63: Prestonlinks Colliery, Cockenzie. With an average workforce of almost 800 miners in the nationalised era, this was East Lothian's largest colliery. Dating from 1899, it was established by Nimmo of Slamannan, and later operated by Edinburgh Collieries Co. After closure in 1964, the site was cleared to make way for Cockenzie Power Station. SMM:1996.3663, SC381293

Figure 5.64: Winton Mine, near Ormiston. Comprising two surface mines, it was opened originally as a ventilation mine in 1943, but was redeveloped by the NCB in 1949, coal production commencing in 1953. It operated for ten years with an average workforce of 85 miners. SMM:1998.659a, SC393053

**Other details:** Built over by the site of Cockenzie Power Station in 1964.
**Figure 5.63**

## Tynemount

NT 4010 6856 (NT46NW/80)
**Parish:** Ormiston
**Region/District:** Lo/Ea
**Council:** East Lothian
**Location:** Ormiston
**Previous owners:** Ormiston Coal Company
**Types of coal:** House and Steam
**Sinking/Production commenced:** 1924
**Year closed:** 1952
**Year abandoned:** 1962
**Average workforce:** 301
**Peak workforce:** 370
**Peak year:** 1948
**Shaft/Mine details:** 1 shaft, 91m deep
**Details in 1948:** Output 335 tons per day, 83,750 tons per annum. 334 employees. Baum-type washer (Blantyre Engineering Company). Canteen, first-aid, ambulance. Electricity bought from SE Scotland Electricity Board. Report dated 15-07-1948.
**Other details:** Washery closed in 1953 and coal sent to new washery at Dalkeith 5.

## Winton

NT 4210 6987 (NT46NW/81)
**Parish:** Ormiston
**Region/District:** Lo/Ea
**Council:** East Lothian
**Previous owners:** National Coal Board
**Sinking commenced:** 1949
**Production commenced:** 1952
**Year closed:** 1962
**Year abandoned:** 1962
**Average workforce:** 85
**Peak workforce:** 85
**Peak year:** 1952
Mine details: 2 surface mines, No. 1 NT 4210 6987, and No. 2 NT 4206 6984.
**Other details:** Opened originally as a ventilation mine in 1943, but redeveloped by the NCB in 1949 to extract coal.
**Figure 5.64**

## Licensed Mines in East Lothian 1947–97

| Name | Licensee | Grid Ref | Opened | Closed |
|------|----------|----------|--------|--------|
| Chancellorville (Ormiston) | Beattie | NT 427 695 | 1971 | 1979 |
| Glencairn (Longniddry) | Glencairn Coal Company | NT 431 750 | 1936 | then NCB |
| Penkaet (Pencaitland) | Alex Gordon | NT 430 673 | pre-1947 | 1965 |
| Policies (Dalkeith) | Daniel Beattie Jnr | NT 333 681 | 1983 | 1987 |

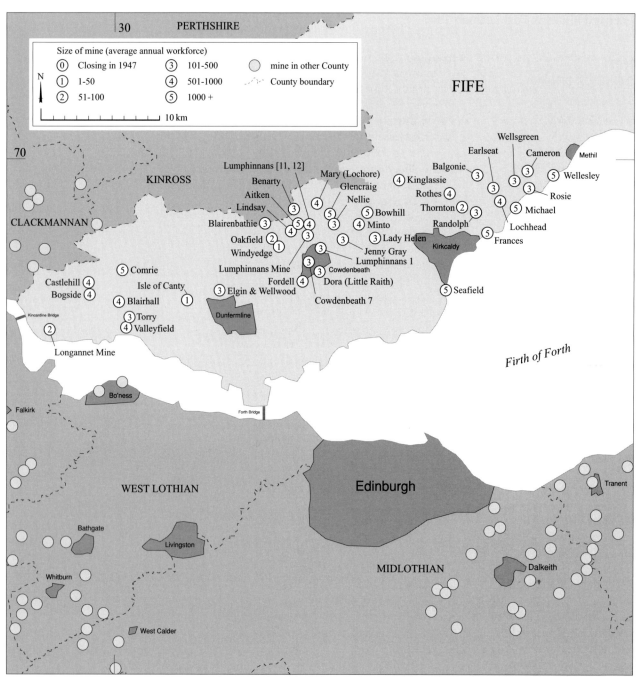

Figure 5.65: Map of National Coal Board (NCB) collieries in Fife.

# Fife

After nationalisation, in terms of manpower and output, the Fife coalfields rapidly grew to become the most important in Scotland, at their peak in 1957 employing over 24,000 miners. This is a reflection of the huge investment by the NCB in new sinkings and reconstructions in Fife, which witnessed some of the most prestigious projects, the most notable being the superpits at Rothes and Seafield. However, the NCB also inherited Michael and Wellesley collieries, two of the largest pits in Scotland, as well as Comrie, a model colliery completed shortly after the outbreak of World War II. In addition, open-cast mining was very important, and for some time one of the largest open-cast mines in Europe was operated by the NCB at Westfield near Cardenden.

The active Fife coalfields extended from the border with Clackmannanshire in the west through the southern half of the county to Buckhaven in the east. In the extreme east and west, coastal locations permitted the exploitation of undersea coals beneath the Firth of Forth, and helped to encourage a large export trade through ports such as Burntisland, and Methil. Indeed, by the early 17th century, Sir George Bruce had pioneered undersea mining with his 'Moat Pit', which emerged through a shaft in an artificial island off the shore from Culross, permitting the export of coal directly by boat. As with the Lothians on the opposite south side of the Forth, much early mining was strongly associated with the salt industry, and one of the finest relics of 19th-century mining is Preston Island near Culross, where the ruins of salt pans and an associated mine complex developed by Sir Robert Preston can still be seen. By the 19th and 20th centuries, Fife coal was fuelling a sophisticated and complex industrial base. In addition to a large ceramics sector, big local coal consumers included the Admiralty (based at Rosyth), the iron, alumina, lime-burning, engineering, linen, jute, floorcloth, linoleum, and paper industries.

Although coal-mining activity had once existed further afield in the north-east half of the county, it had died out by 1947. At that date, the largest concentration of mining was in the central south-eastern part of the county, around Lochgelly, Kelty, Cowdenbeath, and Cardenden, the most important collieries being Aitken, Bowhill, Glencraig and Lindsay. However, the biggest collieries, Michael, Seafield and Wellesley, were in the south-east between Kirkcaldy and Methil. In the far west beyond Dunfermline near Culross, Blairhall, Comrie and Valleyfield collieries were very important, and in the late 20th century, the Longannet complex became the most important producer of coal in Scotland.

The NCB took 38 collieries into state ownership on vesting day, and later established a further eight new mines. Fife was also home to some of the most successful short-term surface drift mines, such as Benarty and Blairenbathie, initially established by the Fife Coal Company on an American model, and designed to boost production during the war years. These were continued and replicated by the NCB in its early years with the aim of aiding the reconstruction of the British economy as a whole, and of enhancing coal exports to pay for international war-induced debt. They also bought time to allow the more ambitious projects to be planned and executed.

Apart from several major reconstructions, the first big project to come on stream was the superpit at Rothes in 1957. From the outset, however, it was plagued with geological and flooding problems and was closed amid great controversy and embarrassment only five years later. In 1965, the NCB ambitiously merged Valleyfield Colliery with a re-constructed Kinneil on the south side of the Forth in West Lothian. A year later, a second superpit at Seafield south of Kirkcaldy began pro-duction, and this proved to be a successful project, operating for over 20 years and becoming one of Scotland's largest pits. Seafield and sev-eral other Fife pits exploited coals from well under the Firth of Forth.

Fife produced a full range of different quality coals. A number of pits such as Valleyfield and Mary produced 'Navigation' coal to Ad-miralty standard, traditionally a Welsh preserve. In addition to house, steam, manufacturing, gas, and coking coals, some anthracite was pro-duced (at Glencraig), and the late 20th-century developments around Longannet extracted low-sulphur upper Hirst coal suitable for burning in the new power station. The power station was specially designed to burn the Hirst coal, the ash from which was refractory in nature, and therefore did not melt and damage the linings of the boilers. Secondary products of the mines included blaes, and some of the earlier pits had also produced ironstone.

Several large companies accounted for most of the collieries, and were amongst the most important mining companies in the UK. These included the Fife Coal Company, which, in addition to sinking many mines, also bought many others. Its collieries included those at Aitken, Benarty, Blairenbathie, Bowhill, Comrie, Cowdenbeath, Kinglassie, Lindsay, Lumphinnans, Mary, Oakfield, Randolph, Thornton, and Wellsgreen. In the east, the Wemyss Coal Company owned Michael, the largest pit in Scotland, and neighbouring Wellesley, Lochhead, Cameron and Rosie. Iron-makers also had a significant interest, Carron Company, Lochgelly Iron Company and Coltness Iron Company at some stage owned collieries such as Blairhall, Minto, Glencraig, Nel-lie, Lady Helen and Dora. Other owners at various times included the Earl of Buckingham (Fordell), the Earl of Elgin, the Earl of Rosslyn's Collieries Limited, the Cowdenbeath Coal Company, the Bowhill Coal Company, and the Balgonie Coal Company.

In terms of workforce, Michael was by far the biggest colliery, peak-ing in 1957 with 3,353 miners, and averaging 2,600 during its 20 years of operation after nationalisation. Other large mines included Seafield with an average of 2,180 miners, Wellesley with 1,900, Bowhill with 1,330, Aitken with 1,250, Comrie with 1,245, Glencraig with 1,105, and Frances with 1,104.

Although Fife, like the other Scottish coalfields, experienced a steady stream of mine closures in the late 1950s and 1960s, there was compensation in the form of the eight new sinkings, and in major re-construction projects. As a result, the number of miners working in the industry grew during the 1950s, reflecting the increased investment and an inward flow of miners from declining coalfields elsewhere. After 1957, however, the workforce began a steep decline which may also be explained in part by the introduction of mechanisation following the major investment. After 1966, the situation stabilised, and just over 6,000 miners were in employment annually until the late 1990s. Ul-timately, only the complex at Longannet survived, and after a period of worsening geological conditions, an underground flood forced its closure in 2002, bringing to an end deep coal mining both in Fife and Scotland as a whole. By 2004, the mine headgear at Frances and Mary (in Lochore meadows) were the only preserved monuments to the coal industry in Fife.

Outside the nationalised sector, 19 small licensed coalmines operated in Fife, some pre-dating World War II. Most originated in the 1950s, and the largest concentration was in the Lassodie area. The last of these ceased renewing its licence in 1991.

An excellent website compiled by Michael Martin, his son Colin, and Chris Sparling providing extensive information on Fife coal mining can be found at www.users.zetnet.co.uk/mmartin/fifepits/.

## Aitken

NT 1551 9481 (NT19SE/33)

**Parish:** Beath

**Region/District:** Fi/Du

**Council:** Fife

**Location:** Kelty

**Previous owners:** Fife Coal Company

**Types of coal:** House, Steam and Navigation

**Sinking/Production commenced:** 1895-9

**Year closed:** 1963

**Year abandoned:** 1969

**Average workforce:** 1,249

**Peak workforce:** 1,431

**Peak year:** 1956

**Shaft/Mine details:** 2 shafts, No. 1 (1896) 370m (NT 1551 9481), No. 2 (1923) 183m (NT 1555 9481) and sunk in 1923, linked to Lindsay Colliery.

**Details in 1948:** Average output 2,000 tons per day, 460,000 tons per annum. 1,368 employees. Baum-type washery, NCB generated electricity at its own power station, the largest at any colliery in Scotland. Baths (1934, for 912 men, with 80 shower cubicles and steam-heated lockers), canteen, first-aid centre. Mechanisation of

Figure 5.66: Aitken Colliery, Kelty, c.1947. A Fife Coal Company pit, sunk in the mid-1890s, and equipped with its own large power station. Pithead baths were built with the assistance of the Miners' Welfare Fund in 1934. On average, 1,249 miners were employed at the colliery in the years after nationalisation until its closure in 1963. SMM:1996.1724, SC393118

Figure 5.67: Balgonie Colliery, Thornton, 1953. Established originally by Balfour and Balgonie Estates in the mid-1880s, in the nationalised era its workforce peaked at 490 in 1952, eight years before its closure in 1960. SMM:1999.297A, SC445856

Figure 5.68: Benarty Mine, near Kelty, c.1947. A successful Fife Coal Company short-term surface mine project, modelled on examples in the USA. Begun originally in 1938, but suspended a year later, the project was revived in 1944. In the NCB era, it worked for a further ten years with an average workforce of 201 miners, closing in 1957. SMM:1997.0444, SC384934

underground haulage already under way in 1947, introducing large mine cars, locomotive haulage and skip winding. Report dated 16-08-1948.

**Other details:** Named after the Fife Coal Company's chairman, Thomas Aitken of Nivingston. The first Fife Coal Company colliery, a showpiece pit, and at the time, said to be the largest coal mine in Scotland.

**Figure 5.66**

## Balgonie (Julian Pit)

NT 3037 9862 (NT39NW/229)

**Parish:** Balgonie

**Region/District:** Fi/Ki

**Council:** Fife

**Location:** Thornton

**Previous owners:** Balfour and Balgonie Estates, later the Balgonie Colliery Company

**Types of coal:** House and Steam

**Sinking/Production commenced:** 1883-5

**Year closed:** 1960

**Year abandoned:** 1960

**Average workforce:** 448

**Peak workforce:** 490

**Peak year:** 1952

**Shaft/Mine details:** 2 shafts, No. 1 147m, No. 2 119m deep

**Details in 1948:** Output 470 tons per day, 123,262 tons per annum. 445 employees. Baum-type washer. DC and AC high tension electricity, none bought from public supply. Report dated 06-08-1948.

**Other details:** Re-organised by NCB in 1954.

**Figure 5. 67**

## Benarty

NT 1518 9594 (NT19NW/43.00)

**Parish:** Ballingry

**Region/District:** Fi/Du

**Council:** Fife

**Location:** Kelty

**Previous owners:** Fife Coal Company

**Types of coal:** Steam

**Sinking/Production commenced:** 1938, suspended 1939, restarted 1944–5

**Year closed:** 1959

**Year abandoned:** 1961

**Average workforce:** 201

**Peak workforce:** 237

**Peak year:** 1957

**Shaft/Mine details:** Surface mine, following seam from surface

**Details in 1948:** Output 470 tons per day, 128,000 tons per annum. 169 employees. Baths. Electricity, generated by NCB. Report dated 16-08-1948.

**Other details:** Built on site of old Benarty Colliery, it was a very successful short-term drift mine, achieving three times the national average productivity. Coal was taken to the washery at Aitken by road.

**Figure 5.68**

## Blairenbathie Mine

NT 1260 9480 (NT19NW/44)

**Parish:** Beath

**Region/District:** Fi/Du

**Council:** Fife

**Location:** Kelty

**Previous owners:** Fife Coal Company

**Types of coal:** House

**Sinking commenced:** 1945

**Production commenced:** 1949

**Year closed:** 1962

**Year abandoned:** 1962

**Average workforce:** 132

**Peak workforce:** 151

**Peak year:** 1957

**Shaft/Mine details:** 2 surface mines, 73m in vertical depth. Older shafts date from 1895.

**Details in 1948:** 120 tons per day, 30,000 tons per annum. 106 employees. Spray baths. Electricity generated by NCB. Report dated 10-08-1948.

**Other details:** A successful short-term drift mine, but not as productive as Benarty because of geological difficulties. 2-ton drop-bottom mine cars introduced with direct-rope and locomotive haulage.

**Figures 5.69 and 5.70**

## Blairhall

NT 0036 8846 (NT08NW/103)

**Parish:** Culross

**Region/District:** Fi/Du

**Council:** Fife

**Location:** East Grange

**Previous owners:** Carron Iron Company, Lochgelly Iron Company (c.1880), then Coltness Iron Company Limited

**Types of coal:** Coking, Gas, House and Steam, and Ironstone

**Sinking/Production commenced:** 1870s, 2 new sinkings 1906–11

**Year closed:** 1969

Figure 5.69: Blairenbathie Mine, near Kelty, 1949. Established close to the site of an earlier colliery dating from 1895, this was another successful Fife Coal Company short-term surface mine project, begun originally in 1945, production commencing in 1949. SMM: 1998.302, SC445859

Figure 5.70: Blairenbathie Mine, near Kelty. View of the mouth of one of the two surface mines. They operated until 1962 with an average workforce of 132 miners. SMM:1996.1588, SC378653

Figure 5.71: Blairhall Colliery, East Grange, Culross, 1952. View underground at the Lord Bruce Pit bottom, showing locomotive haulage (a Hunslet locomotive) installed as a result of the colliery reconstruction in 1952. Morris Allan of Dunfermline, SC382454

**Year abandoned:** 1969
**Average workforce:** 982
**Peak workforce:** 1,053
**Peak year:** 1947
**Shaft/Mine details:** 2 shafts, both 567m by 1957. Lady Veronica shaft (upcast, NT 0038 8854), Lord Bruce shaft (downcast and pumping, NT 0039 8856), named after Lord Elgin's son and daughter, and both used for winding men and coal.
**Details in 1948:** Output 1,100 tons per day, 290,000 tons per annum. 993 employees. Lurig type washer. Baths (1930), canteen, ambulance room. Electricity, all AC and generated at colliery. Report dated 23-08-1948.
**Other details:** Originally established in the 1870s by the Carron Iron Company to develop blackband ironstone, Carron sold the colliery to the Coltness Iron Company, who sank two new shafts in 1911. NCB commenced reconstruction in 1952, involving electrification and underground locomotive haulage, the aim being the increase of production to 1,500 tons per day. A new headframe and coal preparation plant (Baum) were installed in 1956.
**Figures 5.71 and 5.72**

## Bogside 1, 2 & 3
NS 9790 8905 (NS98NE/193)
**Parish:** Culross
**Region/District:** Fi/Du
**Council:** Fife
**Previous owners:** National Coal Board
**Types of coal:** Upper Hirst
**Sinking commenced:** 1957
**Production commenced:** 1959
**Year closed:** 1986
**Year abandoned:** 1987

Figure 5.72: Blairhall Colliery, East Grange, Culross, c.1960. A showpiece colliery established in the 1870s by the Carron Company in search of blackband, and later owned by the Lochgelly and Coltness Iron Companies respectively. In the nationalised era, its workforce averaged 982, and it continued to produce coking and other types of coal until 1969. SMM:1996.02401, SC382528

**Average workforce:** 618

**Peak workforce:** 875

**Peak year:** 1971

**Shaft/Mine details:** 3 surface mines, No .1 NS 9866 8121; No. 2 NS 9789 8902; No. 3 NS 9777 8898

**Other details:** Part of the group of mines designed to access 15 million tons of Upper Hirst coal, and serving initially Kincardine Power Station, and subsequently Longannet, to which it was linked by an underground conveyor running continuously at 10mph. They were highly productive, regularly exceeding 2,000 tons output per day. Flood damage was incurred during an overtime ban in 1983. Final closure was caused by flooding and gas problems during the 1984–5 strike, and the workforce was transferred to new mines at Castlebridge and Solsgirth, and to the established collieries at Frances and Comrie.

**Figures 5.73 and 5.74**

## Bowhill 1, 2 & 3

NT 2115 9567 (NT29NW/54)

**Parish:** Auchterderran

**Region/District:** Fi/Ki

**Council:** Fife

**Location:** Cardenden

**Previous owners:** Bowhill Coal Company, Fife Coal Company from 1909

**Types of coal:** House, Steam, and blaes

**Sinking/Production commenced:** 1895–9

**Year closed:** 1965

**Year abandoned:** 1968

**Average workforce:** 1,331

**Peak workforce:** 1,490

**Peak year:** 1961

Figure 5.73: Bogside Colliery, near Culross, c.1960. A large NCB project comprising three surface mines, begun in 1957, production commencing in 1959. It worked for 27 years producing Upper Hirst coal initially for Kincardine Power Station, and then also for Longannet. Its average workforce was 618, and it finally closed in 1986. SC446737

Figure 5.74: Bogside Colliery, near Culross, c.1960. View of surface arrangement, with one of the three surface mines visible (left), and the pithead baths in the foreground. SC706725

Figure 5.75: Bowhill Colliery, Cardenden. Established in 1895 by the Bowhill Coal Co, but became one of the Fife Coal Co's most important pits after 1911. It was reconstructed in the 1940s, just prior to nationalisation. SC446231

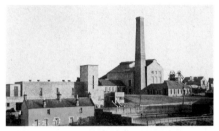

Figure 5.76: Bowhill Colliery, Cardenden, c.1947. This view shows the fine pithead baths, which were built with the support of the Miners' Welfare Fund in 1937. During the nationalised era, the pit employed an average of 1,330 miners, and operated until 1965. SC446220

Figure 5.77: Bowhill Colliery, Cardenden. General view of surface arrangement, clearly showing the headgear of its three shafts. Two were established in the 1890s when the pit was first sunk, and a third in 1952. SMM:1998.0323, SC378658

**Shaft/Mine details:** 2 shafts, No. 1 (NT 2115 9567) 399m, and No. 2 (NT 2116 9565) 304m deep. A third shaft was sunk in 1952.

**Details in 1948:** Output 1,300 tons per day, 350,000 tons per annum. 1,230 employees, Norton washer. Baths (1938, for 1,512 men and 26 women), canteen. Electricity generated at colliery. Report dated 20-08-1948.

**Other details:** Known locally as 'Josephine Pit', after lady Josephine Haig, daughter of the chairman of the company. Noted for a disaster in October 1931 in which a firedamp explosion caused the death of ten men. Reconstructed in 1945, involving a major reorganisation of underground transport including the introduction of large mine cars with direct-rope and locomotive haulage. At the same time, new mine-car handling facilities were installed at the surface, along with a new aerial ropeway.

**Figures 5.75, 5.76 and 5.77**

## Brighills

(see Minto)

## Cameron Mine

NT 3444 9919 (NT39NW/230)

**Parish:** Markinch

**Region/District:** Fi/Ki

**Council:** Fife

**Location:** near Windygates

**Previous owners:** Wemyss Coal Company

**Types of coal:** Steam

**Sinking/Production commenced:** 1934

**Year closed:** 1959

**Year abandoned:** 1959

**Average workforce:** 143

**Peak workforce:** 207

**Peak year:** 1957

**Shaft/Mine details:** 2 surface mines, one downcast and one upcast, 914m long at 1:6 dip (No. 1 NT 3444 9919 and No. 2 NT 3445 9917). Built on site of old Isabella Pit.

Details in 1949: Output 250 tons per day, 62,500 tons per annum. 156 employees. Coal washed at Denbeath central washery. Baths, canteen (snack meals). First-aid. Electricity from Michael Power Station (NCB). Report dated 15-02-1949.

**Other details:** Reconstructed in 1948, including the provision of baths at a cost of £51,000.

**Figure 5.78**

## Castlehill Mine

NS 9782 9002 (NS99SE/05)

**Parish:** Culross

Figure 5.78: Cameron Mine, near Windygates, c.1948. Begun by the Wemyss Coal Company in 1934 on the site of Isabella Colliery, the two surface mines employed an average of 143 miners until closure in 1959. SC378660

**Region/District:** Fi/Du

**Council:** Fife

**Location:** nr. Culross

**Previous owners:** National Coal Board

**Types of coal:** Upper Hirst

**Sinking commenced:** 1965

**Production commenced:** 1969

**Closed:** 1990

**Average workforce:** 749

**Peak workforce:** 770

**Peak year:** 1972

**Shaft/Mine details:** Part of the Longannet complex, for which it was the main ventilation unit and second egress. 2 surface mines, each 4.88m by 3.66m, approximately 1,600m long at an incline of 1 in 4.

**Other details:** Production dedicated to Longannet Power Station, to which it was linked underground via Longannet Mine. Operated as single unit with Castlebridge from September 1986. By 1998, ventilation and access only. The Longannet complex closed in 2002 following the failure of an underground dam.

**Figure 5.79**

Figure 5.79: Castlehill Mine, near Culross, c.1970. Sinking of the two surface mines commenced in 1965, production following in 1969. Part of the Longannet complex, for which it also provided ventilation, it employed an average of 749 miners until its closure in 1990. SC446748

## Comrie

NT 0053 9100 (NT09SW/31.01)

**Parish:** Saline

**Region/District:** Fi/Du

**Council:** Fife

**Location:** Oakley, Dunfermline

**Previous owners:** Fife Coal Company

**Types of coal:** House, and blaes

**Sinking/Production commenced:** 1936–9

**Year closed:** 1986

**Average workforce:** 1,245

**Peak workforce:** 1,498

**Peak year:** 1963

**Shaft/Mine details:** 2 circular shafts, both 130m deep and with concrete lining, 6.1m and 6.7m diameter respectively. No. 1 shaft (NT 0053 9100) utilised a skip-winding system, and No. 2 shaft (NT 0064 9094) carried men and materials in conventional cages, and was also used for downcast ventilation. Main pumping was from No. 2 shaft pit bottom. Both shafts had ground-mounted electric winding engines with dynamic braking and rope guides.

**Details in 1948:** Output 1,600 per day, 425,000 per annum. 1,000 employees. Norton Washer, Baum type. Baths (1942), canteen, first-aid room. All electricity AC, supplied by NCB power station. Report dated 23-08-1948.

**Other details:** At the time of construction, Comrie was the new showpiece pit of the Fife Coal Company, and was described in detail in *The Layout and Equipment of Comrie Colliery in Fifeshire*

Figure 5.80: Comrie Colliery, Oakley, c.1947. A model colliery built by the Fife Coal Co, and incorporating the latest German mining technology. Sinking was completed in 1939, just after the beginning of World War II. The colliery produced coal until 1986, employing an average of 1,245 miners. SC706654

Figure 5.81: Comrie Colliery, Oakley, c.1947. This view shows No. 2 shaft headframe and adjacent surface buildings. The ornamental fountain in the foreground cooled circulation water from No. 2 shaft's braking system. SC706704

Figure 5.82: Cowdenbeath No. 7 Colliery, c.1947. Sunk originally by the Cowdenbeath Coal Co, and taken over by the Fife Coal Company in 1896, the pit employed an average of 435 miners until its closure in 1960. SC446327

by William Reid, Inst. of Mining Engineers, Cardiff, 18-5-1939. Important features included the use of forced-fan ventilation (unusual at the time), and of skip-winding for raising coal from the pit bottom. The skips were brought in from Germany, and one week before the outbreak of World War II, the German engineers responsible for their installation were called home with the job unfinished. Unusual features included the ornamental fountain and pond at the heart of the surface arrangement, the real purpose of which was to provide a cooling system for Shaft No. 2's braking system. Underground transport was fully mechanised, with locomotive haulage and belt conveyors. Surface facilities included a Baum-type washer, and an aerial ropeway to the bing. Scottish Rexco established a smokeless fuel plant at Comrie in 1964. In 1983, 75% of the mine's output was sold to the South of Scotland Electricity Board.
**Figures 5.80 and 5.81**

## Cowdenbeath 7

NT 1634 9171 (NT19SE/35.01)

**Parish:** Beath

**Region/District:** Fi/Du

**Council:** Fife

**Location:** Cowdenbeath

**Previous owners:** Cowdenbeath Coal Company, Fife Coal Company from 1896

**Types of coal:** House and Steam

**Sinking/Production commenced:** 1860

**Year closed:** 1960

**Year abandoned:** 1967

**Average workforce:** 435

**Peak workforce:** 454

**Peak year:** 1950

**Shaft/Mine details:** 1 shaft, 240m deep (also used for pumping and ventilation downcast)

**Details in 1948:** Output 560 tons per day, 151,200 per annum. 440 employees. Foulford Central Washery, Baum-type. Canteen, baths at No. 10 Cowdenbeath. Electricity supplied from NCB Aitken power station. Report dated 11-08-1948.

**Other details:** Reconstruction of underground transport occurred in the mid-1940s, together with the installation of new mine-car handling facilities at the surface, and a new winder for No. 10 shaft.
**Figure 5.82**

## Denbeath

(see Wellesley)

## Dora (Little Raith)

NT 1723 9091 (NT19SE/53.01)

**Parish:** Auchtertool

**Region/District:** Fi/Du

**Council:** Fife

**Location:** Cowdenbeath

**Previous owners:** Lochgelly Iron & Coal Company

**Types of coal:** House and Steam

**Sinking/Production commenced:** 1875

**Year closed:** 1959

**Year abandoned:** 1959

**Average workforce:** 266

**Peak workforce:** 280

**Peak year:** 1957

**Shaft/Mine details:** 1 shaft, 178m deep, but winding from 127m

**Details in 1948:** 263 men employed (204 underground), Campbell Binnie bash-tank feldspar washer, canteen (full meals), no baths, first aid room. Steam winder, all electricity supplied by Nellie Colliery Power Station. Report dated 11-08-1948.

**Other details:** The last of a number of pits on the 'Little Raith' estate.


## The Dubbie

(see Frances)


## Earlseat

NT 3166 9783 (NT39NW/165)

**Parish:** Markinch

**Region/District:** Fi/Ki

**Council:** Fife

**Previous owners:** National Coal Board

**Sinking/Production commenced:** 1950

**Year closed:** 1958

**Year abandoned:** 1959

**Average workforce:** 114

**Peak workforce:** 162

**Peak year:** 1956

**Other Information:** Originally, the site of earlier Earlseat Colliery operated by the Wemyss Coal Company dating from 1903, and closed in 1926.


## Elgin & Wellwood (Leadside)

NT 0876 8942 (NT08NE/177.01)

**Parish:** Dunfermline

**Region/District:** Fi/Du

**Council:** Fife

**Location:** by Dunfermline

**Previous owners:** Thomas Spowart & Company

Figure 5.83: Fordell Colliery (Alice Pit), Crossgates (undated). Sunk originally by the Earl of Buckingham in 1750, this was one of Fife's oldest collieries. In the nationalised era, it employed an average of 509 miners, closing in 1966. SC446720

Figure 5.84: Fordell Colliery, Crossgates (undated). View of William Pit, sunk in 1843, showing Fordell waggons passing under the loading plant. SMM:1997.0469, SC378663

**Types of coal:** House, Manufacturing and Steam

**Sinking/Production commenced:** 1827

**Year closed:** 1950

**Year abandoned:** 1950

**Average workforce:** 214

**Peak workforce:** 214

**Peak year:** 1947

**Shaft/Mine details:** 1 shaft, 73m deep

**Details in 1948:** Output 220 tons per day, 55,750 tons per annum. 242 employees. Dickson and Mann bash tank washer. Canteen (packed meals and teas). Steam and electricity, latter supplied by Fife Electric Power Company. Report dated 11-08-1948.

**Other details:** Elgin merged with Wellwood between 1869 and 1880. Associated brickworks, operating from 1934 to 1981. Closure in 1950 caused by underground fire, many miners being moved to Aitken. Wellwood situated at NT0965 8912 (NT08NE/295).

### Fordell

NT 1584 9013 (NT19SE/32.00)

**Parish:** Dunfermline

**Region/District:** Fi/Du

**Council:** Fife

**Location:** Crossgates

**Previous owners:** Earl of Buckinghamshire

**Types of coal:** Gas, House and Steam

**Sinking/Production commenced:** 1750

**Year closed:** 1966

**Year abandoned:** 1970

**Average workforce:** 509

**Peak workforce:** 622

**Peak year:** 1959

**Shaft/Mine details:** 3 shafts, No. 1 (Alice Pit, NT 1583 9013) 174m, in production by 1894. No. 2 (William Pit, NT 1602 8951) 128m dated from 1843, No. 3 (Lady Anne Pit, NT 1595 8910) 69m deep. Henderson Mine was a drift begun in 1946 and opened in 1948.

**Details in 1948:** Output 540 tons per day, 145,800 tons per annum. 464 employees. Snack canteen, first-aid room, electricity bought from the former Fife Electric Power Company, some steam also used. Report dated 11-08-1948.

**Other details:** Situated on the edge of the Cowdenbeath coalfield, and associated with Sir Robert Henderson's wooden waggon way to St David's Harbour – one of the earliest railways in Scotland.

**Figures 5.83 and Figure 5.84**

### Frances (The Dubbie)

NT 3098 9388 (NT39SW.25)

**Parish:** Kirkcaldy and Dysart

**Region/District:** Fi/Ki

**Council:** Fife

**Location:** Dysart

**Previous owners:** Originally the Early of Rosslyn's Collieries Limited, Fife Coal Company from 1923

**Types of coal:** House and Steam

**Sinking/Production commenced:** *c.*1850

**Year closed:** 1988 (production ceased in 1985)

**Average workforce:** 1,104

**Peak workforce:** 1,482

**Peak year:** 1957

**Shaft/Mine details:** Frances shaft 460m deep (upcast, eliptical for 183m, and circular for remaining 277m (NT 3098 9388) Lady Blanche shaft 73m, Frances surface mine 732m long at 1 in 4 (NT 3096 9392) driven in 1924. New ventilation mine (NT 3091 9405).

**Details in 1948:** Output 1,250 tons per day, 266,000 tons per annum. 916 employees. Baum-type washer, baths (1931), canteen, ambulance room. Reconstruction of underground transport in mid-1940s, including the introduction of large mine cars with direct-rope and locomotive haulage. Electricity supplied by NCB from their Kelty power station. Report dated 10-08-1948.

**Other details:** Frances worked undersea coals from its cliff-top location. It was known locally as 'The Dubbie' because of wet undergound conditions, and was taken over by the Fife Coal Company in 1923 and subsequently equipped with its own washery, built by Simon Carves in 1925. Further redevelopment occurred in the 1930s, and in the 1940s, new headgear and a ground-mounted Robey & Metro Vickers electric 1,600hp winding engine were installed with minimum disruption to production. Underground locomotive haulage (electric Greenbat units) was introduced in 1957. Its washery was closed in 1965, coal being taken to Bowhill for treatment. Linked underground to Seafield by 1981, and drained latterly from a unit retained at Michael (previously closed in 1967). Fires caused by spontaneous combustion broke out during the 1984 strike. Retained on care and maintenance basis after 1985, but planned 'Frances Project' of 1990 never materialised, and the surface buildings were subsequently demolished with the exception of the headframe, which survives as a monument to the Fife coal industry.

**Figure 5.85 and 5.86**

## Glencraig

NT 1812 9557 (NT19NE/59)

**Parish:** Ballingry

**Region/District:** Fi/Du

**Council:** Fife

**Location:** Glencraig

**Previous owners:** Wilsons and Clyde Coal Company – a subsidiary of the Coltness Iron Company

**Types of coal:** Anthracite and Steam

Figure 5.85: Frances Colliery (also known as the Dubbie), Dysart, c.1960. The pit dates from the mid-19th century, and was operated by the Fife Coal Company by 1923. It was reconstructed in the 1940s and 1950s, and linked to Seafield in 1981. SMM:1996.3402, SC379678

Figure 5.86: Frances Colliery, Dysart, 1974. This view shows the colliery's surface arrangement, and its location on the Fife coast. In its 41 years as an NCB colliery, it regularly employed over 1,100 miners, finally closing for good in 1988. The surface remains were subsequently demolished, with the exception of the headframe, which survives as a monument to the Scottish Coal industry. J R Hume, SC444501

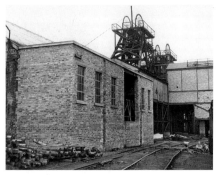

Figure 5.87: Glencraig Colliery, c.1966. Operated by Wilsons & Clyde, a subsidiary of the Coltness Iron Co, this pit produced steam coal and anthracite. It was reconstructed by the NCB, and linked to neighbouring Mary Colliery, after which there was a serious rat infestation. Morris Allan of Dunfermline, SC445867

Figure 5.88: Glencraig Colliery, 1953. In the nationalised era, it employed an average of over 1,100 miners, closing in 1966. SMM:1998.0234, SC378668

**Sinking/Production commenced:** 1896
**Year closed:** 1966
**Year abandoned:** 1968
**Average workforce:** 1,105
**Peak workforce:** 1,316
**Peak year:** 1950
**Shaft/Mine details:** 2 rectangular shafts, No. 1 (downcast) 450m, No. 2 (upcast) 340m deep in 1948, both with steam winders supplied by 9 Lancashire boilers. Drained by a beam engine.
**Details in 1948:** Output 1,050 tons per day, 230,000 per annum. 1,030 employees. 2 washers (Dickson and Mann and Hugh Martin). Baths (1936), canteen, first-aid. DC electricity, none bought from outside. Report dated 10-08-1948.
**Other details:** Mine begun by W H Telfer from Overtown, Lanarkshire. Reconstructed in NCB era, and linked to Mary (Lochore) and Nellie, but suffered serious rat infestation, a problem common in coal mines where pit ponies were used, and animal feed plentiful.
**Figures 5.87 and 5.88**

## Isle of Canty
NT 0617 8868 (NT08NE/292)
**Parish:** Carnock
**Region/District:** Fi/Du
**Council:** Fife
**Location:** Carnock Road, Dunfermline
**Previous owners:** Alloa Coal Company
**Types of coal:** Anthracite (semi-anthracite)
**Sinking/Production commenced:** 1939
**Year closed:** 1948
**Year abandoned:** 1949
**Average workforce:** 41
**Peak workforce:** 41
**Peak year:** 1947
**Shaft/Mine details:** one main mine, and one airway mine from surface
**Details in 1948:** Output 48 tons per day, 13,500 per annum. 36 employees. No washer, no canteen or baths. Electricity AC, supplied by Fife Electric Power Company, all public supply. Report dated 20-08-1948.
**Other details:** The last mine to be opened by the Alloa Coal Company, and the first to be closed by the NCB after nationalisation in 1947.

## Jenny Gray
NT 1917 9352 (NT19SE/47.01)
**Parish:** Auchterderran
**Region/District:** Fi/Du
**Council:** Fife
**Location:** Lochgelly

**Previous owners:** Lochgelly Iron & Coal Company

**Types of coal:** Gas, House, Manufacturing and Steam

**Sinking/Production commenced:** 1854

**Year closed:** 1959

**Average workforce:** 467

**Peak workforce:** 492

**Peak year:** 1950

**Shaft/Mine details:** 3 shafts, No. 1 (possibly 1854) 154m (NT 1917 9352), No. 2 (1889) 154m (NT 1922 9349), and No. 3 (1927) 46m deep (NT 1922 9347)

**Details in 1948:** 610 tons per day, 166,947 tons per annum. 440 employees. (Coal washed at Nellie and Mary pits). Canteen, first-aid. Electricity and steam power. Report dated 06-80-1948.

**Other details:** A very wet mine, resulting in the installation of an underground steam pumping engine. After closure, men moved to Kinglassie, Minto and Valleyfield.

**Figure 5.89**

Figure 5.89: Jenny Gray Colliery, Lochgelly, 1954. Dating from 1854, and owned previously by the Lochgelly Iron and Coal Co, this pit closed in 1959, having employed an average 467 miners in the nationalised era. SMM:1997.0428, SC378669

## Kelty 4 & 5

(see Lindsay)

## Kinglassie

NT 2377 9827 (NT29NW/52)

**Parish:** Kinglassie

**Region/District:** Fi/Ki

**Council:** Fife

**Location:** by Cardenden

**Previous owners:** Fife Coal Company

**Types of coal:** House and Steam

**Sinking/Production commenced:** 1908

**Year closed:** 1966

**Year abandoned:** 1968

**Average workforce:** 652

**Peak workforce:** 764

**Peak year:** 1956

**Shaft/Mine details:** 2 shafts, No. 1 320m (NT 2377 9827), and No. 2 312m deep (NT 2378 9827). Surface mine sunk 1934.

**Details in 1948:** Output 800 tons per day, 210,000 per annum. 657 employees. Baum washer, baths (under construction), canteen (snacks), AC electricity generated by NCB. Report dated 23-08-1948.

**Other details:** Well known for water problems, but these ominously eased with the sinking of the ill-fated Rothes nearby.

**Figure 5.90**

Figure 5.90: Kinglassie Colliery, near Cardenden, c.1945. Sunk by the Fife Coal Company from 1908, this pit employed an average of over 652 miners after nationalisation, closing in 1966. SC446329

## Lady Helen (Dundonald)

NT 2170 9367 (NT29SW/43)

Figure 5.91: Lady Helen, Dundonald, near Cardenden. Operated after 1910 by the Lochgelly Iron & Coal Co, it dates originally from 1892. After 1947, it was retained by the NCB until 1964, employing an average of 342 miners. The National Archives, COAL80/576/5, SC614084

Figure 5.92: Lady Helen, Dundonald, near Cardenden. This view was taken in 1953 and shows the buildings around the new surface mine, which commenced production a year later. SMM:1998.0322(a), SC382526

**Parish:** Auchterderran

**Region/District:** Fi/Ki

**Council:** Fife

**Location:** Cardenden

**Previous owners:** Dundonald Coal Company, Lochgelly Iron & Coal Company from *c*.1910

**Types of coal:** Steam

**Sinking/Production commenced:** 1892

**Year closed:** 1964

**Year abandoned:** *c*.1964

**Average workforce:** 342

**Peak workforce:** 455

**Peak year:** 1957

**Shaft/Mine details:** No. 2 surface mine ('Dothan') at NT 2170 9367, completed in 1954. Previously 1 shaft (Lady Helen), 47m deep, and No. 1 (west) mine (operated 1897–1950)

**Details in 1948:** Output 400 tons per day, 115,476 tons per annum. 369 employees. Campbell Binnie feldspar washer. Canteen (snacks), ambulance room. AC electricity, NCB generated. Report dated 09-08-1948.

**Other details:** Originally named after Lady Helen, Viscountess Novar of Raith. New drift mine operated on a similar basis to others like Benarty and Blairenbathie.

**Figures 5.91 and 5.92**

## Lindsay (Kelty 4 & 5)

NT 1484 9414 (NT19SW/61)

**Parish:** Beath

**Region/District:** Fi/Du

**Council:** Fife

**Location:** Kelty

**Previous owners:** Fife Coal Company

**Types of coal:** House and Steam

**Sinking/Production commenced:** 1873

**Year closed:** 1965

**Year abandoned:** 1967

**Average workforce:** 818

**Peak workforce:** 970

**Peak year:** 1957

**Shaft/Mine details:** 1 shaft, 239m (NT 1484 9414). 2 surface mines (NT 1475 9414), No. 1 214m (sunk in 1924), No. 2 98m deep (sunk in 1945). Lindsay 2, a surface mine sunk in 1939, closed in 1947, at which time it employed 83 people. The pit was reconstructed in the 1950s, and included pithead baths, a canteen and medical centre. An underground explosion on 14 December 1957 killed nine men.

**Details in 1948:** Output 1,000 tons per day, 250,000 tons per annum. 790 employees. Simon Carves washer (1920). Small 'piece-canteen', first-aid. AC electricity, NCB-generated. Report dated 16-08-1948.

**Other details:** Originally Kelty 4 and 5, but renamed Lindsay after the first chairman of the Fife Coal Company, William Lindsay. Connected to Aitken. Reconstruction of underground facilities under way in 1946.

**Figure 5.93**

Figure 5.93: Lindsay Colliery, Kelty, c.1947. Originally known as Kelty 4 & 5, but renamed after the chairman of the Fife Coal Co, William Lindsay. It was reconstructed in the 1950s, and closed in 1965, having employed an average of 818 miners in the nationalised era. SC446271

## Little Raith

(see Dora)

## Lochhead (Lochhead/Victoria)

NT 3224 9664 (NT39NW/149)

**Parish:** Wemyss

**Region/District:** Fi/Ki

**Council:** Fife

**Location:** Coaltown of Wemyss

**Previous owners:** Wemyss Coal Company

**Types of coal:** House and Steam

**Sinking/Production commenced:** 1890

**Year closed:** 1970

**Year abandoned:** 1970

**Average workforce:** 580

**Peak workforce:** 841

**Peak year:** 1957

**Shaft/Mine details:** 4 shafts, Lochhead 116m (originally 'Lady Lilian'), Victoria 133m, Duncan 69m, Earlseat Road 38m deep, and surface mine

**Details in 1948:** Output 660 tons per day, 184,100 tons per annum. 660 employees. Baths (1930), canteen. Steam, and electricity from Michael Colliery. Report dated 16-08-1948.

**Other details:** Also known as 'Wemyss Colliery, Victoria shaft possibly dated from c.1830, and may have been named following the coronation. Production ceased in 1915, but was retained as a ventilation shaft for Lochhead. Very vulnerable to spontaneous combustion. Shared ventilation systems with Michael.

**Figure 5.94**

Figure 5.94: Lochhead Colliery, Coaltown of Wemyss. Operated by the Wemyss Coal Company since 1890, but possibly dating from the 1830s. Well known for spontaneous combustion problems, it shared its ventilation with Michael Colliery nearby. The pit employed an average of 580 miners in the NCB era until its closure in 1970. SMM:1998.643b, SC393138

## Longannet Complex

NS 9455 8635

**Parish:** Tulliallan

**Region/District:** Fi/Du

**Council:** Fife/Clackmannanshire

**Production commenced:** 1969

**Year closed:** 2002

**Type of coal:** Upper Hirst

**Other details:** Comprised Bogside, Castlebridge, Castlehill, Solsgirth

and Longannet Mine, directly serving Longannet Power Station and designed to supply 10,160 tons of coal per day. The 8.8km tunnel between Solsgirth and Longannet contained what was claimed to be the longest underground conveyor belt in the world at the time. It was described as a blueprint for modern mining, and many aspects of the development were duplicated elsewhere in the UK (e.g. Selby), and overseas. Despite geological problems, its miners regularly broke productivity records for output from single faces. The last deep coal mine in Scotland, it closed after a catastrophic flood in April 2002.
**Figure 5.95**

**Longannet Mine**
NS 9454 8622 (NS98NW/65)
**Parish:** Tulliallan
**Region/District:** Fi/Du
**Council:** Fife/ Clackmannanshire
**Previous owners:** National Coal Board
**Types of coal:** Upper Hirst
**Sinking commenced:** 1964
**Production commenced:** 1969
**Year closed:** 2002
**Average workforce:** 56
**Peak workforce:** 60
**Peak year:** 1987
**Shaft/Mine details:** surface mine

Figure 5.95: Longannet Colliery c.1970. The last colliery to operate in Scotland, Longannet mine was part of a large complex of mines in West Fife and Clackmannanshire, extracting low-sulphur Upper Hirst coal for use in the adjacent Longnet Power Station. The mine itself commenced production in 1969, and employed an average of only 56 men until the transfer of miners from other parts of the complex in the 1990s. It closed after a major underground flood in 2002. SMM:1996.02577, SC706720

**Other details:** Directly served Longannet Power Station, and was the main arterial tunnel (containing a cable belt conveyor) connecting the power station with the outstations at Bogside, Castlebridge, Castlehill and Solsgirth. Following the closure of Bogside, Castlebridge and Solsgirth, new drivage began in 1999 to exploit the coals under the River Forth near Airth. After experiencing worsening geological difficulties, the mine was flooded after the failure of a dam in 2002, and subsequently closed.

Figure 5.96: Lumphinnans No. 1 Colliery, Cowdenbeath, c.1947. Originally sunk in the 1850s by the Cowdenbeath Coal Co, but operated by the Fife Coal Company after 1896. In the nationalised era, it operated with an average workforce of 182 miners until its closure in 1957. SC378673

## Lumphinnans Mine

NT 1620 9428 (NT19SE/71)

**Parish:** Beath

**Region/District:** Fi/Du

**Council:** Fife

**Location:** Cowdenbeath

**Previous owners:** Fife Coal Company

**Types of coal:** House

**Sinking/Production commenced:** 1946

**Year closed:** 1966

**Statistics:** Included within Lumphinnans 11 & 12

**Details in 1948:** Output 220 tons per day, 50,000 tons per annum. 147 employees. All electricity supplied by Aitken Colliery power station (NCB). Report dated 11-08-1948.

## Lumphinnans 1

NT 1743 9303 (NT19SE/114)

**Parish:** Beath

**Region/District:** Fi/Du

**Council:** Fife

**Location:** Cowdenbeath

**Previous owners:** Cowdenbeath Coal Company, Fife Coal Company from 1896

**Types of coal:** House

**Sinking/Production commenced:** 1852

**Year closed:** 1957

**Year abandoned:** 1957

**Average workforce:** 182

**Peak workforce:** 202

**Peak year:** 1951

**Shaft/Mine details:** 2 shafts, No. 1 318m, and No. 2 236m deep

**Details in 1948:** Output 220 tons per day, 50,000 tons per annum. 147 employees. All electricity supplied by Aitken Colliery power station (NCB). Report dated 11-08-1948.

**Other details:** Sunk originally as an ironstone pit.

**Figure 5.96**

Figure 5.97: Lumphinnans Nos XI & XII Colliery, Cowdenbeath, c.1947. Taken over by the Fife Coal Company in 1896, and a producer of ironstone as well as Navigation coal for the Admiralty. With an average workforce of almost 600 men in the nationalised era, the pit operated until 1966. SC378674

## Lumphinnans XI & XII (Peewit or Peeweep)

NT 1620 9426 (NT19SE/72)

**Parish:** Beath

**Region/District:** Fi/Du

**Council:** Fife

**Location:** Kelty

**Previous owners:** Cowdenbeath Coal Company, taken over by Fife Coal Company in 1896

**Types of coal:** House, Steam, Navigation, and ironstone

**Sinking/Production commenced:** pre-1896

**Year closed:** 1966

**Year abandoned:** 1968

**Average workforce:** 596

**Peak workforce:** 764

**Peak year:** 1947

**Shaft/Mine details:** 2 shafts, No. 11 shaft 410m (NT 1620 9426), and No. 12 169m deep (NT 1618 9424) and sunk in 1924. New surface mine sunk in 1946 (see Lumphinnans Mine).

**Details in 1948:** Output 1,050 tons per day, 262,500 tons per annum. 794 employees. Baum washer. Baths, canteen, first-aid. Electricity supplied from NCB Aitken power station. Report dated 11-08-1948.

**Other details:** Also known as the 'Peeweep'. Washery built in 1928, and baths in 1933. Transport reconstruction underground in progress in 1946, including the introduction of 2-ton drop-bottom mine cars and locomotive and direct-rope haulage. In addition, a new surface mine was being developed to replace No. XI shaft. Surface improvements included new car handling arrangements and a conveyor direct to the screening plant. Reconstructed continued by the NCB in the 1950s.

**Figure 5.97**

## Kelty 4 & 5

(see Lindsay)

## Mary (Lochore)

NT 1717 9636 (NT19NE/27)

**Parish:** Ballingry

**Region/District:** Fi/Du

**Council:** Fife

**Location:** Lochore

**Previous owners:** Fife Coal Company

**Types of coal:** House, Navigation

**Sinking/Production commenced:** 1904, No. 2 shaft in 1923

**Year closed:** 1966

**Year abandoned:** 1968

**Average workforce:** 614

**Peak workforce:** 780

**Peak year:** 1957

**Shaft/Mine details:** 2 shafts, No. 1 (ventilation) 201m (NT 1717 9636), and No. 2 521m deep (NT 1704 9619), built in 1923 with a reinforced concrete headframe. Original sinking in 1904 took it to 613m, at the time the deepest mine in Scotland.

**Details in 1948:** Output 800 tons per day, 211,355 per annum. 647 employees. Baum-type washer. Canteen (packed meal), first-aid room. AC electricity. Report dated 10-08-1948.

**Other details:** Reconstruction programme under way in 1946 included the introduction underground of large mine cars and locomotive haulage. New car handling plant was added at the surface, as was a redd disposal plant. Reinforced concrete headframe for No. 2 shaft now survives as part of Lochore Meadows Country Park.

**Figures 5.98 and 5.99**

## Michael

NT 3356 9611 (NT39NW/23)

**Parish:** Wemyss

**Region/District:** Fi/Ki

**Council:** Fife

**Location:** East Wemyss

**Previous owners:** Wemyss Coal Company

**Types of coal:** House and Steam

**Sinking/Production commenced:** 1895/1898

**Year closed:** 1967

**Average workforce:** 2,598

**Peak workforce:** 3,353

**Peak year:** 1957

**Shaft/Mine details:** 2 original shafts, both 549m in 1948 (No. 1 downcast, NT 3357 9612, and No. 2 upcast, NT 3356 9610). Closed because of fire in 1967. Pumping only from 1968. Resinking in 1926 and deepening and enlargements of shafts in 1930s to aid ventilation and prepare for the exploitation of undersea coals. Third shaft (also downcast) added at NT 3364 9624 in 1928. Steam winder supplied by 13 Lancashire boilers built on site of old power station in 1932.

**Details in 1948:** Output 2,330 tons per day, 633,037 tons per annum. 2,027 employees. (No washer). Baths (1937) designed by J A Dempster for 2,552 men (with 122 cubicles) and 96 women (10 cubicles), canteen, first-aid room. Electricity and steam, own supply. Report dated 06-08-1948.

**Other details:** Michael became the largest producer of coal in Scotland, and the Wemyss Coal Company's showpiece pit, despite continuous problems of gas and spontaneous combustion. Whilst still Scotland's biggest pit, and after massive investment, a disastrous fire broke out on 9 September 1967, destroying the new reserves. Although 302 men escaped, nine were killed. The disaster highlighted many safety issues, including the dangers of using polyurethane foam in underground workings, and the lack of portable respiratory equipment. A resulting campaign by Scottish NUM leader Michael

Figure 5.98: Mary Colliery, Lochore, 1924. A Fife Coal Company colliery, sunk originally in 1904. Its first shaft was 613 metres, and said to be the deepest shaft in Scotland. SMM:1997 0434, SC378675

Figure 5.99: Mary Colliery, Lochore, c.1947. Shaft No. 2 was sunk in 1923, and equipped with a reinforced concrete headframe, which survives as a monument in Lochore Meadows Country Park. An average of 614 miners were employed at Mary in the NCB era, the colliery closing in 1966. SC446331

Figure 5.100: Michael Colliery, East Wemyss, 1926. Sunk by the Wemyss Coal Company in the 1890s, Michael was Scotland's largest colliery. In the nationalised era, it employed an average of 2,598 miners, peaking in 1957 with a workforce of 3,353. SMM:1996 0678, SC378677

Figure 5.101: Michael Colliery, East Wemyss, c.1950. The shafts were re-sunk in 1926, and the pit was regularly modernised and redeveloped both before and after nationalisation. The last major investment occurred in the mid-1960s by the NCB, but an underground fire in 1967 killed nine miners and caused the closure of the colliery. SMM:1996 0540, SC446421

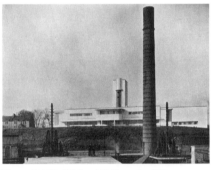

Figure 5.102: Michael Colliery, East Wemyss, c.1937. Michael was equipped with one of the finest pithead baths in the UK, designed by J A Webster of the Miners' Welfare Fund and completed in 1937. The tops of the headgear can be seen in the foreground. SC675062

McGahey resulted in the mandatory issue of self-rescuers (breathing aids) to all personnel working underground, and the installation of new emergency telephone systems.
**Figure 5.100, 5.101 and 5.102**

## Minto (Brighills)

NT 2051 9475 (NT29SW/44)
**Parish:** Auchterderran
**Region/District:** Fi/Ki
**Council:** Fife
**Location:** Cardenden
**Previous owners:** Lochgelly Iron & Coal Company
**Types of coal:** House and Steam
**Sinking/Production commenced:** 1903
**Year closed:** 1967
**Year abandoned:** 1968
**Average workforce:** 640
**Peak workforce:** 730
**Peak year:** 1957
**Shaft/Mine details:** 2 shafts, No. 1 184m (NT 2051 9475), and No. 2 302m deep (NT 2050 9477)
**Details in 1948:** Output 910 tons per day, 230,000 tons per annum. 684 employees. Campbell Binnie washer. Baths (1935), canteen, first-aid. Electricity, steam and compressed air. Report dated 07-08-1948.
**Other details:** Prone to spontaneous combustion and sudden roof falls, it suffered increasing water problems as neighbouring pits closed. An excellent model of the surface arrangement is on display at the Royal Museum of Scotland in Chambers Street, Edinburgh.
**Figure 5.103**

## Mossbeath

NT 1569 9102 (NT19SE/37)
**Parish:** Beath
**Region/District:** Fi/Du
**Council:** Fife
**Location:** Cowdenbeath
**Previous owners:**
**Types of coal:** not produced since 1947
**Sinking/Production commenced:** 1895
**Year closed:** pre-1947
**Shaft/Mine details:** 2 shafts, reorganised in 1930s and linked to Cowdenbeath No. 10
**Other details:** Retained by NCB as a training pit

## Muircockhall

NT 1130 8980 (NT18NW/33)

**Parish:** Dunfermline
**Region/District:** Fi/Du
**Council:** Fife
**Location:** Townhill
**Previous owners:** Henry Ness & Company
**Sinking/Production commenced:** unknown
**Year closed:** 1944
**Year abandoned:** 1969
**Other details:** Former colliery, retained by the NCB as a training centre until 1969

## Nellie

NT 1845 9477 (NT19SE/42.01)
**Parish:** Auchterderran
**Region/District:** Fi/Du
**Council:** Fife
**Location:** Lochgelly
**Previous owners:** Lochgelly Iron & Coal Company
**Types of coal:** House and Steam
**Sinking/Production commenced:** 1880
**Year closed:** 1965
**Year abandoned:** 1965
**Average workforce:** 428
**Peak workforce:** 508
**Peak year:** 1958
**Shaft/Mine details:** 2 shafts, No. 1 (Main Winding) 318m (NT 1845 9477), and No. 2 169m deep (NT 1846 9467)
**Details in 1948:** Output 450 tons per day, 125,000 tons per annum. 357 employees. Campbell Binnie washer. Canteen, ambulance facility. Electricity generated at colliery's own power station, and exported to Rosie. Report dated 06-08-1948

Figure 5.103: Minto Colliery, Cardenden, also known as 'Brighills'. A Lochgelly Iron and Coal Company pit, dating from 1903, it operated until 1967, employing an average of 640 miners in the nationalised era. A very fine model of its surface layout can be seen on display at the National Museums of Scotland in Edinburgh. SMM:1996 0688, SC378678

**Other details:** Closed in 1894, re-opened in 1903. Said to have had the last operating steam winder in Fife.

## Oakfield

NT 1333 9355 (NT19SW/62)

**Parish:** Beath

**Region/District:** Fi/Du

**Council:** Fife

**Location:** Kelty

**Previous owners:** Fife Coal Company

**Types of coal:** Steam

**Sinking/Production commenced:** 1944

**Year closed:** 1949

**Year abandoned:** 1950

**Average workforce:** 70

**Peak workforce:** 70

**Peak year:** 1947

**Shaft/Mine details:** two surface mines, main mine 503m, return mine 375m long

**Details in 1948:** Output 60 tons per day, 15,000 tons per annum. 62 employees. All electricity from NCB. Report dated 10-08-1948.

**Other details:** With Benarty and Blairenbathie, one of a group of successful post-war short-term drift mines commenced by the Fife Coal Company and continued by the NCB.

## Peeweep or Peewip

(see Lumphinnans XI and XII)

## Randolph

NT 3027 9577 (NT39NW/226)

**Parish:** Wemyss

**Region/District:** Fi/Ki

**Council:** Fife

**Location:** near Dysart

**Previous owners:** Earl of Rosslyn's Collieries Limited, Fife Coal Company from 1923

**Types of coal:** Steam and Navigation

**Sinking/Production commenced:** c.1850

**Year closed:** 1968

**Year abandoned:** 1968

**Average workforce:** 208

**Peak workforce:** 333

**Peak year:** 1947

**Shaft/Mine details:** 2 shafts, No. 1 115m (NT 3027 9577), and No. 2 152m deep (NT 3026 9573), sunk in 1921

**Details in 1948:** Output 430 tons per day, 112,000 tons per annum.

333 employees. British Baum-type washer, canteen (disused), no baths, ambulance room. 3-phase electricity, own supply. Report dated 07-08-1948.

**Other details:** very wet pit

**Figure 5.104**

## Rosie

NT 3460 9769 (NT39NW/227)

**Parish:** Wemyss

**Region/District:** Fi/Ki

**Council:** Fife

**Location:** by Buckhaven

**Previous owners:** Bowman & Company, Wemyss Coal Company from 1905

**Types of coal:** House

**Sinking/Production commenced:** 1880

**Year closed:** 1953

**Year abandoned:** 1953

**Average workforce:** 250

**Peak workforce:** 306

**Peak year:** 1947

**Shaft/Mine details:** No. 1 shaft NT 3460 9769, 183m, outlet to Wellesley Colliery. Shaft No. 2 NT 3026 9573.

**Details in 1948:** 291 workers (248 underground), no washery, no baths, canteen (snacks and meals), basic first-aid room. Electric power DC from Wellesley and Michael Collieries. None from public supply. Report dated 19-08-1948.

**Other details:** Bevin Boys hostel nearby

**Figure 5.105**

## Rothes

NT 2811 9727 (NT29NE/08)

**Parish:** Kirkcaldy and Dysart

**Region/District:** Fi/Ki

**Council:** Fife

**Location:** Thornton

**Previous owners:** National Coal Board

**Sinking commenced:** 1946

**Production commenced:** 1957

**Year closed:** 1962

**Year abandoned:** 1969

**Average workforce:** 960

**Peak workforce:** 1,235

**Peak year:** 1960

**Shaft Details:** Two shafts with Koepe electric winders mounted in huge concrete towers (No. 1 at NT 2811 9727 and No. 2 at NT 2796

Figure 5.104: Randolph Colliery, near Dysart, c.1947. Like Frances nearby, it was sunk in the mid-19th century by the Earl of Rosslyn's Collieries Limited. By 1923, it was operated by the Fife Coal Co, and produced steam and Navigation coal. The pit closed in 1968, having been operated by the NCB with an average workforce of 208 miners. SC382471

Figure 5.105: Rosie Colliery, near Buckhaven, c.1947. Founded by Bowman and Company in 1880, Rosie was operated by the Wemyss Coal Company from 1905. After nationalisation, it employed an average of 250 miners, closing in 1953. SC706494

Figure 5.106: Rothes Colliery, Thornton, 1957. Designed by the NCB Scottish Division's architect, Egon Riss, this was Scotland's first 'superpit', and was formally opened by Her Majesty Queen Elizabeth II in 1958. Originally a Fife Coal Company scheme, it ran into insurmountable geological and drainage problems, closing after only five years of operation in 1962. A hugely embarrassing failure, on average it employed 960 miners in its few years of operation. Its winding towers, which were a useful landmark in the Glenrothes landscape, were demolished in 1993. SC446437

9726). The winders lifted conventional cages with minecars (not skips), which were ejected onto a minecar circuit at the pithead.

**Other details:** The first and perhaps visually the finest of the new NCB superpits, it was designed by NCB architect Egon Riss, although the plans for the development dated back to the Fife Coal Company era. It was opened amid great optimism by HM Queen Elizabeth II on 30th June 1958, and was described by Lord Forbes (Minister for State for Scotland) as having a lifespan of 100 years. The new town of Glenrothes was built in anticipation of its longevity, but it became the most spectacular failure of the NCB era in Scotland. Its estimated cost was £1.65 million, and it was expected to produce 5,000 tons per day. In reality, it cost £20 million, and geological problems and severe flooding resulted in only a fraction of the anticipated productivity, and closure after only five years. It became a huge political embarrassment, and the associated records in the Public Record Office were therefore closed and inaccessible until 2000. Meanwhile, the shafts were subsequently used to test underwater equipment, some of which was used offshore in the development of the North Sea oil fields. In 1993, the two huge winding towers, which had become an important navigational aid in Glenrothes, were blown up, leaving other parts of the large surface facilities occupied by the local Fire Brigade, and a number of other tenants.

**Figure 5.106**

## Seafield

NT 2782 8964 (NT28NE/45)
**Parish:** Kinghorn
**Region/District:** Fi/Ki
**Council:** Fife
**Location:** South of Kirkcaldy
**Previous owners:** National Coal Board
**Sinking commenced:** 1954
**Production commenced:** 1966

Figure 5.107: Seafield Colliery, near Kirkcaldy, 1967. Built to exploit undersea coals, this was one of the NCB's most successful superpits. Production commenced after 11 years of development in 1967, and it was later linked underground with Frances on the other side of Kirkcaldy. It operated for 22 years with an average workforce of 2,178. The colliery buildings were demolished after its closure in 1988, to make way for a housing estate. SMM:1997.0441, SC378683

**Year closed:** 1988

**Year abandoned:** 1988

**Average workforce:** 2,178

**Peak workforce:** 2,466

**Peak year:** 1970

**Shaft/Mine details:** 2 shafts, each concrete lined and 7.3m diameter, with multi-rope electric friction winders mounted in concrete towers, each equipped with passenger lifts. No. 1 upcast, 642m deep and used for skip coal winding (NT 2782 8964), No. 2 downcast, 592m deep, and used for winding men and materials (NT 2782 8961), both being equipped by ASEA of Sweden. No. 1 Shaft had two winding engines with 'Ward Leonard' control systems, each having two 15-ton skips. No. 2 shaft had a double-decked cage and counterweight, each deck accommodating up to 55 men (i.e. 110 men at a time).

**Other details:** One of Scotland's successful superpits, and famous for its prowess at incline mining (1 in 0.8). Spontaneous combustion problems were tackled by nitrogen injection into the 'waste' extracted area. Equipped with its own dense-medium coal preparation plant. Tragic accident in 1973 when a face collapse killed five and injured four men. Linked underground with Frances by 1981. Just survived the miners' strike of 1984, but closed four years later.

Two towers blown up and surface buildings cleared after closure.

**Figure 5.107**

Figure 5.108: Thornton Mine, Balbeggie, near Kirkcaldy, c.1947. Sinking commenced in 1945 by the Fife Coal Co, and one of several short-term surface mines in Fife, it operated with a workforce averaging at 84 for six years in the NCB era, closing in 1953. SC378685

## Thornton Mine

NT 2895 9615 (NT29NE/30)

**Parish:** Kirkcaldy and Dysart

**Region/District:** Fi/Ki

**Council:** Fife

**Location:** Balbeggie, by Kirkcaldy

**Previous owners:** Fife Coal Company

**Types of coal:** House and Steam

**Sinking/Production commenced:** 1945

**Year closed:** 1953

**Year abandoned:** 1953

**Average workforce:** 84

**Peak workforce:** 113

**Peak year:** 1950

**Shaft/Mine details:** 2 surface mines, each 76m. No. 1 at NT 2895 9615, No. 2 at NT 2899 9610

**Details in 1948:** Output 300 tons per day, 80,000 tons per annum. 93 employees. Screened and washed at Frances Colliery. Baths. Electricity from NCB Aitken Colliery power station (AC). Report dated 09-08-1948

**Other details:** A Fife Coal Company post-War drift mine taken on by the NCB, equipped with the latest 2-ton drop-bottom mine cars

**Figure 5.108**

Figure 5.109: Valleyfield Colliery, near Culross. A large Fife Coal Company colliery, sunk originally between 1908 and 1911 which, in addition to producing Navigation coal for the Admiralty, yielded house and steam coals. SC446334

Figure 5.110: Valleyfield Colliery, near Culross. After major reconstructions in the 1940s and 1950s, it was merged with Kinneil Colliery on the opposite bank of the Firth of Forth in 1965. SC446338

### Torry

NT 0123 8707 (NT08NW/104)

**Parish:** Torryburn

**Region/District:** Fi/Du

**Council:** Fife

**Previous owners:** National Coal Board

**Sinking commenced:** 1951

**Production commenced:** 1952

**Year closed:** 1965

**Year abandoned:** 1966

**Average workforce:** 147

**Peak workforce:** 198

**Peak year:** 1962

**Mine Details:** No. 1 mine NT 0123 8707, No. 2 mine NT 0122 8712

**Other details:** A short-term NCB drift developed to satisfy demand until the new sinkings came on tap. Linked to Valleyfield by a purpose-built railway, including an 800m long tunnel. Designed to produce 400 tons a day, all of which was sent to be screened and washed at Valleyfield.

### Valleyfield 1 & 2

NT 0095 8641 (NT08NW/43)

**Parish:** Culross

**Region/District:** Fi/Du

**Council:** Fife

**Location:** Newmills

**Previous owners:** Fife Coal Company

**Types of coal:** House, Steam, and Navigation

**Sinking/Production commenced:** 1908–11

**Year closed:** 1978

**Year abandoned:** 1979

**Average workforce:** 982

**Peak workforce:** 1,052

**Peak year:** 1959

**Shaft/Mine details:** 2 oval brick-lined shafts, No. 1 393m (NT 0095 8641), and No. 2 384m deep (NT 0097 8641), deepened to 686m in 1950s reconstruction. Originally equipped with steam winders built by Douglas & Grant of Kirkcaldy. Surface mine at NT 0093 8622 driven in 1945 as an airway. Colliery merged under the Forth with Kinneil in 1965. Kinneil was closed in 1982 and abandoned in 1983.

**Details in 1948:** Output, 930 tons per day, 240,000 tons per annum. 843 employees. Baum-type washery. New baths opened 25-06-1949, accommodation for 1,044 men, modern canteen and medical centre attached. Steam for winding, electricity for other purposes, supplied from NCB Central Power Station. Original report dated 18-08-1948.

**Other details:** Said to produce the best coking and navigation coal in Scotland. Reconstructed in 1930s, at which time ponies were replaced by mechanised haulage. Valleyfield House was demolished in 1941 to build the miners' village of High Valleyfield. Reconstruction in progress in 1946, involving a major reorganisation of underground transport including the introduction of large mine cars with direct-rope and locomotive haulage. At the same time, new mine-car handling facilities were installed at the surface, along with a new surface drift mine for ventilation. Augmented by NCB's addition of Valleyfield 3 in 1954 (see below). Prone to methane problems, and the gas was extracted to be added to local town gas supplies. Gas explosions killed three men in 1911, and 35 in 1939, shortly after the outbreak of World War II. Conveniently positioned to supply navigation coal to the Royal Navy at nearby Rosyth.

**Figures 5.109, 5.110, and 5.111**

Figure 5.111: Valleyfield 1& 2 Colliery, 1974. In the nationalised era, Valleyfield operated with a workforce of on average 982 miners, closing in 1978. Kinneil, with which it had been linked seven years earlier, closed four years later. J R Hume, SC451803

## Valleyfield 3

NT 0095 8629 (NT08NW/43.1)

**Parish:** Culross

**Region/District:** Fi/Du

**Council:** Fife

**Location:** Newmills

**Previous owners:** Fife Coal Company

**Types of coal:** House and Steam (including navigation)

**Sinking/Production commenced:** 1954

**Year closed:** *c.*1964

**Year abandoned:** 1964

**Other details:** A major NCB new sinking designed to raise Valleyfield output, topped by a re-inforced concrete winding tower. Rendered redundant by plans to merge Valleyfield with Kinneil Colliery.

**Figure 5.112**

Figure 5.112: Valleyfield No. 3 Colliery, near Culross. Built as part of a major reconstruction project in 1954, it was rendered redundant by the merger under the Firth of Forth of Valleyfield with Kinneil Colliery in 1965. SMM:1998 0384, SC378688

## Wellesley (also known as Denbeath)

NT 3666 9875 (NT39NE/59)

**Parish:** Wemyss

**Region/District:** Fi/Ki

**Council:** Fife

**Location:** Methil

**Previous owners:** Bowman & Company, Wemyss Coal Company from 1905

**Types of coal:** House and Steam

**Sinking/Production commenced:** 1883–5

**Year closed:** 1967

**Year abandoned:** 1972

**Average workforce:** 1,908

**Peak workforce:** 2,603

**Peak year:** 1957

**Shaft/Mine details:** 2 circular shafts, both 481m. No. 1 shaft NT 3660 9877 (1883), and No. 2 shaft NT 3662 9879 (1885). Third eliptical

Figure 5.113: Wellesley Colliery, Buckhaven. Previously known as Denbeath Colliery, the Wellesley was one of the Wemyss Coal Co's largest pits, in the nationalised era continuing to employ on average almost 2,000 miners. SMM:1996 2524, SC378692

shaft added in 1907. In 1938, two steam winders were in operation, fed by 14 Lancashire boilers.

**Details in 1948:** Output 1,700 tons per day, 440,300 tons per annum. 1,543 employees. First washer installed in 1904, later replaced by large Baum-type central washery (largest in Europe in 1938), which also served Randolph, Lochhead, Frances and Michael. Baths (extension completed by MWF in 1938), canteen (under construction), first-aid (hospital close by colliery). Steam and electricity, all generated at mine. Report dated 12-08-1948.

**Other details:** Originally known as Denbeath Colliery, but the name was changed at the onset of the Wemyss Coal Company era in 1905. Major redevelopment in 1907, with addition of third shaft, production coming back on line in 1910. Pithead baths built by the Wemyss Coal Company in 1915 were the first to be built in Scotland, and the second in the UK. In 1947, second only to Michael Colliery in Scotland in terms of annual output. New washery added in 1960. Closed in 1967 during period of major closures, within two months of another of Scotland's largest pits, Minto, and only two months prior to the disastrous fire that closed neighbouring Michael.
**Figures 5.113 and 5.114**

**Wellsgreen**
NT 3352 9835 (NT39NW/164)
**Parish:** Markinch
**Region/District:** Fi/Ki
**Council:** Fife
**Location:** by Windygates
**Previous owners:** Fife Coal Company
**Types of coal:** House, Steam, and Manufacturing
**Sinking/Production commenced:** 1888

Figure 5.114: Wellesley Colliery, Buckhaven, May 1959. Begun in 1883, Wellesley operated until 1967, and was one of the biggest collieries in Scotland. It is also famous for having the first pithead baths in Scotland, built in 1915 by the Wemyss Coal Co. Simmons Aerofilms, R35976

**Year closed:** 1959

**Year abandoned:** 1959

**Average workforce:** 268

**Peak workforce:** 407

**Peak year:** 1949

**Shaft/Mine details:** 2 shafts, each 153m. No. 1 shaft NT 3352 9835, and No. 2 shaft NT3350 9835. New surface mine opened in 1946, but closed in 1948.

**Details in 1948:** Output 320 tons per day, 86,000 tons per annum. 243 employees. Norton washery. Canteen, first-aid. 3-phase AC used, all own supply. Report dated 07-08-1948.

**Other details:** Closed temporarily in 1927.

**Figure 5.115**

Figure 5.115: Wellsgreen Colliery, near Windygates, c.1947. A Fife Coal Company colliery, and dating from 1888, this pit closed in 1959, having been operated by a workforce averaging over 260 miners in the nationalised era. SC446325

## Westfield Opencast

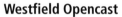
NT 195 982 (NT19NE/20)

**Parish:** Auchterderran

**Region/District:** Fi/Ki

**Council:** Fife

**Location:** near Cardenden

**Other details:** Begun in 1955 on the site of the former Kirkness Colliery, production commencing in 1956. Produced 20,000 tons per week, 40% of Scottish opencast capacity at the time, said to be the biggest opencast site in UK, and claimed to be Europe's biggest hole. Had its own coal preparation plant. Experimental Lurgi plant opened nearby in 1960 at Westfield.

**Figure 5.116**

Figure 5.116: Westfield Opencast, near Cardenden, c.1980. Thought at the time to be one of the biggest open-cast projects in Europe, extraction commenced in 1955 on the site of Kirkness Colliery, and the yield of coal was sufficient to merit its own dedicated coal preparation plant. SMM:1997 0807, SC378696

**Windyedge**
NT 1373 9297 (NT19SW/12)
**Parish:** Beath
**Region/District:** Fi/Du
**Council:** Fife
**Location:** Lassodie Mill, near Kelty
**Previous owners:** Lassodie Coal Company
**Types of coal:** House and Steam
**Sinking commenced:** 1949
**Production commenced:** 1950
**Year closed:** 1951
**Year abandoned:** 1951
**Average workforce:** 26
**Peak workforce:** 26
**Peak year:** 1950
**Shaft/Mine details:** 2 surface mines, No. 1 at NT 1373 9297, and No. 2 at NT 1365 9256
**Details in 1948:** Output 25 tons per day, 7,000 tons per annum. 20 employees. Electricity bought from Central Electricity Board. Report dated 15-02-1948.

## Licensed Mines in Fife 1947–97

| Name | Licensee | Grid Ref | Opened | Closed |
|------|----------|----------|--------|--------|
| Buchanan Mine (Saline Mine) | N B Buchanan | NS 810 270 | 1979 | 1991 |
| Cartmore (Lochgelly) | W S Rosland | NT 170 940 | c. 1981 | 1987 |
| Easter Clune | Easter Clune Coal Company | NT 067 894 | 1957 | 1966 |
| Killernie (Saline) | J Payne | NT 027 928 | 1966 | 1974 |
| Lassodie | John M Heeps | NT 139 934 | 1961 | 1991 |
| Lassodie 2 | J Methven & Sons | NT 113 929 | pre-1947 | c. 1950 |
| Lassodie 3 | J Methven & Sons | NT 113 929 | 1950 | c. 1952 |
| Lassodie 4 | J Methven & Sons | NT 113 929 | 1950 | 1952 |
| Lassodie 5 | J Methven & Sons | NT 113 929 | 1952 | 1953 |
| Lassodie 6 | J Methven & Sons | NT 117 926 | 1952 | c. 1959 |
| Lassodie 7 | J Methven & Sons | NT 113 929 | 1955 | c. 1957 |
| Lassodie 8 | J Methven & Sons | NT 113 929 | 1955 | c. 1956 |
| Lassodie 9 | J Methven & Sons | NT 119 925 | 1956 | c. 1959 |
| Lethan's Mine 1 (Dunfermline) | John M Heeps | NT 061 944 | 1955 | 1958 |
| Lethan's Mine 2 (Dunfermline) | John M Heeps | NT 055 942 | 1957 | 1959 |
| Lochhead (Dunfermline) | Lochside Coal & Fireclay Company Limited | NT 077 903 | c. 1957 | c. 1977 |
| Lochside 1/2/3 (Dunfermline) | Lochside Coal & Fireclay Company | NT 102 892 | pre-1947 | c. 1963 |
| North Steelend (Dunfermline) | J Payne | NT 058 936 | 1958 | c. 1965 |
| Windyedge (Lassodie) | J Summerville | NT 137 929 | pre-1947 | 1949 |

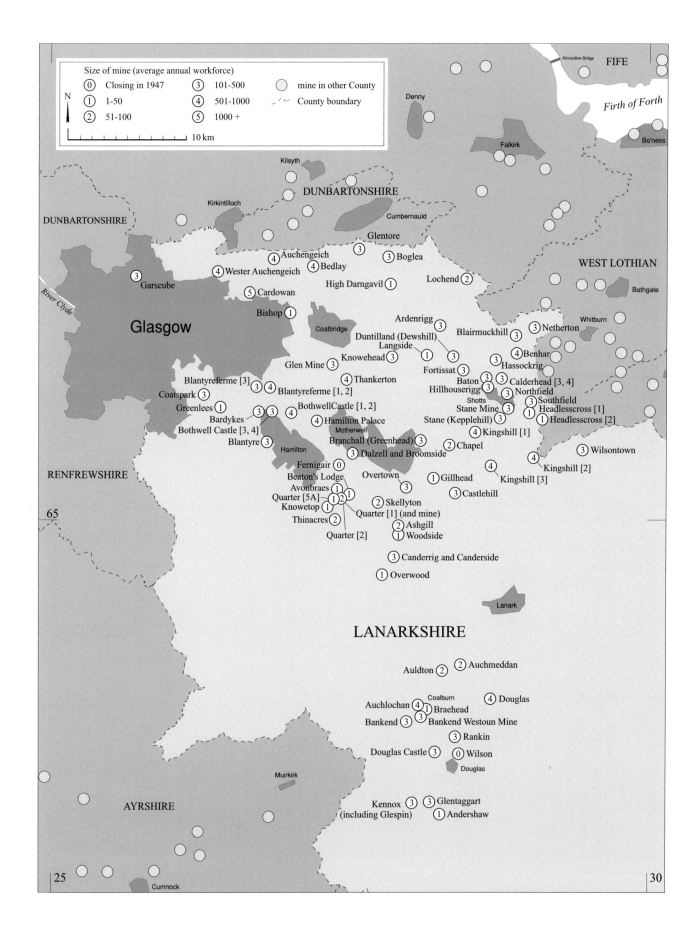

# Lanarkshire

In its heyday in the early 20th century, Lanarkshire was the most important mining county in Scotland, and one of the most significant in the UK. Situated to the south and east of Glasgow, and including much of the city itself, the county was host to major technological advances in coal mining, including early electrification and mechanisation, which was driven by the presence of advanced engineering industries, led by companies such as Mavor & Coulson of Glasgow, and Anderson Boyes of Motherwell. Also important was the huge market for coal in the Glasgow conurbation, in particular from the iron and steel industry, the greatest concentration of which had developed during the 19th century in the north-east of the county, especially in the Monklands and around Shotts, where blackband ironstone ores were also to be found.

The geographical distribution of mines in Lanarkshire reflects this pattern, most collieries being located in the north of the county immediately to the east and south of Glasgow. The city had itself witnessed substantial coal-mining activity, but by 1947, only Garscube in the north and a cluster of mines around Cambuslang to the south had survived. Further afield, the greatest concentrations were in the Glenboig, Coatbridge and Airdrie areas of the Monklands, around Shotts, Newmains and Forth on the border with West Lothian, and on either side of the Clyde valley in the Hamilton, Motherwell and Wishaw areas. A detached coalfield further south, centred around Douglas, Coalburn and Lesmahagow, was also significant, with mining activity stretching through Glentaggart and Glespin south-west towards Muirkirk in Ayrshire.

The depression and lack of investment that followed World War I had damaged the coal industry throughout the UK, but many of Lanarkshire's surviving pits were already showing signs of exhaustion during the inter-war years. At the time of nationalisation in 1947, just over 21,000 miners were employed in the 60 collieries that were taken into state ownership by the NCB. These figures fell steadily as the most unproductive mines were closed down, only a few being selected for long-term reconstruction. A number of new projects were established to maintain output, such as those at Andershaw, Avonbraes, Greenlees, Knowehead, Knowetop and Stane, but these were relatively small compared to the major new sinkings elsewhere in Fife, Ayrshire and the Lothians. The only substantial project was the sinking of Kingshill No. 3 near Newmains. Mining activity continued to decline rapidly during the 1950s and 1960s, and was reduced to only ten mines by 1967. Five years later, two collieries, Cardowan and Bedlay, were the only survivors, Cardowan being the last to close in 1983 just prior to the miners' strike.

The mines produced several qualities of coal, including anthracite (at pits such as Ardenrigg, Duntilland, Hassockrigg and Lochend 5), and in most cases, house, gas, steam and manufacturing coals. However, given the needs of the once massive iron and steel industries, coking coal was perhaps most important, significant providers being the collieries at Auchengeich, Cardowan, Kingshill 1, 2 and 3, Overtown and Wester Auchengeich. In addition, fireclay and blaes were often mined from adjacent seams, and were an important source of raw materials

Figure 5.117: Map of National Coal Board (NCB) collieries in Lanarkshire.

for the brick and refractory industries throughout the region, but most famously in the Glenboig area.

Although there were many independent small mining businesses, the industry was dominated by a group of large companies. By 1947, these included the Coltness Iron Company (Kingshill 1 & 2, Douglas, Hassockrigg, Branchall, Overtown, and Duntilland), Wm Baird (Bedlay, Bothwell Castle 1/2 & 3/4), Wm Dixon Limited (Blantyre, Wilsontown, Auchlochan), A G Moore (Blantyreferme 1, 2 & 3), James Nimmo (Cardowan, Auchengeich, W Auchengeich, Canderigg & Canderside), the Shotts Iron Company (Castlehill 6, Southfield, Northfield & Hall, Baton, Calderhead, Fortissat, Kepplehill/Stane), the Summerlee Iron Company (Benhar, Bardykes, Garscube), United Collieries (Quarter, Netherton), and Wilson & Clyde (Douglas Castle, Woodside, Skellyton, and Ashgill).

After nationalisation, Cardowan was by far the largest colliery, averaging almost 1,500 miners over its 37 years of operation despite serious methane problems. Other large collieries included Kingshill 1 (with an average of almost 900 miners for over 20 years), Douglas and Bedlay (annually each with over 800 miners over 20 and 34 years, respectively), and Auchengeich (employing an average of 650 miners over a period of 18 years). Others routinely with over 500 miners on the books included Kingshill 3, and Hamilton Palace.

Symbolic though it was, the closure of Cardowan in 1983 did not signal the end of deep coal mining in Lanarkshire. During the nationalised era, almost 160 small licensed mines operated within the county, mostly in areas where mining had prospered previously such as in the Monklands, and in the upland plateau around Shotts and Forth. Several had originated in the years before nationalisation, but most were subsequent new developments, some as recently as in the 1980s. The last of the licences, which was for a mine at Rashiehill near Forth, had lapsed by 1995.

## Andershaw 1, 2, 3, 5 & 6

NS 8269 2606 (NS82NW/10)

**Parish:** Douglas

**Region/District:** St/CY

**Council:** South Lanarkshire

**Previous owners:** National Coal Board

**Sinking commenced:** 1948

**Production commenced:** 1949

**Year closed:** 1959

**Average workforce:** 49

**Peak workforce:** 55

**Peak year:** 1952

**Shaft/Mine details:** Two surface mines

## Ardenrigg 6

NS 8275 6565 (NS86NW/19)

**Parish:** Shotts

**Region/District:** St/Mo

**Council:** North Lanarkshire

**Location:** Plains, by Airdrie

**Previous owners:** Ardenrigg Coal Company Limited

**Types of coal:** Anthracite

**Sinking/Production commenced:** 1926

**Year closed:** 1963

**Average workforce:** 282

**Peak workforce:** 304

**Peak year:** 1951

**Shaft/Mine details:** Main mine, 1 in 8, 61m long to upper levels, and 134m at 1 in 5 to lower levels

**Details in 1948:** Output 280 tons per day, 77,000 tons per annum, longwall working. 249 employees. Screening plant and access to railway 4km away to the north-west at Stepends, supplied by aerial ropeway. Baum-type washery (Blantyre Engineering Company). Baths (1938, for 238 men), canteen (not operating), first-aid services. All electricity from public supply. Report dated 28-07-1948.

**Figure 5.118**

## Ashgill

NS 7797 4990 (NS74NE/24)

**Parish:** Dalserf

**Region/District:** St/Ha

**Council:** South Lanarkshire

**Location:** Larkhall

**Previous owners:** Wilsons and Clyde Coal Company

**Types of coal:** House

**Sinking/Production commenced:** 1945

Figure 5.118: Ardenrigg Colliery, Plains, near Airdrie, 1961. Developed from 1926 by the Ardenrigg Coal Co, this mine was noted for its anthracite. It continued after nationalisation with an average workforce of 282 miners until its closure in 1963. SMM:1996.0453B, SC380996

**Year closed:** 1951
**Year abandoned:** 1951
**Average workforce:** 60
**Peak workforce:** 60
**Peak year:** 1947
**Shaft/Mine details:** 2 mines, both 38m deep
**Details in 1948:** Output 60 tons per day, 15,000 tons per annum. 59 employees. Bar screen and bagging platform. No washery, baths or canteen, and no medical services. All electricity supplied from outside. Report dated 10-08-1948.

## Auchengeich ('The Geich')

NS 6860 7114 (NS67SE/28)
**Parish:** Cadder
**Region/District:** St/St
**Council:** North Lanarkshire
**Location:** Bridgend, Chryston
**Previous owners:** James Nimmo & Company
**Types of coal:** Gas, House, Steam and Coking
**Sinking/Production commenced:** 1908
**Year closed:** 1965
**Year abandoned:** 1982
**Average workforce:** 650
**Peak workforce:** 860
**Peak year:** 1956
**Shaft/Mine details:** 2 shafts, both 320m deep, No. 1 rectangular and timber-lined, No. 2 round and brick-lined. Pumping continued after closure for neighbouring Bedlay Colliery.
**Details in 1948:** Output 880 tons per day, 237,600 tons per annum. 797 employees. Baum-type washery (Blantyre Engineering Company). Baths (1931), canteen, first-aid room. Electricity supplied by Clyde Valley. Report dated 19-08-1948.

Figure 5.119: Auchengeich Colliery, Bridgend, Chryston. Known as 'The Geich', production began in 1908 by James Nimmo and Co. A significant producer of gas and coking coals, the colliery had its own coke ovens. Well known for the disaster in 1959 which killed 47 miners, it operated in the nationalised era with an average workforce of 650 miners until closure in 1965. SC446200

**Other details:** Disastrous fires in 1931 and 1959 killed 6 and 47 men, respectively.
**Figure 5.119**

## Auchlochan 6 & 7

NS 8102 3523 (NS83SW/17)
**Parish:** Lesmahagow
**Region/District:** St/CY
**Council:** South Lanarkshire
**Location:** Coalburn
**Previous owners:** Wm Dixon Limited
**Types of coal:** Manufacturing and Steam, and fireclay
**Sinking/Production commenced:** 1894
**Year closed:** 1968
**Average workforce:** 534
**Peak workforce:** 687
**Peak year:** 1952
**Shaft/Mine details:** 2 shafts, No. 6 being 91m, and No. 7 88m deep. (Auchlochan 7 at NS 8090 3399)
**Details in 1948:** Output 600 tons per day, 138,500 tons per annum, stoop and room and longwall working. 457 employees. 4 screens for dry coal (2 at No. 6 and 2 at No. 9), Jig washer (Campbell, Binnie and Reid). Baths, canteen. Steam and electricity, none from public supply. Report dated 12-08-1948.

## Auchlochan 9 & 10

NS 8133 3630 (NS83NW/35.01)
**Parish:** Lesmahagow
**Region/District:** St/CY
**Council:** South Lanarkshire
**Location:** Coalburn
**Previous owners:** Caprington & Auchlochan Collieries Limited, Wm Dixon Limited from 1930s
**Sinking/Production commenced:** 1894
**Year closed:** 1968
**Statistics:** see Auchlochan 6 and 7
**Shaft/Mine details:** 2 brick-lined shafts, No. 10 108m and No. 9 99m deep
**Other details:** Washer handled coal from other local mines. Baths built 1930.
**Figure 5.120**

## Auchmeddan 1, 2 & 3

NS 8441 3826 (NS83NW/40)
**Parish:** Lesmahagow
**Region/District:** St/CY

Figure 5.120: Auchlochan 9 & 10 Colliery, Coalburn, c.1950. Sunk originally in the 1890s by Caprington and Aucholochan Collieries Limited, it was taken over by William Dixon Limited in the 1930s. Together with Auchlochan 6, it regularly employed over 500 miners in the NCB era, finally closing in 1968. Harrison Collection, SC446067

**Council:** South Lanarkshire

**Location:** Auchmeddan Mains

**Previous owners:** Daniel Beattie & Company

**Types of coal:** Steam

**Sinking/Production commenced:** 1945

**Year closed:** 1968

**Average workforce:** 85

**Peak workforce:** 101

**Peak year:** 1967

**Shaft/Mine details:** 3 surface mines, No. 1 NS 845 383, No. 2 845 386, and No. 3 846 381. Auchmeddan No. 4 added in 1955, also NS 846 381

**Details in 1949:** Output 160 tons per day, 40,000 tons per annum, stoop and room working. 53 employees. No screening, washing, baths or canteen. Electricity all from public supply. Report dated 22-02-1949.

**Other details:** Transferred to NCB in 1949.


### Auldton 1, 2 & 3

NS 8272 3792 (NS83NW/38)

**Parish:** Lesmahagow

**Region/District:** St/CY

**Council:** South Lanarkshire

**Location:** Auldton Mains

**Previous owners:** Auldton Colliery (Boyd Bros.)

**Types of coal:** Steam

**Production commenced:** 1941

**Year closed:** 1963

**Average workforce:** 84

**Peak workforce:** 96

**Peak year:** 1960

**Shaft/Mine details:** 3 surface mines, first sunk in 1942, closing in 1949, the others being developed 1951–4

**Details in 1949:** Output 40 tons per day from Mines 2 & 3, 10,000 tons per annum, longwall working. 34 employees. Fixed bar screen, no washer. No baths or canteen. Electricity all public supply. Report dated 22-02-1949.


### Avonbraes

NS 7406 5222 (NS75SW/28)

**Parish:** Hamilton

**Region/District:** St/Ha

**Council:** South Lanarkshire

**Location:** Larkhall

**Previous owners:** National Coal Board

**Sinking commenced:** 1949

**Production commenced:** 1952
**Year closed:** 1965
**Year abandoned:** 1965
**Average workforce:** 32
**Peak workforce:** 39
**Peak year:** 1952


## Bankend Colliery

NS 7995 3365 (NS73SE/ 7)
**Parish:** Lesmahagow
**Region/District:** St/CY
**Council:** South Lanarkshire
**Location:** Coalburn
**Previous owners:** Arden Coal Company Limited
**Types of coal:** Gas, House, Manufacturing and Steam
**Sinking/Production commenced:** 1890
**Year closed:** 1958
**Year abandoned:** 1958
**Average workforce:** 167
**Peak workforce:** 226
**Peak year:** 1947
**Shaft/Mine details:** Several surface mines, including Nos. 3, 8, 9, 11, 12, 13, 15 and 16, and Hagshaw South. Bankend No. 3 was situated at NS 7923 3309 (site No. NS73SE/8), and No. 13, which had a shaft, was at NS 7988 3351 (site No. NS73SE/7)
**Details in 1948:** Output 270 tons per day, 70,000 tons per annum, stoop and room and longwall working. 221 employees. 2 screens for dry coal, Jig (Dickson and Mann washer). No baths, small canteen serving packed meals. 100% electricity from public supply. Report dated 12-08-1948.
**Other details:** Sinking of No. 12 (NS 792 339) by the NCB commenced in 1951, but the resulting mine may never have produced significantly, being abandoned in 1954. The entire area was subsequently engulfed by a very large opencast mine (named Dalquhandy).


## Bankend (Westoun Mine)

(see Westoun)


## Bardykes

NS 6754 5882 (NS65NE/44)
**Parish:** Cambuslang
**Region/District:** St/Gl
**Council:** South Lanarkshire
**Location:** Cambuslang
**Previous owners:** Summerlee Coal and Iron Company

Figure 5.121: Bardykes Colliery, Cambuslang. Previously owned by the Summerlee Coal & Iron Co, and dating from 1874, this colliery continued to operate after nationalisation, despite the collapse of No. 2 shaft in 1949. It closed in 1962 having been operated by a workforce of on average 479 miners. SMM:1996.02594, SC381039

**Types of coal:** House and Steam

**Sinking/Production commenced:** 1874

**Year closed:** 1962

**Average workforce:** 479

**Peak workforce:** 775

**Peak year:** 1947

**Shaft/Mine details:** 2 shafts, No. 1 393m, and No. 2 497m deep

**Details in 1948:** Output 750 tons per day, 202,500 tons per annum. 750 employees. 4 picking tables. Dickson and Mann washer. Canteen. Ambulance man (day shift). Steam for winding. Electricity all generated at colliery. Report dated 11-08-1948.

**Other details:** No. 2 shaft collapsed in 1949, but new mine road driven from Blantyreferme 3 (Newton) nearby.

**Figure 5.121**


**Baton**

NS 8651 6146 (NS86SE/30)

**Parish:** Shotts

**Region/District:** St/MW

**Council:** North Lanarkshire

**Location:** Shotts

**Previous owners:** Shotts Iron Company Limited

**Types of coal:** House and Steam

**Sinking/Production commenced:** *c.*1850

**Year closed:** 1950

**Year abandoned:** 1960

**Average workforce:** 445

**Peak workforce:** 772

**Peak year:** 1950

**Shaft/Mine details:** 1 shaft, 64m deep, later shaft added

**Details in 1948:** Output 286 tons per day, 91,069 tons per annum, longwall working. 278 employees. Three picking tables, Baum-type washer. No baths, no canteen (snacks from adjacent colliery), first-aid services. Electricity from Central NCB power station at Shotts. Report dated 28-07-1948.


**Beaton's Lodge**

NS 7514 5204 (NS75SE/32)

**Parish:** Dalserf

**Region/District:** St/Ha

**Council:** South Lanarkshire

**Location:** Larkhall

**Previous owners:** Currie Brothers

**Types of coal:** House

**Sinking/Production commenced:** 1940

**Year closed:** 1960

**Year abandoned:** 1961

**Average workforce:** 15

**Peak workforce:** 31

**Peak year:** 1952

**Shaft/Mine details:** surface mine

**Details in 1949:** Output 20 tons per day, 1,000 tons per annum. 23 employees. Bar screen. No washer, canteen, or medical facilities. Electricity supplied by Electricity Board. Report dated 15-02-1949.

**Other details:** Transferred to NCB in 1949.

**Figure 5.122**

Figure 5.122: Beaton's Lodge Mine, near Larkhall, 1950. A small private surface mine operated by Currie Brothers, opened originally in 1940. It was taken into public ownership in 1949, thereafter being operated by a workforce on average of 15 miners until its closure in 1960. SMM:1998.0226, SC381000

## Bedlay

NS 7205 7056 (NS77SW/22)

**Parish:** Cadder

**Region/District:** St/St

**Council:** North Lanarkshire

**Location:** by Glenboig

**Previous owners:** William Baird & Company Limited, later Bairds & Scottish Steel Limited

**Types of coal:** Coking, Manufacturing and House

**Sinking/Production commenced:** 1905

**Year closed:** 1981

**Year abandoned:** 1982

**Average workforce:** 792

**Peak workforce:** 870

**Peak year:** 1959

**Shaft/Mine details:** 3 shafts, No. 1 372m (NS 7205 7056), No. 2 372m (NS 7204 7058), and No. 3 362m deep (NS 7208 7049). Coal lifted in Nos.1 & 2, and men up and down No. 3.

**Details in 1948:** Output 540 tons per day, 145,800 tons per annum. 683 employees. Campbell Binnie jig washer. No baths. Canteen, first-aid room. Electricity supplied by overhead power lines from Gartsherrie Iron Works, but 90% bought from Clyde Valley Supply. Report dated 19-08-1948.

Figure 5.123: Bedlay Colliery, near Glenboig, c.1960. Sunk by William Baird and Company Limited, which later became Baird and Scottish Steels Limited, this was one of Lanarkshire's most important collieries in the nationalised era, and a significant producer of coking coal. SMM:1996.1090, SC445857

Figure 5.124: Bedlay Colliery, near Glenboig, c.1980. A major reconstruction in the late 1950s led to the addition of the concrete-framed winding tower, and a new coal preparation plant. East Dunbartonshire Information and Archives P5277, SC706475

Figure 5.125: Bedlay Colliery, near Glenboig, 1981. In the NCB era, Bedlay's workforce averaged almost 800 miners. It closed in 1981, and this view shows the last coal train leaving the colliery. East Dunbartonshire Information and Archives P5250, SC706487

**Other details:** New coal preparation plant installed by Simon Carves in 1957 as part of a £1.25 million reconstruction project completed in 1958, which included the construction of a new shaft and concrete winding tower, and the introduction of battery-locomotive underground haulage. A gassy pit with a methane drainage system, but important producer of coking coal for the iron and steel industry, particularly Ravenscraig. Latterly the coal preparation plant processed output from Auchengeich.
**Figures 5.123, 5.124 and 5.125**

## Benhar
NS 8899 6340 (NS86SE/04.00)
**Parish:** Shotts
**Region/District:** St/MW
**Council:** North Lanarkshire
**Location:** Harthill
**Previous owners:** Summerlee Iron Company from 1936
**Types of coal:** House and Steam
**Sinking/Production commenced:** 1914
**Year closed:** 1962
**Average workforce:** 545
**Peak workforce:** 772
**Peak year:** 1947
**Shaft/Mine details:** 2 shafts: No. 19 209m, and No. 3 210 deep

**Details in 1948:** Output 750 tons per day, 222,000 tons per annum, longwall working. 755 employees. 4 bar tables, Norton Washer. No baths, but full-meal canteen, first-aid services. All electricity AC, generated at mine. Report dated 28-07-1948.

**Other details:** Baths opened in 1950

**Figure 5.126**

Figure 5.126: Benhar Colliery, near Harthill, c.1960. Dating from 1914, and operated by the Summerlee Iron Company after 1936, Benhar employed on average 545 miners in the nationalised era until its closure in 1962. SMM:1996.02593, SC446544

**Bishop No. 3 Mine**

NS 7010 6652 (NS76NW/25)

**Parish:** Old Monkland

**Region/District:** St/Mo

**Council:** North Lanarkshire

**Location:** Gartcosh

**Previous owners:** Mosside Coal and Fireclay Company

**Types of coal:** House, Steam, and fireclay

**Sinking/Production commenced:** 1939

**Year closed:** 1948

**Year abandoned:** 1948

**Average workforce:** 21

**Peak workforce:** 21

**Peak year:** 1947

**Shaft/Mine details:** 1 shaft, 30m deep, surface mine 121m in length, dipping 1 in 3.75

**Details in 1948:** Output 26 tons per day, 7,020 tons per annum. 16 employees. No washer, no canteen, no baths. First-aid room. Electricity from Clyde Valley. Report dated 19-08-1948.

**Blairmuckhill**

NS 8897 6492 (NS86SE/08)

**Parish:** Shotts

**Region/District:** St/MW

**Council:** North Lanarkshire

**Location:** Harthill

**Previous owners:** A and G Anderson

**Types of coal:** House, Manufacturing and Steam

**Sinking/Production commenced:** 1910

**Year closed:** 1959

**Year abandoned:** 1959

**Average workforce:** 321

**Peak workforce:** 347

**Peak year:** 1952

**Shaft/Mine details:** 2 shafts: No. 11 112m, and No. 10 37m deep (ventilation)

**Details in 1948:** Output 300 tons per day, 86,000 tons per annum, longwall working. 309 employees. 2 bar tables, Norton-Harty washer (obsolete). No baths, but full-meal canteen, first-aid services. Electricity all from public supply. Report dated 28-07-1948.

## Blantyre 1 & 2 (and 5) (Dixons)

NS 6827 5624 (NS65NE/46)

**Parish:** Blantyre

**Region/District:** St/Ha

**Council:** South Lanarkshire

**Location:** High Blantyre

**Previous owners:** Wm Dixon Limited

**Types of coal:** House and Manufacturing

**Sinking/Production commenced:** 1865

**Year closed:** 1957

**Year abandoned:** 1957

**Average workforce:** 460

**Peak workforce:** 533

**Peak year:** 1947

**Shaft/Mine details:** 3 shafts, No. 1 256m, No. 2 226m, and No. 5 also 226m deep

**Details in 1948:** Output 450 tons per day, 109,210 tons per annum. 533 employees. Screening plant with 2 tables at each pit (Nos. 1 and 2). Lurig washer. No baths. Canteen (pieces only), only local doctors as medical service. Steam power for surface. Electricity underground. All electricity supplied by Clyde Valley Power Company Report dated 11-08-1948.

**Other details:** Famous for Scotland's worst mining disaster on 22nd October 1877 when a firedamp explosion killed at least 207 men. The final death toll remained uncertain because at the time there was no reliable record of the number of men working underground.

## Blantyreferme 1 & 2

NS 6848 6070 (NS66SE/34)

**Parish:** Blantyre

**Region/District:** St/Ha

**Council:** South Lanarkshire

**Location:** Uddingston

**Previous owners:** A G Moore & Company

**Types of coal:** House and Steam

**Sinking/Production commenced:** 1894

**Year closed:** 1962

**Average workforce:** 507

**Peak workforce:** 656

**Peak year:** 1950

**Shaft/Mine details:** 2 shafts, each 244m deep

**Details in 1948:** Output 450 tons per day, 110,00 tons per annum. 457 employees. 2 travelling bar tables. Dickson and Mann washer.

Baths and canteen, but no medical services. All electricity bought from public supply. Report dated 12-08-1948.

**Figure 5.127**

## Blantyreferme 3 (Newton)

NS 6740 6079 (NS66SE/35)

**Parish:** Cambuslang

**Region/District:** St/Ha

**Council:** South Lanarkshire

**Location:** Newton, Cambuslang

**Previous owners:** A G Moore & Company

**Types of coal:** House and Steam

**Sinking/Production commenced:** 1850

**Year closed:** 1964

**Average workforce:** 394

**Peak workforce:** 430

**Peak year:** 1956

**Shaft/Mine details:** 1 shaft, winding from 165m, further 24m to pumping level. Originally 249m deep

**Details in 1948:** Output 300 tons per day, 82,000 tons per annum. 300 employees. Screening: two-bar type. Washing done at Blantyreferme No. 1. Baths (396 men), canteen, first-aid room. All electricity AC, supplied by Clyde Valley Power Company Report dated 10-08-1948.

**Other details:** Linked to Bardykes *c.*1950. Associated brickworks operated between 1924 and 1974, and was supplied by a drift mine sunk to extract fireclay.

**Figure 5.128**

## Boglea

NS 7832 7124 (NS77SE/16)

**Parish:** New Monkland

**Region/District:** St/Mo

**Council:** North Lanarkshire

**Location:** Greengairs, by Airdrie

**Previous owners:** Brownieside Coal Company Limited

**Types of coal:** House

**Sinking/Production commenced:** 1942

**Year closed:** 1962

**Average workforce:** 123

**Peak workforce:** 160

**Peak year:** 1956

**Shaft/Mine details:** 3 surface mines, No. 1 (downcast, NS 7832 7124) 193m long, 1 in 3; No. 2 (upcast, NS 7832 7122) also 193m long, 1 in 3; No. 6 new mine (NS 7841 7127). All adjacent to earlier colliery.

Figure 5.127: Blantyreferme 1 & 2 Colliery, Uddingston. Sunk in 1894 and owned by A & G Moore and Co. In the nationalised era, it operated with an average workforce of over 500 miners, closing in 1962. The National Archives, COAL80/135/4, SC613968

Figure 5.128: Blantyreferme 3 Colliery (Newton), Cambuslang. An A & G Moore and Company colliery, dating from 1850. Well known also for its adjacent brickworks, which operated between 1924 and 1974. The colliery itself had a workforce averaging almost 400 miners after 1947, closing in 1964. The National Archives, COAL80/135/2, SC614080

Figure 5.129: Bothwell Castle 1 & 2 Colliery, 1920. Situated adjacent to Bothwell Station, it was established by William Baird and Company Limited in 1875. After nationalisation, its workforce averaged 623 miners until its closure in 1950. Rokeby Collection, SC706247

Figure 5.130: Bothwell Castle 3 & 4, near Blantyre, c.1910. Also known as 'Priory', sinking was commenced in the 1880s by William Baird & Company Limited. Having been worked by a workforce averaging 345 miners after nationalisation, the colliery was closed in 1959, but continued pumping for several years. SMM:1998.665, SC393232

**Details in 1948:** Output 147 tons per day, 39,690 tons per annum. 121 employees. No washer, no baths. Canteen (pieces only), first-aid room. Electricity supplied by Clyde Valley. Report dated 19-08-1948.

### Bothwell Castle 1 & 2

NS 7025 5867 (NS75NW/63)

**Parish:** Bothwell

**Region/District:** St/Ha

**Council:** South Lanarkshire

**Location:** Bothwell

**Previous owners:** William Baird & Company, later Bairds & Scottish Steel Limited

**Types of coal:** House and Steam

**Sinking/Production commenced:** 1875

**Year closed:** 1950

**Year abandoned:** 1953

**Average workforce:** 623

**Peak workforce:** 751

**Peak year:** 1947

**Shaft/Mine details:** 2 shafts, No. 1 293m, and No. 2 346m deep

**Details in 1948:** Output 480 tons per day, 129,360 per annum. 646 employees. Screening: 3 tables and tipplers. Campbell Binnie Reid washer. Canteen (pieces), no baths. Ambulance room. All electricity AC, and generated at colliery. Report dated 19-08-1948.

**Figure 5.129**

### Bothwell Castle 3 & 4 (Priory)

NS 6862 5876 (NS65NE/45)

**Parish:** Bothwell

**Region/District:** St/Ha

**Council:** South Lanarkshire

**Location:** Blantyre

**Previous owners:** Bairds & Scottish Steel Limited

**Types of coal:** House, Steam, Manufacturing and Gas

**Sinking/Production commenced:** 1889

**Year closed:** 1959, but remained pumping for some time

**Average workforce:** 345

**Peak workforce:** 655

**Peak year:** 1947

**Shaft/Mine details:** 2 shafts, each 410m deep

**Details in 1948:** Output 550 tons per day, 165,000 tons per annum. 630 employees. 3 bar screens. Bash-tank washer. Baths (1932), canteen. No medical services. All electricity self-generated. Report dated 16-08-1948.

**Figure 5.130**

## Braehead

NS 8135 3435 (NS83SW/26)

**Parish:** Lesmahagow

**Region/District:** St/CY

**Council:** South Lanarkshire

**Location:** Coalburn

**Previous owners:** Daniel Beattie & Company

**Types of coal:** no data

**Sinking/Production commenced:** 1938

**Year closed:** 1948

**Year abandoned:** 1951

**Statistics:** no data

**Shaft/Mine details:** 3 surface mines

**Details in 1949:** Longwall and stoop and room working. No further data. Report dated 22-02-1949.

**Other details:** Transferred to NCB in 1949, and appears not to have been re-opened.

Figure 5.131: Branchal Colliery, near Wishaw, 1955. Originally an ironstone mine developed by the Coltness Iron Co, the surface mine was begun 1924. The colliery also incorporated Greenhead mines, and in the nationalised era, employed on average 315 miners until its closure in 1959. SMM:1996.02479, SC381012

## Branchal (Greenhead & Branchal)

NS 8107 5639 (NS85NW/12)

**Parish:** Cambusnethan

**Region/District:** St/MW

**Council:** North Lanarkshire

**Location:** Branchal, Wishaw

**Previous owners:** Coltness Iron Company Limited

**Types of coal:** House

**Sinking/Production commenced:** 1924

**Year closed:** 1959

**Average workforce:** 315

**Peak workforce:** 366

**Peak year:** 1952

**Shaft/Mine details:** 2 operating mines in 1948, to depth of 165m. Also known as Greenhead and Branchal, incorporating Greenhead Mines. Older shaft (No. 1) at NS 8095 5630, and Mine No. 3 at 8095 5624.

**Details in 1948:** Output 280 tons per day, 66,500 tons per annum. 280 employees. Screening: 2 tables. No washer (dry dross goes straight to Royal George Central Washery at Newmains). No baths. Canteen under construction. All electricity AC, supplied by Clyde Valley Supply Company Report dated 16-08-1948.

**Other details:** Partly closed in 1949. Previously an ironstone mine.

**Figure 5.131**

## Broomfield

(see Canderigg 4 & 5)

Figure 5.132: Calderhead 3 & 4 Colliery, near Shotts, c.1947. A Shotts Iron Company colliery dating from c.1850. In the nationalised era, its workforce of, on average, 437 miners, produced mostly house and steam coal until its closure in 1958. The National Archives COAL80/197/1, SC614055

## Broomside
(see Dalziel and Broomside)

## Calderhead 3 & 4 (Muiracre)
NS 8777 6141 (NS86SE/14)

**Parish:** Shotts

**Region/District:** St/MW

**Council:** North Lanarkshire

**Location:** Shotts

**Previous owners:** Shotts Iron Company Limited

**Types of coal:** House and Steam

**Sinking/Production commenced:** c.1850

**Year closed:** 1958

**Average workforce:** 437

**Peak workforce:** 617

**Peak year:** 1947

**Shaft/Mine details:** 2 shafts, No. 3 75m, and No. 4 138m deep

**Details in 1948:** Output 490 tons per day, 144,410 tons per annum, longwall working. 396 employees. 3 picking tables, Baum-type washer. Baths under construction. Snack canteen available. First-aid services. Electricity (both AC and DC) from Central NCB power station at Shotts. Report dated 28-07-1948.

**Other details:** also known as Muiracre

**Figure 5.132**

## Canderigg 4 & 5 (Broomfield)
NS 7876 4696 (NS74NE/25)

**Parish:** Dalserf

**Region/District:** St/Ha

**Council:** South Lanarkshire

**Location:** Larkhall

**Previous owners:** James Nimmo & Company

**Types of coal:** Steam, House and Gas

**Sinking/Production commenced:** 1902

**Year closed:** 1954

**Year abandoned:** 1954

**Average workforce:** 379

**Peak workforce:** 582

**Peak year:** 1947

**Shaft/Mine details:** 3 shafts, No. 4 shaft 62m, No. 5 shaft 84m, and Draffan shaft 59m deep. No. 6 mine (NS7766 4679) and No. 7 mine (NS 7749 4669) were subsequently referred to as Canderside 6 & 7.

**Details in 1948:** Output 190 tons per day, 51,300 tons per annum. 267 employees. Screening: 3 picking belts. Bash-tank type washer, by Inglis Engineers, Airdrie. Canteen (pieces only), no baths or medical services.

All electricity DC, generated at colliery. Report dated 11-08-1948.
**Figure 5.133**

Figure 5.133: Canderigg 4 & 5 Colliery, near Larkhall. Dating from 1902, this was a James Nimmo and Company pit. After 1947, it produced gas, steam and house coals, and supported an average workforce of 379 miners, closing in 1954. Richard Stenlake, SC706298

## Canderside 6 & 7

NS 7766 4679 (NS74NE/26)

**Parish:** Dalserf

**Region/District:** St/Ha

**Council:** South Lanarkshire

**Location:** Larkhall

**Previous owners:** James Nimmo & Company

**Types of coal:** Gas, Steam and House

**Sinking/Production commenced:** 1939

**Year closed:** 1964

**Statistics:** combined with Canderigg 4 and 5

**Shaft/Mine details:** 2 surface mines (originally part of Canderigg), No. 6 mine (at NS 7766 4679) 168m long, dipping 1 in 6, No. 7 mine (at NS 7749 4669) 104m long, dipping at 1 in 2.24

**Details in 1948:** Output 300 tons per day, 81,000 tons per annum. 311 employees. All output screened at No. 4 pit plant, as is washing of dross. No baths, canteen or medical services. DC electricity all generated at No. 4 pit. Report dated 11-08-1948.

## Cardowan (The Stepps)

NS 6661 6834 (NS66NE/07)

**Parish:** Cadder

**Region/District:** St/St

**Council:** North Lanarkshire

**Location:** Stepps, by Glasgow

**Previous owners:** Nimmo and Dunlop

**Types of coal:** Gas, Manufacturing, House and Coking

**Sinking/Production:** 1924–28, production 1929

**Year closed:** 1983

**Average workforce:** 1,493

**Peak workforce:** 1,970

**Peak year:** 1959

**Shaft/Mine details:** 3 shafts, No. 1 (upcast) 626m (NS 6661 6834), No. 2 (upcast) 626m deep (NS 6664 6835), No. 3 (downcast) new shaft sunk in 1958 (NS 6682 6823) with steel-framed tower 23m high, built by Redpath & Brown to house overhead electric winder. Shafts 1 & 2 powered by steam winders. Pneumatic mine-car handling circuit at the surface.

**Details in 1948:** Output 1,550 tons per day, 418,500 tons per annum. 1,346 employees. Simon Carves Baum-type washer, and flotation and filter plant for fines recovery. Baths (1935 for 792 men, with 65 cubicles), canteen, first-aid room. 50% electricity self-generated, 50% from Clyde Valley. Report dated 19-08-1948.

Figure 5.134: Cardowan Colliery, Stepps, 1983. Also known as 'The Stepps', this colliery was operated by Nimmo and Dunlop, production commencing in 1929. A notoriously gassy pit, an explosion in 1982 injured 42 men. After nationalisation and subsequent reconstructions, a workforce averaging almost 1,500 miners continued to produce coking, gas, house and manufacturing coal, closure occurring in 1983. SC376874

**Other details:** Reconstructed by the NCB in the 1950s involving third shaft in a plan to increase output from 1,500 to 2,750 tons per day. The scheme also included a new coal preparation plant, and underground locomotive haulage. A very gassy pit, restricting underground power to compressed air until 1940s, when improved conditions permitted the introduction of electricity. Methane was subsequently tapped to supply the colliery's 11 Lancashire boilers from 1958, and later to supply the neighbouring Black & White whisky bottling plant. An explosion in 1982 injured 42 men. This led to changes in national ventilation regulations, and the automatic isolation of electricity in facelines when methane levels rose above 1.25% by volume. Following closure in 1983, the surviving Murray and Paterson of Coatbridge steam winder was dismantled and is now on display not far away at the industrial museum in Summerlee Heritage Park, Coatbridge.

**Figure 5.134**

**Castlehill No. 6**
NS 8399 5217 (NS85SW/25)
**Parish:** Carluke
**Region/District:** St/CY
**Council:** South Lanarkshire
**Location:** Carluke
**Previous owners:** Shotts Iron Company Limited
**Types of coal:** House and Steam (locomotive)
**Sinking/Production commenced:** 1916
**Year closed:** 1954
**Year abandoned:** 1954
**Average workforce:** 422
**Peak workforce:** 465
**Peak year:** 1947

**Shaft/Mine details:** 1 shaft, 91m deep

**Details in 1948:** Output 400 tons per day, 100,000 tons per annum. 420 employees. Screening: 3 Bar picking tables. Central washery, Baum-type. Baths (1935, for 450 men, with 37 shower cubicles), canteen. Medical service on request to local GP. Electricity previously DC, but now AC and bought from SWS Electricity Board. Report dated 17-08-1948.

**Figure 5.135**

Figure 5.135: Castlehill No. 6 Colliery, near Carluke, c.1934. A Shotts Iron Company pit which commenced production in 1916. This view shows the colliery prior to the construction of new pithead baths in 1935. The colliery only survived for seven years after nationalisation, operating with an average workforce of 422 miners. The National Archives, COAL80/224/4, SC614065

## Chapel 1 & 2

NS 8405 5480 (NS85SW/73)

**Parish:** Cambusnethan

**Region/District:** St/MW

**Council:** North Lanarkshire

**Location:** Cathburn Rd, Newmains

**Previous owners:** Haywood Coal Company

**Types of coal:** House

**Sinking/Production commenced:** 1935

**Year closed:** 1949

**Year abandoned:** 1949

**Average workforce:** 61

**Peak workforce:** 61

**Peak year:** 1947

**Shaft/Mine details:** 1 surviving surface mine. Air shaft 9m deep.

**Details in 1948:** Output 60 tons per day, 16,000 tons per annum. 65 employees. Screening: 3 fixed-bar perforated plate screens. No washer, baths, canteen or medical facilities. All electricity bought from Clyde Valley Electrical Power Company. Report dated 12-08-1948.

## Coatspark

NS 6282 6012 (NS66SW/115)

**Parish:** Cambuslang

**Region/District:** St/Gl

**Council:** South Lanarkshire

**Location:** Rutherglen, Burnside, Glasgow

**Previous owners:** Flemington Coal Company Limited

**Types of coal:** House and Steam

**Sinking/Production commenced:** 1937

**Year closed:** 1958

**Year abandoned:** 1958

**Average workforce:** 104

**Peak workforce:** 112

**Peak year:** 1948

**Shaft/Mine details:** Steep inclined surface mine 305m long, vertical depth 82m. Return shaft 12m deep.

**Details in 1948:** Output 110 tons per day, 26,000 tons per annum,

Figure 5.136: Dalziel & Broomside Colliery, Motherwell. Originating in 1869, and operated by the Glasgow Iron and Steel Co, the colliery was closed in 1948, only a year after nationalisation. At the time, there were 234 miners employed at the pit. This view shows the pithead baths, which were completed in 1928. North Lanarkshire Council Museums, SC706291

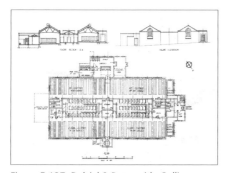

Figure 5.137: Dalziel & Broomside Colliery, Motherwell. Plan of the pithead baths built with the support of the Miners' Welfare Fund in 1928 – one of the Fund's earliest projects in Scotland. SC674992

Figure 5.138: Dewshill Colliery, near Shotts. This view of shows the surface arrangement prior to closure in 1943, after which it was retained only for pumping purposes. SMM:1996.02587, SC381015

longwall working. 112 employees. Single plate table. No washer. No baths, canteen or medical facilities. All electricity from public supply. Report dated 10-08-1948.

## Dalziel & Broomside

NS 7548 5529 (NS75NE/23)

**Parish:** Dalziel

**Region/District:** St/MW

**Council:** North Lanarkshire

**Location:** Motherwell

**Previous owners:** Glasgow Iron and Steel Company Limited

**Types of coal:** House and Steam

**Sinking/Production commenced:** 1869

**Year closed:** 1948

**Year abandoned:** 1950

**Average workforce:** 234

**Peak workforce:** 234

**Peak year:** 1947

**Shaft/Mine details:** 2 shafts, each 30m deep

**Details in 1948:** Output 120 tons per day, 33,000 tons per annum. 155 employees. Screening: 3 bar tables and 1 plate table. Boon-type washer (Blantyre Engineering Co). Baths (1928), canteen, and ambulance room. DC electricity supplied by Burgh of Motherwell and Wishaw. 5 Lancashire boilers generate steam for winding engines, surface fan, picking tables and baths. Report dated 10-08-1948.

**Figures 5.136 and 5.137**

## Dewshill

NS 8538 6390 (NS86SE/26)

**Parish:** Shotts

**Region/District:** St/ Mo

**Council:** North Lanarkshire

**Other details:** Redeveloped 1923, closed 1943. New shaft sunk by NCB in 1957 for pumping only.

**Figure 5.138**

## Dixons

(see Blantyre)

## Douglas (Ponfeigh)

NS 8689 3543 (NS83NE/20)

**Parish:** Douglas/Douglasdale

**Region/District:** St/CY

**Council:** South Lanarkshire

**Location:** Douglas

**Previous owners:** Coltness Iron Company Limited

**Types of coal:** Coking, House, Manufacturing and Steam

**Sinking/Production commenced:** 1893–8

**Year closed:** 1967

**Year abandoned:** 1968

**Average workforce:** 842

**Peak workforce:** 1,031

**Peak year:** 1952

**Shaft/Mine details:** 2 shafts (Lady Mary Pit and Lord Dunglass Pit), each 238m deep. NCB erected a reinforced concrete tower winder.

**Details in 1948:** Output 1,000 tons per day, 260,000 tons per annum, stoop and room working. 886 employees. 3 screens for dry coal. Jig washer (Campbell, Binnie and Reid). Baths (1932), canteen (full meal), morphia administration scheme, and physiotherapy. Steam and electricity, none from public supply. Report dated 12-08-1948.

**Other details:** Closed in 1967 after fire. Coal said to be well suited to express railway locomotives.

**Figure 5.139**

Figure 5.139: Douglas Colliery, Douglas, also known as Ponfeigh. Opened by the Coltness Iron Company in the late 1890s, and a source of coking as well as gas and other coals. In the NCB era, it employed on average over 840 miners, and closed in 1967 after a fire. SMM:1996.02591, SC381018

## Douglas Castle

NS 8242 3113 (NS83SW/27)

**Parish:** Douglas

**Region/District:** St/CY

**Council:** South Lanarkshire

**Location:** Douglas

**Previous owners:** Wilsons and Clyde Coal Company Limited

**Types of coal:** House and Steam

**Sinking/Production commenced:** 1912

**Year closed:** 1959

**Year abandoned:** 1960

**Average workforce:** 364

**Peak workforce:** 416

**Peak year:** 1948

**Shaft/Mine details:** 3 surface mines, Wilson Mine, and Rankin (main mine) 1:1, Douglas (main mine) 1:1

**Details in 1948:** Output 400 tons per day, 100,000 tons per annum, stoop and room. 416 employees. 2 screens for dry coal, Baum-type washer (Norton). Baths (extension completed 1938, for 308 men) and canteen. Steam and electricity, 6% from public supply. Report dated 12-08-1948.

**Other details:** In 1919, Wilsons built pithead baths, the second earliest example in Scotland (after Denbeath Colliery (Wellesley) in Fife.

**Figures 5.140 and 5.141**

Figure 5.140: Douglas Castle Colliery, near Douglas. A Wilsons and Clyde Coal Company pit dating from 1912, and notable because in 1919, the company provided the second pithead baths to be built at a colliery in Scotland. The National Archives, COAL 80/361/2, SC614056

Figure 5.141: Douglas Castle Colliery, near Douglas. The earlier baths of 1919 were replaced with the assistance of the Miners' Welfare Fund in 1938. In the nationalised era, the colliery operated with a workforce of on average 364 miners until closure in 1959. The National Archives, COAL80/361/3, SC614102

## Duntilland

NS 8382 6329 (NS86SW/64)

**Parish:** Shotts

Figure 5.142: Fernigair Colliery, Hamilton. Owned by Archibald Russell Limited and dating from c.1850, this colliery was closed down in 1947 immediately after nationalisation. South Lanarkshire Libraries SC706610

**Region/District:** St/Mo
**Council:** North Lanarkshire
**Location:** Salsburgh
**Previous owners:** Coltness Iron Company Limited
**Types of coal:** Manufacturing, Anthracite and Steam
**Sinking/Production commenced:** 1943
**Year closed:** 1951
**Year abandoned:** 1951
**Average workforce:** 181
**Peak workforce:** 182
**Peak year:** 1947
**Shaft/Mine details:** Surface mine 1 in 5, to 15m depth. New shaft sunk in 1957 by NCB for pumping, known as Dewshill (sometimes confusingly referred to as Dewshill, NS86SE/26, NS 8538 6390)
**Details in 1948:** Output 120 tons per day, 34,400 tons per annum, longwall working. 166 employees. 1 plate table, washer in Coltness. No baths, canteen (pieces only), first-aid services. All electricity bought in from public supply. Report dated 28-07-1948.

## Fernigair

NS 7421 5434 (NS75SW/33)
**Parish:** Hamilton
**Region/District:** St/Ha
**Council:** South Lanarkshire
**Location:** Hamilton
**Year Opened:** *c.*1850
**Previous owners:** Archibald Russell Limited, co-owned by Colvilles
**Types of coal:** House, Manufacturing and Steam
**Year closed:** 1947
**Year abandoned:** 1947
**Other details:** Probably responsible for much of the subsidence beneath Chatelherault. Baths built in 1931.
**Figure 5.142**

## Fortissat

NS 8461 6207 (NS86SW/46)
**Parish:** Shotts
**Region/District:** St/MW
**Council:** North Lanarkshire
**Location:** Shotts
**Previous owners:** Shotts Iron Company Limited
**Types of coal:** House and Steam
**Sinking/Production commenced:** 1870
**Year closed:** 1949
**Year abandoned:** 1949
**Average workforce:** 278

**Peak workforce:** 283

**Peak year:** 1948

**Shaft/Mine details:** 1 shaft, 87m deep

**Details in 1948:** Output 210 tons per day, 60,507 tons per annum, longwall working. 242 employees. 2 picking tables. No washer. Baths (built in 1930 for 196 men), canteen (snacks), first-aid services. All electricity from central NCB power station in Shotts. Report dated 28-07-1948.

## Garscube (Gilshie)

NS 5720 6963 (NS56NE/233)

**Parish:** Glasgow

**Region/District:** St/Gl

**Council:** City of Glasgow

**Location:** Maryhill, Glasgow

**Previous owners:** Summerlee Iron Company until 1897. Re-opened in 1916 by Messrs. Dron & Company until 1921. Thereafter, the Garscube Coal Company Limited

**Types of coal:** originally Ironstone, Steam, House and Manufacturing

**Sinking/Production commenced:** *c.*1850

**Year closed:** 1966

**Year abandoned:** 1967

**Average workforce:** 203

**Peak workforce:** 311

**Peak year:** 1963

**Shaft/Mine details:** 2 earlier shafts, 27m (No. 2 NS 5720 6963, No. 3 NS 5724 6962), surface mine (NS 5727 6968). No. 1 (downcast) 1 in 4, 116m, brick and concrete reinforced circle girders, 10' x 8', direct rope haulage. No. 2 (upcast) same dimensions and construction as No. 1, but 110m and 1 in 3, balanced double-drum direct haulage.

**Details in 1948:** Output: nil. Under reconstruction in 1948, mine driving and development in progress. 110 employees. No washer, no baths. Canteen, first-aid room. Electricity supplied by Glasgow Corporation. Report dated 19-08-1948.

**Figure 5.143**

Figure 5.143: Garscube Colliery, Glasgow. A surface mine complex originally producing ironstone for the Summerlee Iron Co. It was reconstructed in 1948 by the NCB, and supported an average workforce of 203 miners until its closure in 1966. Its closure brought to an end deep coal mining in Glasgow. T Weir of Glasgow, SMM:1998.0222, SC382480

**The Geich**
(see Auchengeich)

**Gillhead**
NS 8205 5326 (NS85SW/26)
**Parish:** Cambusnethan
**Region/District:** St/MW
**Council:** North Lanarkshire
**Location:** Waterloo, Wishaw
**Previous owners:** Coltness Iron Company Limited
**Types of coal:** House and Manufacturing
**Sinking/Production commenced:** 1945
**Year closed:** 1955
**Year abandoned:** 1955
**Average workforce:** 35
**Peak workforce:** 41
**Peak year:** 1947
**Shaft/Mine details:** 2 surface mines and 1 shaft
**Details in 1948:** Output 65 tons per day, 17,000 tons per annum. 40 employees. Static screening plant. No washer, no baths, no canteen. First-aid room. All electricity bought from Clyde Valley Electric Company Report dated 11-08-1948.

**Gilshie**
(see Garscube)

**Glen Mine**
NS 7384 6277 (NS76SW/32)
**Parish:** Old Monkland
**Region/District:** St/Mo
**Council:** North Lanarkshire
**Location:** Coatbridge
**Previous owners:** Robert Addie and Sons Collieries Limited
**Types of coal:** House and Steam (including locomotive)
**Sinking/Production commenced:** 1940
**Year closed:** 1954
**Year abandoned:** 1954
**Average workforce:** 117
**Peak workforce:** 126
**Peak year:** 1948
**Shaft/Mine details:** surface mine 305m long, 1 in 3
**Details in 1948:** Output 90 tons per day, 27,000 tons per annum. 119 employees. Screening: 1 bar table. No washer, no baths, packed meals supplied. First-aid room. AC electricity, all supplied by British Electrical Authority. Report dated 11-08-1948.

## Glentaggart

NS 8189 2717 (NS82NW/03)

**Parish:** Douglas

**Region/District:** St/CY

**Council:** South Lanarkshire

**Location:** Douglas

**Previous owners:** Glentaggart Coal Company Limited

**Types of coal:** House

**Sinking/Production commenced:** 1943

**Year closed:** 1969

**Year abandoned:** 1969

**Average workforce:** 120

**Peak workforce:** 130

**Peak year:** 1952

**Shaft/Mine details:** Originally a surface mine: second mine sunk 1947, closed and abandoned 1948

**Details in 1948:** Output 135 tons per day, 35,000 tons per annum, stoop and room working. 94 employees. 1 screen for dry coal, no washing. No baths. Canteen in Douglas village for packed meals. Steam and electricity, none from public supply. Report dated 12-08-1948.

**Figure 5.144**

Figure 5.144: Glentaggart Colliery, near Douglas, 1974. One of a cluster of mines in the Douglas coalfield, it was sunk in 1943 by the Glentaggart Coal Company Limited, and was operated subsequently by the NCB with an average workforce of 120 miners until it closed in 1969. John R Hume, SC451969

## Glentore

NS 7956 7190 (NS77SE/17)

**Parish:** New Monkland

**Region/District:** St/Mo

**Council:** North Lanarkshire

**Previous owners:** National Coal Board

**Sinking commenced:** 1954

**Production commenced:** 1957

**Year closed:** 1964

**Average workforce:** 112

**Peak workforce:** 160

**Peak year:** 1957

**Other information:** Wester Glentore Colliery was at NS7864 7141

## Glespin

NS 8043 2790 (NS82NW/11)

**Parish:** Douglas

**Region/District:** St/CY

**Council:** South Lanarkshire

**Location:** Glespin

**Previous owners:** possibly Kennox Coal Company Limited

**Sinking/Production commenced:** *c.*1908

**Year closed:** 1964

**Shaft/Mine details:** Part of Kennox. Thought to be a shallow shaft, and associated with 'Carmacoupe Mine', and to have operated a longwall system.

**Other details:** Earlier mine operated by Glespin Coal Company said to have closed by 1917 (derived from letter sent by A S Martin of Ilkeston, Derbyshire, 2/1/1999).

### Greenlees
NS 6427 5911 (NS65NW/79)

**Parish:** Cadder

**Region/District:** St/Gl

**Council:** South Lanarkshire

**Location:** Cambuslang

**Previous owners:** National Coal Board

**Types of coal:** House

**Sinking commenced:** 1947

**Production commenced:** 1948

**Year closed:** 1957

**Year abandoned:** 1957

**Average workforce:** 94

**Peak workforce:** 179

**Peak year:** 1956

**Shaft/Mine details:** 2 surface mines

**Details in 1949:** Output 30 tons per day, 8,500 tons per annum (planned 25,000 tons). 45 employees. No screening, washer, canteen, baths or medical services. All electricity supplied by Clyde Valley AC. Report dated 31-03-1949.

### Hamilton Palace (The Pailice)
NS 7240 5799 (NS75NW/64)

**Parish:** Bothwell

**Region/District:** St/MW

**Council:** North Lanarkshire

**Location:** Bothwellhaugh

**Previous owners:** Bent Colliery Company

**Types of coal:** House, Gas, Manufacturing and Steam (including locomotive)

**Sinking/Production commenced:** 1884

**Year closed:** 1959

**Year abandoned:** 1959

**Average workforce:** 552

**Peak workforce:** 605

**Peak year:** 1948

**Shaft/Mine details:** 2 shafts each 291m deep, No. 1 winding from 168m

**Details in 1948:** Output 500 tons per day, 137,500 tons per annum.

605 employees. Screening: 5 bar travelling tables. Campbell Binnie Reid washer. No baths. Canteen (pieces only). No medical services. DC electricity generated at colliery. AC bought from Central Electricity Board to drive 2 main pumps. Report dated 10-08-1948.

**Other details:** No. 1 pit dependent on ponies for underground haulage, No. 2 pit more advanced. In 1913, one of Scotland's biggest producers, but increasing problems from the 1920s. At this time, permission was given to mine under the Duke of Hamilton's mausoleum, which had been structurally designed to withstand ground movement. Splint coals especially suited to railway steam locomotives, and exported widely (e.g. to Argentina).

**Figures 5.145, see also 2.3**

Figure 5.145: Hamilton Palace Colliery, Bothwellhaugh. Sunk by the Bent Colliery Company in 1884 at a time when the Lanarkshire coal industry was flourishing. After nationalisation, the NCB retained the pit with a workforce of on average 552 miners until closure in 1959. SMM:1997.0504, SC381020

## Hassockrig

NS 8733 6292 (NS86SE/12)

**Parish:** Shotts

**Region/District:** St/Mo

**Council:** North Lanarkshire

**Location:** Harthill

**Previous owners:** Coltness Iron Company Limited

**Types of coal:** Anthracite, House and Steam

**Sinking/Production commenced:** 1885

**Year closed:** 1962

**Average workforce:** 362

**Peak workforce:** 379

**Peak year:** 1958

**Shaft/Mine details:** 2 shafts, No. 1 97m, and No. 2 165m deep

**Details in 1948:** Output 400 tons per day, 108,000 tons per annum, longwall working. 359 employees. 1 bar and 1 plate table. Washer at Coltness. No baths, but full-meal canteen. First-aid services. All electricity bought from public supply. Report dated 28-07-1948.

**Other details:** Redeveloped in 1920s

**Figure 5.146**

Figure 5.146: Hassockrig Colliery, near Harthill. Notable for its double headframe, this pit was sunk by the Coltness Iron Company in 1885, and was an important source of anthracite and steam coal. After 1947, it was operated by the NCB with a workforce of on average 362 miners, closing in 1962. SMM:1996.0466, SC446518

## Headlesscross 1

NS 9010 5850 (NS95NW/34.00)

**Parish:** Cambusnethan

**Region/District:** St/MW

**Council:** North Lanarkshire

**Location:** Fauldhouse

**Previous owners:** J Hunter and Sons

**Types of coal:** House and Manufacturing

**Sinking/Production commenced:** 1938

**Year closed:** 1949

**Year abandoned:** 1949

**Average workforce:** 50

**Peak workforce:** 52

**Peak year:** 1948

**Shaft/Mine details:** 2 shafts: No. 1 29m, and No. 2 22m vertical, then mine 1 in 0.93 for 29m

**Details in 1948:** Output 55 tons per day, 13,750 tons per annum, longwall working. 52 employees. 3 bar screens, no washer. No baths, canteen or medical facilities. All electricity from public supply. Report dated 30-07-1948.

**Other details:** Site of mine consumed by opencast mining 1979–89

## Headlesscross 2

NS 9106 5814 (NS95NW/74)

**Parish:** Cambusnethan

**Region/District:** St/MW

**Council:** North Lanarkshire

**Location:** Fauldhouse

**Previous owners:** J Hunter and Sons

**Types of coal:** Steam, House and Manufacturing

**Sinking/Production commenced:** 1946

**Year closed:** 1953

**Year abandoned:** 1953

**Average workforce:** 49

**Peak workforce:** 91

**Peak year:** 1951

**Shaft/Mine details:** 2 surface mines: No. 1 1 in 4 for 66m. No. 2 9m vertical, and then 1 in 1 for 6m.

**Details in 1948:** Output 40 tons per day, 10,000 tons per annum, longwall working. 34 employees. 2 screens. No baths, canteen or medical facilities. All power from public supply. Report dated 30-07-1948.

**Other details:** Site of mine consumed by opencast mining 1979–89

## High Darngavil

NS 7891 6919 (NS76NE/26)

**Parish:** New Monkland

**Region/District:** St/Mo

**Council:** North Lanarkshire

**Location:** Greengairs

**Previous owners:** J Crawford & Company

**Types of coal:** House (landsale)

**Sinking commenced:** pre-1947

**Year closed:** 1949

**Year abandoned:** 1949

**Statistics:** appears to have worked only for a short time in 1949

**Shaft/Mine details:** 2 surface mines, main mine 27m (1 in 3), companion mine 18m (1 in 2.5)

**Details in 1949:** Output 10 tons per day, 2,500 tons per annum. 14

employees. No washer, no baths, no canteen. Steam power, generated from small vertical boiler. Report dated 12-02-1949.

**Other details:** Transferred to NCB in 1949 and subsequently closed. Situated close to earlier Darngavil collieries. The area has since been erased by opencast coal mining.

## Hillhouserigg (Greystonelee)

NS 8665 6083 (NS86SE/18)

**Parish:** Shotts

**Region/District:** St/MW

**Council:** North Lanarkshire

**Location:** Shotts

**Previous owners:** Shotts Iron Company Limited

**Types of coal:** House and Steam

**Sinking/Production commenced:** *c.*1850

**Year closed:** 1949

**Year abandoned:** 1949

**Average workforce:** 265

**Peak workforce:** 265

**Peak year:** 1947

**Shaft/Mine details:** 1 shaft, 112m deep

**Details in 1948:** Output 259 tons per day, 59,060 tons per annum, longwall working. 241 employees. 2 screens. No washer. Small snack canteen, no baths. First-aid services. All DC and AC electricity from NCB Central Power Station in Shotts. Report dated 28-07-1948.

## Kennox Mines

NS 8039 2709 (NS82NW/12)

**Parish:** Douglas

**Region/District:** St/CY

**Council:** South Lanarkshire

**Location:** Douglas

**Previous owners:** Kennox Coal Company Limited

**Types of coal:** Steam

**Sinking/Production commenced:** 1908

**Year closed:** 1972

**Year abandoned:** 1972

**Average workforce:** 190

**Peak workforce:** 227

**Peak year:** 1960

**Shaft/Mine details:** includes No. 6 (1 in 2), No. 6a (1 in 2.25), and No. 7 (1 in 3) (see also Kennox Bridge Mine)

**Details in 1948:** Output 270 tons per day, 70,000 tons per annum, stoop and room and longwall working. 170 employees. 2 screens for dry coal, Baum-type (Blantyre Engineering Company). No baths. No canteen, but packed meals. Steam, and 100% electricity supplied from public supply. Report dated 12-08-1948.

**Other details:** Three men killed in tragedy caused by inrush of water in 1943.

## Kennox Bridge

NS 7997 2732 (NS72NE/11)

**Parish:** Douglas

**Region/District:** St/CY

**Council:** South Lanarkshire

**Sinking commenced:** pre-1948

**Production commenced:** no data

**Year closed:** 1968

**Year abandoned:** 1975

**Statistics:** see Kennox 6 & 7

**Shaft/Mine details:** surface mine (1 in 2.25), part of Kennox 6 & 7 complex

## Kepplehill 1 & 2 (also known as Stane)

NS 8765 5816 (NS85NE/37)

**Parish:** Cambusnethan

**Region/District:** St/MW

**Council:** North Lanarkshire

**Location:** Shotts

**Previous owners:** Kepplehill Coal Company, Shotts Iron Company from 1918

**Types of coal:** Steam and House

**Sinking/Production commenced:** 1897

**Year closed:** 1951

**Year abandoned:** 1955

**Average workforce:** 279

**Peak workforce:** 303

**Peak year:** 1948

**Shaft/Mine details:** 2 shafts, No. 1 108m, and No. 2 57m deep (ventilation)

**Details in 1948:** Output 250 tons per day, 69,052 tons per annum, longwall working. 303 employees. 3 picking tables, Lurig washer. No baths, snack canteen, first-aid services. AC and DC electricity all from NCB Central Power Station at Shotts. Report dated 28-07-1948.

**Other details:** Opencast mine at site between 1989 and 1999

**Figure 5.147**

Figure 5.147: Kepplehill Colliery, near Shotts, also known as Stane. Sunk in 1897 by the Kepplehill Coal Co, the colliery was operated by the Shotts Iron Company after 1918. It survived only four years into the NCB era, with an average workforce of 279 miners. This view shows a civil defence exercise under way at the colliery during World War II. SMM:1997.1049, SC381032

## Kingshill 1

NS 8562 5703 (NS85NE/27.00)

**Parish:** Cambusnethan

**Region/District:** St/MW

**Council:** North Lanarkshire

**Location:** Allanton, Shotts

**Previous owners:** Coltness Iron Company Limited

**Types of coal:** Coking, Gas, House and Steam

**Sinking/Production commenced:** 1919

**Year closed:** 1968

**Year abandoned:** 1975

**Average workforce:** 876

**Peak workforce:** 1,496

**Peak year:** 1951

**Shaft/Mine details:** 2 shafts, No. 1 344m, and No. 2 371m

**Details in 1948:** Output 1,350 tons per day, 335,000 tons per annum, longwall working. 1,323 employees. 4 screen tables. Baum-type washer (Simon Carves). Baths (1937), canteen, and well-equipped first-aid room. 40% of electricity purchased from public supply. Report dated 30-07-1948.

**Other details:** One of the first uses of underground diesel-locomotive haulage occurred at Kingshill 1 in 1935. Much of its output was later diverted up Kingshill 3 (from 1951), but was brought back by surface haulage (a tramway) for treatment at the washer. Following the Auchengeich disaster in 1962, self-rescuers were introduced for the first time in Scotland at the Kingshill pits. After the Michael disaster five years later, they became mandatory for all underground personnel.

**Figures 5.148 and 5.149**

## Kingshill 2 (also known as Queenshill)

NS 9044 5501 (NS95SW/117)

**Parish:** Cambusnethan

**Region/District:** St/MW

**Council:** North Lanarkshire

**Location:** Forth, by Lanark

**Previous owners:** Coltness Steel Company Limited

**Types of coal:** Coking, Gas, House and Steam

**Sinking commenced:** 1931

**Year closed:** 1963

**Year abandoned:** 1975

**Average workforce:** 624

**Peak workforce:** 686

**Peak year:** 1954

**Shaft/Mine details:** 1 shaft, 296m deep (downcast with forcing fan), return mine at Climpy, 1 in 1.2 for 128m

**Details in 1948:** Output 640 tons per day, 160,000 tons per annum, longwall working. 589 employees. 3 screening tables. Baum-type washer (Simon Carves). Baths (1929), canteen (packed meals), and first-aid room. All electricity from public supply. Report dated 30-07-1948.

**Other details:** connected to Kingshill No. 1

**Figure 5.150**

Figure 5.148: Kingshill No. 1 Colliery, Allanton, c.1947. Sunk in 1919 by the Coltness Iron Co, it produced steam, gas, house and coking coals. Guthrie Hutton, SC706294

Figure 5.149: Kingshill No. 1 Colliery, Allanton, 1954. From 1951, coal was taken up the shaft of the new Kingshill No. 3 Colliery, but was then brought back by surface haulage to be cleaned at this washery. During the NCB era, Kingshill No. 1 maintained a workforce of on average 876 miners, eventually closing in 1968. SMM:1996.2442, SC381025

Figure 5.150: Kingshill No. 2 Colliery, near Forth, c.1945. The second of the three Kingshill pits, it was sunk by the Coltness Iron Company in 1931, and also produced coking, gas, steam and house coals. It continued in production for 16 years after nationalisation, with an average workforce of 624 miners. SC706275

Figure 5.151: Kingshill No. 3 Colliery, Allanton Moor, c.1950. The first major new sinking in Scotland by the NCB, and one of the few new projects to occur in Lanarkshire. Production commenced in 1951, and continued with an average workforce of 600 miners until 1974. SC446725

## Kingshill 3

NS 8690 5434 (NS85SE/36)
**Parish:** Cambusnethan
**Region/District:** St/MW
**Council:** North Lanarkshire
**Location:** South-east of Newmains, Allanton Moor
**Previous owners:** National Coal Board
**Types of coal:** Coking, House and Steam
**Sinking commenced:** 1946–50
**Production commenced:** 1951
**Year closed:** 1974
**Year abandoned:** 1975
**Average workforce:** 600
**Peak workforce:** 769
**Peak year:** 1958
**Shaft/Mine details:** Single new shaft, 234m deep, 4.57m diameter, concrete lined, designed to wind 1200 tons per shift, driven by electric 550hp winder with 'bi-cylindro' drum.
**Other details:** The first major NCB sinking in Scotland at a time when coking coal was desperately needed, and the first to wind large-capacity mine cars (2.5 ton) to the surface. It was designed to take half of Kingshill 1's production to the surface, but the coal was then sent by endless-rope tramway across the moor to be washed at Kingshill No. 1.
**Figure 5.151**

## Knowehead 2 (New Mine)

NS 7876 6303 (NS76SE/23)
**Parish:** Shotts
**Region/District:** St/Mo
**Council:** North Lanarkshire
**Location:** Chapelhall
**Previous owners:** National Coal Board
**Types of coal:** House, (bituminous, possibly semi-anthracite in parts)
**Sinking commenced:** 1949
**Production commenced:** 1952
**Year closed:** 1962
**Average workforce:** 152
**Peak workforce:** 203
**Peak year:** 1959
**Shaft/Mine details:** surface mine, estimated 305m long, gradient 1 in 4
Details in 1949: Planned output of 150 tons per day, 37,500 tons per annum. 92 employees. Screening: jig screen, picking band. No washer. Baths, canteen. AC electricity. Report dated 05-04-1949.

## Knowetop

NS 7326 5099 (NS75SW/29)

**Parish:** Hamilton

**Region/District:** St/Ha

**Council:** South Lanarkshire

**Location:** Quarter, Hamilton

**Previous owners:** National Coal Board

**Types of coal:** House and Steam

**Sinking commenced:** 1947

**Production commenced:** 1948

**Year closed:** 1966

**Average workforce:** 31

**Peak workforce:** 42

**Peak year:** 1952

**Shaft/Mine details:** 2 surface mines

**Details in 1949:** Output 30 tons per day, 7,500 tons per annum. 27 employees. Bar screens. No washer, baths or canteen. Ambulance room. All electricity from public supply. Report dated 31-03-1949.

**Figure 5.152**

Figure 5.152: Knowetop Colliery, near Quarter, c.1950. An NCB scheme commenced in 1947, comprising two surface mines which employed on average 31 miners during the colliery's 19 years of operation. Willie Brown, SC706296

## Langside Mine

NS 8158 6309 (NS86SW/34)

**Parish:** Shotts

**Region/District:** St/Mo

**Council:** North Lanarkshire

**Location:** Salsburgh

**Previous owners:** Langside Coal Company Limited

**Types of coal:** House and Manufacturing

**Sinking commenced:** 1945

**Year closed:** 1949

**Year abandoned:** 1949

**Average workforce:** 17

**Peak workforce:** 17

**Peak year:** 1948

**Shaft/Mine details:** surface mine, dipping 1 in 3

Details in 1949: Output 20–25 tons per day, 5,600–6,250 per annum, longwall working. 17 employees. Fixed-plate screen. No baths or canteen, first-aid equipment only. All electricity from public supply. Report dated 11-02-1949

**Other details:** Transferred to NCB in 1948.

## Lochend 5

NS 8487 6945 (NS86NW/54)

**Parish:** New Monkland

**Region/District:** St/Mo

**Council:** North Lanarkshire

Figure 5.153: Lochend No. 5 Colliery, Caldercruix, 1998. Aerial view of the colliery's extraordinary bing, known locally as the 'Mexican's Hat'. The colliery was operated from the 1880s by the Brownieside Coal Company Limited. It closed only a year after nationalisation in 1948, when it had a workforce of 89 miners. SC354310

**Location:** Caldercruix

**Previous owners:** Brownieside Coal Company Limited

**Types of coal:** Anthracite

**Sinking/Production commenced:** *c.*1880

**Year closed:** 1948

**Year abandoned:** 1948

**Average workforce:** 89

**Peak workforce:** 101

**Peak year:** 1948

**Shaft/Mine details:** No. 5 shaft 70m (to Coxrod or Lower Drumgray). Air Pit 46m to Upper Drumgray.

**Details in 1948:** 19,250 tons per annum. 101 employees. Screening facilities: bar screen only, dross hand fed into washer. Bash-type washer made partly by Dickson and Mann. Method of working longwall, hand-filled into tubs, with aid of coal cutting machinery. No conveying machinery. Canteen, but no baths. All electricity AC, and supplied from outside by Clyde Valley Power Company Estimated future life of colliery only 3 years. Report dated 28-08-1948.

**Other details:** Notable for its extraordinary bing, known locally as 'the Mexican's Hat'.

**Figure 5.153**


**Lumloch**

(see Wester Auchengeich)


**Muiracre**

(see Calderhead 3 & 4)


**Netherton Mine**

NS 9070 6559 (NS96NW/21)

**Parish:** Shotts

**Region/District:** St/MW

**Council:** North Lanarkshire
**Location:** Harthill
**Previous owners:** United Collieries Limited
**Types of coal:** House
**Sinking/Production commenced:** 1938
**Year closed:** 1950
**Year abandoned:** 1950
**Average workforce:** 106
**Peak workforce:** 110
**Peak year:** 1948
**Shaft/Mine details:** surface mine, to 33m
**Details in 1948:** Output 90 tons per day, 28,000 tons per annum, longwall working. 110 employees. 2 plate tables. No washer. No baths, piece canteen run by workmen, first-aid services. All electricity AC, supplied from NCB Westrigg Power Station. Report dated 28-07-1948.

## Newton

(see Blantyreferme 3)

## Northfield and Hall (The Pennypit)

NS 8822 6018 (NS86SE/40)
**Parish:** Shotts
**Region/District:** St/MW
**Council:** North Lanarkshire
**Location:** Shotts
**Previous owners:** Shotts Iron Company Limited
**Types of coal:** House and Steam, and blaes
**Sinking/Production commenced:** 1917
**Year closed:** 1961
**Year abandoned:** 1961
**Average workforce:** 443
**Peak workforce:** 484
**Peak year:** 1958
**Shaft/Mine details:** 2 shafts, Northfield 73m, and Hall 84m deep
**Details in 1948:** Output 490 tons per day, 129,524 tons per annum, longwall working. 395 employees. 3 picking tables, Baum-type washer. Baths (1932), canteen, first-aid services. AC and DC electricity, 60% from public supply. Report dated 28-07-1948.
**Other details:** Supplied blaes to local NCB brickworks, which operated from *c.*1920 until the mid-1980s.

## Overtown

NS 7990 5278 (NS75SE/33)
**Parish:** Cambusnethan
**Region/District:** St/MW

Figure 5.154: Overtown Colliery, near Wishaw. A Coltness Iron Company colliery sunk in 1931, and closed by the NCB in 1954 to permit a major redevelopment facilitating access to coking coals. Production resumed in 1959, and the colliery operated until 1968 with an average workforce of 284 miners. SMM:1996.0435, SC381027

**Council:** North Lanarkshire

**Location:** Overtown, Wishaw

**Previous owners:** Coltness Iron Company Limited

**Types of coal:** House, and later Coking

**Sinking/Production commenced:** 1931, re-opened 1956

**Year closed:** 1954, and again in 1968

**Year abandoned:** 1968

**Average workforce:** 284

**Peak workforce:** 466

**Peak year:** 1960

**Shaft/Mine details:** 2 shafts, each 305m deep

**Details in 1948:** Output 180 tons per day, 45,000 tons per annum, longwall working. 160 employees. 2 bar travelling screens. Baths, canteen (packed meals), ambulance room. Electricity all from public supply. Report dated 11-08-1948.

**Other details:** Closed in 1954 prior to redevelopment involving the deepening of the shafts (from 305 to 430m) to reach coking coal measures, re-opening in 1959.

**Figure 5.154**


**Overwood Mine**

NS 7745 4530 (NS74NE/27)

**Parish:** Dalserf

**Region/District:** St/Ha

**Council:** South Lanarkshire

**Location:** Carlisle Road, Stonehouse

**Previous owners:** Overwood Coal Company

**Types of coal:** House, Manufacturing

**Sinking/Production commenced:** 1937

**Year closed:** 1955

**Average workforce:** 39

**Peak workforce:** 39

**Peak year:** 1952

**Shaft/Mine details:** 2 surface mines 91m long, dipping 1 in 2.5

Details in 1949: Output 35 tons per day, 9,000 tons per year, longwall working. 34 employees. 1 small bar screen. No washer, no baths, canteen or medical facilities. All electricity from public supply. Report dated 15-02-1949.

**Other details:** Transferred to NCB in 1949. May have been linked with Canderside.


**The Pailice**

(see Hamilton Palace)


**The Pennypit**

(see Northfield)

## Ponfeigh

(see Douglas)

## Priory

(see Bothwell Castle 3 & 4)

## Quarter

NS 7385 5173 (NS75SW/30)

**Parish:** Hamilton

**Region/District:** St/Ha

**Council:** South Lanarkshire

**Location:** Quarter, Hamilton

**Previous owners:** United Collieries Limited

**Types of coal:** Steam and House

**Sinking/Production commenced:** 1815

**Year closed:** 1951

**Year abandoned:** 1960

**Average workforce:** 106

**Peak workforce:** 120

**Peak year:** 1947

**Shaft/Mine details:** 2 shafts, Quarter1 at NS 7386 5173 (NS75SW/30), and Quarter 2 at NS 7447 5172 (NS75SW/31). Surface mine at Quarter 5 (NS 7413 5148), known as 5A, sunk in 1947, but closed in 1950.

**Details in 1948:** Output 115 tons per day, 27,445 tons per annum, pillar and stall working. 98 employees. Plate tables for screening. No washer. No baths, canteen or medical facilities, but nurse in village. DC electricity, none from public supply. Report dated 19-08-1948.

**Other details:** Quarter 1, 2 and 5 were linked by a triangular arrangement of surface mineral railways (between 1 and 5), and a tramway between 5 and 2.

**Figure 5.155**

## Queenshill

(see Kingshill 2)

## Rankin

NS 8394 3235 (NS83SW/28)

**Parish:** Douglas

**Region/District:** St/CY

**Council:** South Lanarkshire

**Sinking/Production commenced:** unknown

**Year closed:** *c.*1947

**Average workforce:** 110

**Peak workforce:** 120

Figure 5.155: Quarter Colliery, near Hamilton, c.1900. The colliery, which was operated by United Collieries Limited and sunk in 1815, comprised several shafts and surface mines. It survived for only four years after nationalisation, with a workforce never exceeding 120 miners. SMM:1996.2354, SC446578

Figure 5.156: Rankin Colliery, near Douglas. A small colliery in the Douglas coalfield which seems to have operated for only one year in the NCB era, at which time it had a workforce of 110 miners. The National Archives, COAL80/831/11, SC614048

**Peak year:** 1947
**Figure 5.156**

## Skellyton 2 & 3
NS 7690 5182 (NS75SE/34)
**Parish:** Dalserf
**Region/District:** St/Ha
**Council:** South Lanarkshire
**Location:** Larkhall
**Previous owners:** Wilsons and Clyde Coal Company
**Types of coal:** House
**Sinking/Production commenced:** 1944
**Year closed:** 1951
**Year abandoned:** 1951
**Average workforce:** 90
**Peak workforce:** 92
**Peak year:** 1948
**Shaft/Mine details:** 2 mines, to 14m (No. 2) and 50m (No. 3)
**Details in 1948:** Output 120 tons per day, 30,000 tons per annum, stoop and room and longwall working. 92 employees. 2-bar screens and bagging platform. No washer. No baths, canteen or medical facilities. All electricity from public supply. Report dated 11-08-1948.

## Southfield
NS 9015 5970 (NS95NW/30)
**Parish:** Cambusnethan
**Region/District:** St/MW
**Council:** North Lanarkshire
**Location:** Shotts
**Previous owners:** Shotts Iron Company Limited
**Types of coal:** House and Steam, and blaes
**Sinking/Production commenced:** 1923
**Year closed:** 1959
**Average workforce:** 492
**Peak workforce:** 556
**Peak year:** 1947
**Shaft/Mine details:** 2 shafts, No. 1 140m, and No. 2 74m deep
**Details in 1948:** Output 560 tons per day, 140,238 tons per annum, longwall working. 548 employees. 3 tables, no washer. Baths (1930), snack canteen, first-aid services. All electricity from public supply. Report dated 28-07-1948.

## Stane
(see Kepplehill 1 & 2)

## Stane Mines

NS 8865 5862 (NS85NE/163)

**Parish:** Cambusnethan

**Region/District:** St/MW

**Council:** North Lanarkshire

**Location:** Shotts

**Previous owners:** National Coal Board

**Types of coal:** Steam and House

**Sinking/Production commenced:** 1948

**Year closed:** 1955

**Year abandoned:** 1955

**Average workforce:** 155

**Peak workforce:** 210

**Peak year:** 1953

**Shaft/Mine details:** surface mines

**Details in 1949:** Output 250 tons per day, 62,500 tons per annum, longwall working. 158 employees. Coal screened at Stane Colliery. 3 picking tables. Small baths, snack canteen, first-aid services to be erected. AC and DC electricity all from NCB Central Power Station at Shotts. Report dated 14-02-1949.


## The Stepps

(see Cardowan)


## Thankerton 1, 3, 6 & 6a

NS 7494 6120 (NS76SW/33)

**Parish:** Bothwell

**Region/District:** St/MW

**Council:** North Lanarkshire

**Location:** Holytown, Motherwell

**Previous owners:** John McAndrew & Company Limited

**Types of coal:** Steam and House, and blaes

**Sinking/Production commenced:** 1850

**Year closed:** 1953

**Year abandoned:** 1954

**Average workforce:** 636

**Peak workforce:** 766

**Peak year:** 1947

**Shaft/Mine details:** 4 shafts, No. 1 (NS 7494 6120, NS76SW/33) 59m, No. 3 (NS 7558 6103, NS76SE/24) 37m, and No. 6 (NS 7613 6068, NS76SE/25) 165m deep. Thankerton 6A (NS 761 606, NS76SE/26) was sunk in 1949.

**Details in 1948:** Output 600 tons per day, 160-180,000 tons per annum, longwall working (hand driving). 736 employees. No. 1 Pit: 2 screening tables. No. 6 Pit: 2 screening tables. Bash washer, now obsolete. No baths or canteen, but ambulance room at each pit. All

Figure 5.157: Thankerton No. 3 Colliery, Holytown, c.1900. One of a group of collieries dating from 1850, and established by John McAndrew and Company Limited. Together, they employed 636 miners at nationalisation in 1947. The last of the pits closed in 1953. SMM:1998.0207, SC381033

electricity from public supply. Report date 11-08-1948.
**Other details:** Thankerton 2 was at NS 7515 6056 (NS76SE/21).
**Figure 5.157**

## Thinacre Mine

NS 7393 5005 (NS75SW/32)
**Parish:** Hamilton
**Region/District:** St/Ha
**Council:** South Lanarkshire
**Location:** Quarter
**Previous owners:** Overwood Coal Company
**Types of coal:** Steam and House
**Sinking/Production commenced:** 1939
**Year closed:** 1963
**Average workforce:** 91
**Peak workforce:** 99
**Peak year:** 1956
**Shaft/Mine details:** surface mine
**Details in 1949:** Output 60 tons per day, 3,000 tons per annum, longwall working. 41 employees. Screens: shaker riddle. No washer. No baths, canteen or medical facilities. All electricity from public supply. Report dated 15-02-1949.
**Other details:** Transferred to NCB in 1949.

## Wester Auchengeich (Lumloch)

NS 6407 7005 (NS67SW/56)
**Parish:** Cadder
**Region/District:** St/St
**Council:** North Lanarkshire
**Location:** Bishopbriggs
**Previous owners:** Nimmo & Dunlop
**Types of coal:** Gas, House, Steam and Coking
**Sinking/Production commenced:** 1928
**Year closed:** 1968
**Year abandoned:** 1968
**Average workforce:** 601
**Peak workforce:** 850
**Peak year:** 1959
**Shaft/Mine details:** 2 shafts, No. 3 315m (NS 6407 7005), and No. 4 314m deep (NS 6407 7004)
**Details in 1948:** Output 920 tons per day, 248,400 tons per annum. 612 employees. Simon Carves Baum-washer. Baths (1939, for 504 men), canteen, first-aid room. Electricity supply from Clyde Valley. Report dated 19-08-1948.
**Other details:** Originally an ironstone pit. Connected to Auchengeich.
**Figure 5.158**

## Westoun Mine (Bankend)

NS 8115 3405 (NS83SW/25)

**Parish:** Lesmahagow

**Region/District:** St/CY

**Council:** South Lanarkshire

**Location:** Coalburn

**Previous owners:** Arden Coal Company Limited

**Types of coal:** Manufacturing

**Sinking/Production commenced:** *c.*1890, re-opened by NCB 1948

**Year closed:** 1961

**Average workforce:** 134

**Peak workforce:** 175

**Peak year:** 1956

**Shaft/Mine details:** surface mines, no further details

**Details in 1948:** Being re-opened in 1948, no employee data. Screening at Auchlochan No. 6. No baths or canteen. 100% electricity from public supply. Report dated 12-08-1948.

**Other details:** also known as Bankend Westoun (and Westown)

## Wilsontown

NS 9410 5562 (NS95NW/120)

**Parish:** Carnwath

**Region/District:** St/CY

**Council:** South Lanarkshire

**Location:** Forth

**Previous owners:** William Dixon Limited

**Types of coal:** Steam, Manufacturing and Coking

**Sinking/Production commenced:** 1898

**Year closed:** 1955

**Year abandoned:** 1955

**Average workforce:** 304

**Peak workforce:** 378

**Peak year:** 1948

**Shaft/Mine details:** 2 shafts, No. 3 104m, and No. 9 73m deep

**Details in 1948:** Output 320 tons per day, 80,000 tons per annum, longwall working. 390 employees. 2 screen tables. Luhrigg (Grantown Engineering Company) washer. No baths. Canteen for packed meals, first-aid centre. DC electricity, all generated at mine. Report dated 30-07-1948.

**Figure 5.159**

## Woodside

NS 7924 4883 (NS74NE/28)

**Parish:** Dalserf

**Region/District:** St/Ha

**Council:** South Lanarkshire

Figure 5.158: Wester Auchengeich Colliery, near Bishopbriggs, also known as 'Lumloch'. An important source of coking coal for Nimmo and Dunlop, the colliery was sunk in 1928 when it also produced ironstone. After nationalisation, it continued to produce coal until 1968, with a workforce which averaged over 600 miners. SMM:1996:02590, SC381037

**Location:** Larkhall

**Previous owners:** Wilsons and Clyde Coal Company Limited

**Types of coal:** House and Steam

**Sinking/Production commenced:** 1848

**Year closed:** 1955

**Year abandoned:** 1955

**Average workforce:** 38

**Peak workforce:** 40

**Peak year:** 1954

**Shaft/Mine details:** surface mines, 1 in 3 to 110m

**Details in 1948:** Output 40 tons per day, 11,000 tons per annum, hand-pick and machine-cutting working. 37 employees. 5-cm bar screens and dross conveyor to wagons. No washer. No baths or canteen. Ambulance room in office. All electricity from public supply. Report dated 11-08-1948.

Figure 5.159: Wilsontown Colliery, Forth. Established in 1898 by William Dixon Limited to produce coking and steam coals, its two shafts operated until 1955, employing on average 304 miners in the years following nationalisation. SMM:1997.0992, SC381038

## Licensed Mines in Lanarkshire 1947–97

| Name | Licensee | Grid Ref | Opened | Closed |
|------|----------|----------|--------|--------|
| Abbey (Forth) | M T & P Stark | NS 934 543 | 1954 | pre-1997 |
| Amberly (Plains) | Shankland & Company | NS 793 678 | 1954 | 1957 |
| Amosknowes (Hartwood) | J McCutcheon | NS 837 593 | 1955 | 1959 |
| Arbuckle (Plains) | R & J Dow | NS 797 680 | pre-1947 | 1949 |
| Arbuckle 13/14 (Plains) | R & J Dow | NS 795 680 | 1946 | c.1949 |
| Arbuckle 15 (Plains) | Charles Tobin | NS 798 684 | 1950 | c.1967 |
| Arbuckle 16 (Plains) | Charles Tobin | NS 794 680 | 1959 | c.1962 |
| Arbuckle 17 (Plains) | Charles Tobin | NS 796 682 | unknown | c.1967 |
| Ashlea 4 (Calderbank) | John Ferguson Limited | NS 768 617 | pre-1947 | 1947 |
| Ashlea 5 (Calderbank) | John Ferguson Limited | NS 768 617 | 1946 | 1956 |
| Ashlea 6 (Calderbank) | John Ferguson Limited | NS 765 623 | 1956 | 1959 |
| Ashlea 7 (Calderbank) | John Ferguson Limited | NS 768 617 | 1958 | c.1963 |
| Auchenheath (N of Lesmahagow) | Auchenheath Mining Company Limited | NS 805 440 | pre-1947 | 1951 |
| Auchmeddan 1 (Lesmahagow) | Daniel Beattie & Company | NS 844 383 | 1945 | NCB |
| Auchmeddan 2 (Lesmahagow) | Daniel Beattie & Company | NS 843 386 | 1945 | NCB |
| Auchmeddan 3 (Lesmahagow) | Daniel Beattie & Company | NS 845 380 | 1945 | NCB |
| Auldton (Lesmahagow) | Joseph Boyd | NS 828 377 | 1942 | NCB |
| Avonhead (Caldercruix) | Headrigg Coal Company | NS 805 697 | pre-1947 | 1948 |
| Avonhead 3 (Caldercruix)) | Headrigg Coal Company | NS 805 697 | pre-1947 | c.1947 |
| Avonhead 4 & 4a (Caldercruix) | Headrigg Coal Company | NS 809 702 | 1947 | 1948 |
| Backshot 4 (Forth) | A Maxwell & Company | NS 935 538 | 1964 | 1995 |
| Ballochney 1 (Airdrie) | J & J Somerville Limited | NS 788 675 | 1950 | 1951 |
| Ballochney 2 (Airdrie) | J & J Somerville Limited | NS 796 673 | 1951 | 1951 |
| Ballochney 3 (Airdrie) | J & J Somerville Limited | NS 796 678 | 1951 | 1951 |
| Balmoral (Greengairs) | Palace Coal Company | NS 775 703 | 1947 | 1948 |
| Beaton's Lodge (Larkhall) | Currie Brothers | NS 751 520 | 1940 | NCB |
| Berryhill (Righead, Airdrie) | T McDonach | NS 769670 | 1962 | 1967 |
| Bethel (Chapelhall) | J G McCracken | NS 791 621 | 1967 | 1973 |
| Biggar 2 (Gorehill, Airdrie) | J & J McCracken | NS 815 691 | 1953 | c.1967 |
| Biggar Ford 2 (Cleland) | R Moffat & Sons | NS 799 595 | pre-1947 | 1948 |
| Biggar/Gorehill (Airdrie) | J & J McCracken | NS 814 688 | 1947 | 1948 |
| Birkrigg (Stonehouse) | Jackson & Tweedie | NS 777 495 | 1952 | 1953 |
| Blackhall (Allanton) | J B Milliken | NS 868 577 | 1957 | 1970 |
| Blairmuckhole (Harthill) | Henderson Brothers | NS 863 644 | pre-1947 | 1947 |
| Blueknowe 1 (Carluke) | Shawfield Coal Company | NS 813 524 | 1953 | 1953 |
| Blueknowe 2 (Carluke) | Shawfield Coal Company | NS 813 521 | 1954 | 1955 |
| Bogside (Wishaw) | Peter Hamil | NS 781 493 | 1953 | 1955 |
| Bogside 1 (Law) | A J & A Smillie | NS 828 536 | 1958 | c.1959 |
| Braehead Farm (Coalburn) | Daniel Beattie & Company | NS 813 343 | 1938 | NCB |
| Bridgend 1 (Wishaw) | Deerpark Coal Company Limited | NS 791 564 | 1954 | c.1963 |
| Broomknowe (Cleland) | R & W Henderson | NS 808 604 | pre-1947 | 1948 |
| Broomknowe 2 (Cleland) | R & W Henderson | NS 807 604 | 1949 | 1950 |

| Name | Licensee | Grid Ref | Opened | Closed |
|---|---|---|---|---|
| Burnfoot (Forth) | M T & P Stark | NS 295 653 | c. 1947 | unknown |
| Burnside 3 (Newarthill) | R Moffat & Sons | NS 799 601 | 1948 | 1951 |
| Burnside 4 (Newarthill) | R Moffat & Sons | NS 802 600 | 1952 | 1961 |
| Calder 1 (Clarkston, Airdrie) | Wester Moffat Coal Company | NS 790 658 | 1957 | 1959 |
| Calder 2 (Clarkston, Airdrie) | Wester Moffat Coal Company | NS 790 658 | 1958 | 1960 |
| Castleview (Chapelhall) | J J Kane | NS 795 622 | 1960 | c. 1970 |
| Catcraig (Braidwood, Carluke) | Catcraig Coal Company | NS 833 472 | 1955 | 1956 |
| Chapelrigg (Chapelhall) | Bonnybridge Company Limited | NS 791 629 | pre-1947 | 1953 |
| Claremount (Coatbridge) | R D Poore | NS 745 635 | 1949 | 1950 |
| Clarkston 2 (Parkhead, Airdrie) | T Reid & Company | NS 781 648 | pre-1947 | 1947 |
| Coalhall 1 (Jerviston, Airdrie) | T & J Campbell | NS 762 586 | pre-1947 | 1951 |
| Coalhall 2 (Jerviston, Airdrie) | T & J Campbell | NS 762 586 | pre-1947 | 1950 |
| Crindledyke (Cleland) | Auchinlea Quarries Limited | NS 820 572 | pre-1947 | 1947 |
| Deerpark (Wishaw) | Deerpark Coal Company | NS 793 569 | 1941 | 1954 |
| Drossyhill (Airdrie) | Armstrong Bros | NS 784 653 | 1958 | 1959 |
| Drumbreck (Dumbreck) | Beveridge Brose | NS 829 686 | 1946 | c. 1963 |
| Drumshangie 4 (Airdrie) | Drumlangie Coal Company | NS 775 685 | pre-1947 | 1947 |
| Drumshangie 5 (Airdrie) | Drumlangie Coal Company | NS 775 685 | pre-1947 | 1947 |
| Dryflats (Airdrie) | Glenrigg Coal Company | NS 748 676 | pre-1947 | 1950 |
| Dryflats 4 (Airdrie) | M Wooton, Airdrie | NS 748 675 | pre-1947 | 1950 |
| Dunrobin (Airdrie) | Browshot Coal Company | NS 783 653 | pre-1947 | 1947 |
| Dunrobin 2 (Airdrie) | Armstrong Coal Company | NS 784 655 | 1948 | 1951 |
| Dunrobin 3 (Airdrie) | Browshot Coal Company | NS 785 651 | pre-1947 | 1947 |
| Dunrobin 4 (Airdrie) | Browshot Coal Company | NS 783 653 | 1948 | 1948 |
| Dunrobin 5 (Airdrie) | Browshot Coal Company | NS 785 651 | 1950 | 1950 |
| Dykehead (Larkhall) | Hillfarm Coal Company | NS 753 435 | 1968 | 1971 |
| Easter Windyedge (Cleland) | R Henderson & Company | NS 801 592 | pre-1947 | 1948 |
| Fairview (Forth) | M T & P Stark | NS 930 545 | 1953 | 1954 |
| Fairybank (Salsburgh) | Greenshields & Company | NS 807 624 | 1950 | 1951 |
| Fence (Kirkmuirhill) | Thomas Sinclair & Sons | NS 803 458 | 1944 | 1961 |
| Gartmillan 1 (Glenboig) | W Wooton | NS 747 695 | 1951 | 1963 |
| Gartness (Chapelhall) | Barr & Sinclair | NS 787 636 | 1936 | 1960 |
| Gillfoot (Braidwood, Carluke) | Gillfoot Coal Company/Currie Bros | NS 832 470 | 1957 | 1969 |
| Glenbank (Dykehead, Shotts) | R Henderson | NS 866 599 | 1951 | 1964 |
| Glenview/Law Mine (Carluke) | Wright & Moffat | NS 817 527 | 1970 | 1991 |
| Goodockhill 3 (Salsburgh) | Greenshields & Company | NS 815 613 | 1951 | 1958 |
| Harestonehill (Newmains) | Blyth Bros | NS 818 543 | 1967 | 1983 |
| Hazelside Mine (Coalburn) | Thomas Sinclair & Sons | NS 814 288 | 1966 | 1968 |
| Heatheryknowe A (Birkenshaw) | Glenrigg Coal Company | NS 701 622 | 1950 | 1951 |
| High Darngavil (Airdrie) | J Crawford & Sons | NS 786 690 | pre-1947 | NCB |
| Hill (Newmains) | R Moffat & Sons | NS 822 584 | 1960 | 1962 |
| Hill of Drumgray 9 (Greengairs) | Darngavil Brickworks Limited | NS 789 699 | pre-1947 | 1951 |
| Hills of Murdostoun 1 (Cleland) | Auchinlea Quarries Limited | NS 820 586 | c. 1941 | 1951 |
| Hills of Murdostoun 3 (Cleland) | Auchinlea Quarries Limited | NS 818 596 | 1950 | 1952 |

| Name | Licensee | Grid Ref | Opened | Closed |
|---|---|---|---|---|
| Hillside 1 (Darngavil, Airdrie) | Patrick McGrady | NS 786 688 | 1951 | 1956 |
| Hillside 2 (Darngavil, Airdrie) | Patrick McGrady | NS 785 686 | 1956 | c. 1958 |
| Hillside 3 (Darngavil, Airdrie) | Patrick McGrady | NS 787 687 | 1958 | 1968 |
| Hopefield (Wishaw) | Hopefield Coal Company | NS 792 531 | pre-1947 | 1947 |
| Hyndshaw 1 (Carluke) | Dr Arthur | NS 842 532 | 1947 | c. 1950 |
| Hyndshaw 2 (Carluke) | Dr Arthur | NS 842 533 | 1948 | 1953 |
| Hyndshaw 3 (Carluke) | Hyndshaw Coal Company Limited | NS 845 528 | 1957 | 1963 |
| Isabella (Larkhall) | William Fisher & Sons | NS 777 506 | 1959 | 1961 |
| Junction (Blackhall J'n, Shotts) | T McDonagh | NS 884 577 | 1973 | 1980 |
| Kippsbyre (Airdrie) | Francis McLean & Company | NS 748 669 | 1939 | 1960 |
| Kittymuir 3 (Stonehouse) | Matthew McCulloch | NS 762 480 | 1953 | 1953 |
| Kittymuir 4 (Stonehouse) | Matthew McCulloch | NS 762 480 | 1954 | 1954 |
| Kittymuirhill (Stonehouse) | Beattie Bros | NS 758 488 | 1951 | 1951 |
| Langside Mine (Salsburgh) | Langside Coal Company | NS 815 632 | 1945 | NCB |
| Langside Farm (Salsburgh) | Union Coal Company | NS 814 630 | pre-1947 | 1948 |
| Little Whitehill 1 (Cleland) | Waddel Hawthorn | NS 812 609 | 1947 | 1950 |
| Little Whitehill 2 (Cleland) | Waddel Hawthorn | NS 812 606 | 1956 | 1956 |
| Manse (Forth) | M T & P Stark | NS 976 554 | 1951 | 1951 |
| Marlage 1 (Larkhall) | J & J McPhee | NS 787 485 | 1951 | 1953 |
| Marlage 2 (Larkhall) | J & J McPhee | NS 783 485 | 1954 | 1959 |
| Marlage 3 (Larkhall) | J & J McPhee | NS 782 485 | 1960 | c. 1963 |
| Mauldslie (Carluke) | Andrew Maxwell & Company | NS 824 508 | pre-1947 | 1949 |
| Mauldslie 7 (Carluke) | Andrew Maxwell & Company | NS 824 508 | 1955 | c. 1968 |
| Mauldslie Upper Ell (Carluke) | Andrew Maxwell & Company | NS 824 508 | 1949 | c. 1955 |
| Meadowhill (Airdrie) | Meadowhill Coal Company Limited | NS 783 674 | 1951 | 1970 |
| Meikle Drumgray 1 (Darngavil) | Felix Travers | NS 782 691 | pre-1947 | 1958 |
| Mossband 10 (Newhouse) | A Allan & Sons Limited | NS 798 606 | 1957 | 1958 |
| Mossband 7 (Newhouse) | A Allan & Sons Limited | NS 796 607 | 1948 | 1950 |
| Mossband 8 (Newhouse) | A Allan & Sons Limited | NS 795 607 | 1947 | 1954 |
| Mossband 9 (Newhouse) | A Allan & Sons Limited | NS 801 603 | 1952 | 1953 |
| New Biggarford (Cleland) | R Moffat & Sons | NS 799 595 | pre-1947 | 1948 |
| North Linrigg 1 (Salsburgh) | Greenshields & Company | NS 807 616 | 1959 | c. 1963 |
| North Linrigg 3 (Salsburgh) | Greenshields & Company | NS 813 617 | pre-1947 | 1950 |
| North Linrigg 5/4 (Salsburgh) | Greenshields & Company | NS 807 614 | pre-1947 | 1947 |
| North Shaws (Cleland) | Auchinlea Quarries | NS 818 586 | 1941 | 1949 |
| O'Wood (Motherwell) | John Dow jnr | NS 773 612 | pre-1947 | 1948 |
| Overwood (Stonehouse) | Overwood Coal Company | NS 778 454 | 1937 | NCB |
| Palace 2 (Airdrie) | Palace Coal Company Limited | NS 742 671 | pre-1947 | 1950 |
| Palace 3 (Airdrie) | Palace Coal Company Limited | NS 742 671 | 1948 | 1948 |
| Parkhead 1 (Cleland) | Auchinlea Quarries/Brickworks Limited | NS 806 599 | 1952 | 1954 |
| Printfield (Airdrie) | Armstrong Bros | NS 789 653 | 1952 | 1959 |
| Purdielodge (Ashgill) | J Sorbie | NS 809 491 | 1959 | 1959 |
| Raebog 6 (Airdrie) | Drumslangie Coal Company | NS 770 686 | pre-1947 | 1947 |
| Rashiehill (Forth) | Hillfarm Coal Company Limited | NS 966 579 | 1987 | 1995 |

| Name | Licensee | Grid Ref | Opened | Closed |
|---|---|---|---|---|
| Shotlinn (Hamilton) | Overwood Coal Company | NS 707 492 | pre-1947 | 1948 |
| South Lanridge 2 (Newarthill) | James Somerville | NS 801 611 | pre-1947 | 1948 |
| South Lanridge 3 (Newarthill) | James Somerville | NS 804 614 | pre-1947 | 1948 |
| South Lanridge 6 (Newarthill) | A Allan & Sons | NS 807 607 | pre-1947 | 1948 |
| Spalehall 3 (Newarthill) | Whittagreen Coal Company | NS 796 597 | 1951 | 1952 |
| Spalehall 5 (Newarthill) | Whittagreen Coal Company | NS 793 596 | pre-1947 | 1948 |
| Spalehall 6 (Newarthill) | Whittagreen Coal Company | NS 791 596 | pre-1947 | 1948 |
| Spalehall 7 (Newarthill) | Whittagreen Coal Company | NS 790 595 | 1948 | 1952 |
| Spoutcroft 2 (Cleland) | E & J Speirs | NS 821 602 | pre-1947 | 1975 |
| Spoutcroft 3 (Cleland) | E & J Speirs | NS 820 603 | 1958 | 1975 |
| Stanrigg 4 (Airdrie) | R & J Dow | NS 782 684 | 1944 | 1946 |
| Stanrigg 5 (Airdrie) | R & J Dow | NS 782 687 | 1945 | 1946 |
| Stanrigg 6 (Airdrie) | R & J Dow | NS 782 687 | 1948 | c. 1969 |
| Staylea (Airdrie) | J Jack (briefly NCB 1948) | NS 771 679 | 1948 | 1955 |
| Thinacre (Hamilton) | Overwood Coal Company | NS 739 500 | pre-1947 | then NCB |
| Thorn (Carluke) | John Rankin | NS 872 516 | 1957 | 1957 |
| Torrance (East Kilbride) | Torrance Coal Company | NS 657 539 | pre-1947 | 1947 |
| Victor (Airdrie) | J O Kane | NS 789 680 | 1954 | 1966 |
| West Machan (Larkhall) | J Burns & Sons | NS 766 511 | pre-1947 | 1960 |
| Wester Dunsyston 1 (Salsburgh) | James Shields | NS 798 628 | 1951 | 1952 |
| Wester Dunsyston 2 (Salsburgh) | James Shields | NS 800 626 | 1952 | 1954 |
| Wester Dunsyston 3 (Salsburgh) | William Hendrie | NS 800 628 | 1955 | 1956 |
| Wester Dunsyston 4 (Salsburgh) | William Hendrie | NS 799 627 | 1956 | 1970 |
| Whiteside (Caldercruix) | Meadowhead Coal Company | NS 858 688 | pre-1947 | 1948 |
| Windsor (Newhouse) | Rigsmuir Coal Company | NS 798 613 | pre-1947 | 1949 |
| Windsor 2 (Newhouse) | Rigsmuir Coal Company | NS 798 613 | pre-1947 | 1950 |
| Windsor 3 (Newhouse) | Rigsmuir Coal Company | NS 798 613 | 1950 | 1950 |
| Windsor 4 (Newhouse) | Rigsmuir Coal Company | NS 797 610 | 1950 | 1954 |
| Woodhall (Carluke) | William Crossan | NS 846 472 | 1951 | 1969 |

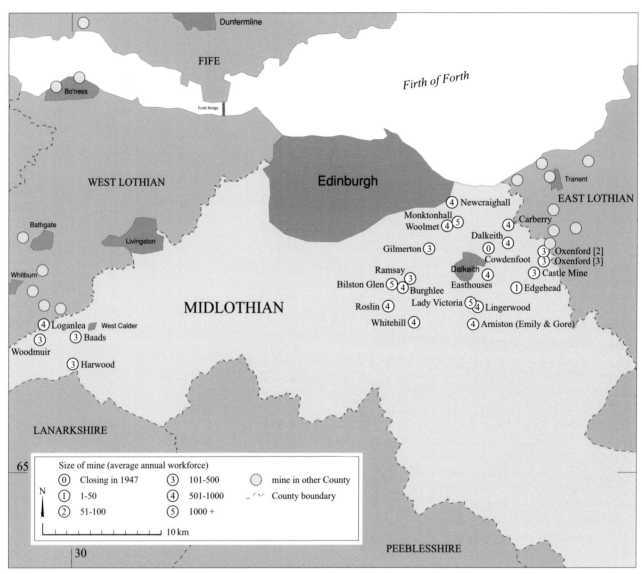

Figure 5:160: Map of National Coal Board
(NCB) collieries in Midlothian.

# Midlothian

Formerly the county of Edinburgh, and including the capital city itself, Midlothian was the fourth most important coal-mining area in Scotland after Fife, Lanarkshire and Ayrshire. Situated directly to the south of Edinburgh, and extending west towards the oil-shale fields of West Lothian, south towards the Pentland Hills and Peeblesshire, and east towards East Lothian (formerly Haddingtonshire), the county contained 26 collieries, and peaked in 1957 when approximately 11,000 miners were employed in its collieries. Two of these collieries, Gilmerton and Newcraighall, fell within the boundaries of the city of Edinburgh, confirming the perhaps surprising fact that in this period, there was more coal-mining activity in Edinburgh than in Glasgow. Indeed, most of the collieries were situated immediately to the south-east of the city around Dalkeith and towards the border with East Lothian. A second cluster of collieries was grouped in the far west near to West Calder in what is now West Lothian Council's domain.

As was the case elsewhere in Scotland, Midlothian has a long history of coal mining stretching back many centuries, some of the earliest extraction having monastic origins relating to Newbattle Abbey. The county is also associated with one of Scotland's worst mining disasters at Mauricewood near Penicuik, which claimed the lives of 63 miners in an underground fire on 5th September 1889. More recently, Midlothian witnessed some of the bitterest scenes in the miners' strike in 1984–5, when Bilston Glen Colliery became the focus of sustained unrest. Monktonhall is also worthy of note because it survived the privatisation of the industry in 1994, being rescued by a miners' co-operative buyout which generated great public sympathy. Unfortunately, it survived for no more than five years because the co-operative was unable to raise enough capital to keep more than one coal face in production, thereby failing to generate enough output to cover costs.

Back in 1947, 20 collieries were taken into state ownership on vesting day, six new mines being sunk subsequently by the NCB. Of these, four were relatively small short-lived surface drifts at Oxenford 3, Cowdenfoot, Castle and Harwood, but two were the prestigious superpits at Bilston Glen and Monktonhall. The former was one of Egon Riss's most successful creations, and the latter had the distinction of clinging on for longer than any of Scotland's other surviving collieries after the 1984 miners' strike (with the exception of the Longannet complex).

In its heyday, Midlothian's coalfield produced a wide range of coal, including gas, house, steam and manufacturing coals. In addition, fireclay, blaes and ironstone were produced in significant quantities by some pits. The most important companies included the Lothian Coal Company (Lady Victoria Colliery, Easthouses, Lingerwood and Whitehill), the Shotts Iron Company (Burghlee, Ramsay and Roslin), and the Niddrie and Benhar Coal Company (Newcraighall and Woolmet). The largest collieries in terms of manpower were Bilston Glen and Monktonhall, which averaged 2,150 and 1,600 miners respectively. Elsewhere, easily the largest of the older pits was Lady Victoria Colliery, with an average workforce of 1,340 miners during 35 years of operation after 1947. Other large pits included Arniston with almost 1,000 miners, Woolmet with over 850, Easthouses with over 800, Burghlee, Newcraighall and Roslin each with over 750, Lingerwood with over 700, Dalkeith with over 600, and Carberry and Whitehill each with over 550. Also of note was Gilmerton Colliery, which was famous for mining almost vertical seams of 'edge coals'.

Midlothian's mining industry appears to have prospered in the early years of nationalisation, the workforce rising from 9,000 in 1947 to almost 11,000 for much of the 1950s. However, decline had set in by 1960 with a sequence of closures, but these were more than offset in 1963 by the opening of Bilston Glen, and again when Monktonhall commenced production in 1967. Ten years later, only these two super-pits survived, sustaining a combined workforce of no more than 1,500 miners. Bilston Glen was closed in 1989, and Monktonhall 12 years later in 1997, both complexes being demolished shortly after closure. Outside the public sector, 11 private licensed mines also operated in the county after 1947, the last of which, at Blinkbonny near Gorebridge, ceased to maintain its licence in 1995.

Despite the total cessation of mining, Midlothian remains hugely significant because of the survival of much of the surface arrangement of Lady Victoria Colliery, which, with great foresight, was saved from destruction after its closure in 1981, later becoming (initially with Prestongrange Colliery in East Lothian) the Scottish Mining Museum. This complex now represents the only recognisable extant deep coal mine in Scotland, and as a late Victorian model colliery, is a fine survivor from the golden era of coal mining in the UK. Within its collections is a superb archive and library relating to coal mining in general, and the Scottish coalfields in particular.

## Arniston (Emily & Gore)

NT 3358 6198 & NS 3391 6142 (NT36SW/42)

**Parish:** Cockpen

**Region/District:** Lo/Mi

**Council:** Midlothian

**Location:** Gorebridge

**Previous owners:** Arniston Coal Company

**Types of coal:** Gas, House and Steam

**Sinking/Production commenced:** 1858

**Year closed:** 1962

**Average workforce:** 980

**Peak workforce:** 990

**Peak year:** 1951

**Shaft/Mine details:** 2 shafts, Emily (NT 3358 6198, NT36SW/42) 302m deep, sunk 1858, and Gore (NS 3391 6142, NT36SW/42.1) 214m deep, sunk 1878.

**Details in 1948:** Output 950 tons per day, 237,500 tons per annum. 964 employees. Norton washer. Baths, first-aid room, canteen. Electricity from SE Scotland Electricity Board. Report dated 15-07-1948.

**Other details:** Emily dated originally from the 1860s, and its timber-lined shaft was 293m deep, said to be the deepest in Scotland at the time. Arniston Coal Company was established in 1874, and Gore was its first pit, having a rectangular brick-lined shaft 213m

deep, completed in 1878. The two pits were subsequently known as Arniston Colliery. Sixty six men died in a fire in 1899. At one stage, Emily had its own power station. One of the most famous features of the pit was the circular pithead baths at Gore, designed by J A Dempster, completed at a cost of £12,000 in 1936, with facilities for 570 men, including 52 shower cubicles. A re-organisation was planned for the pit in 1948.

**Figures 5.161 and 5.162**

## Baads

NT 0017 6100 (NT06SW/39)

**Parish:** West Calder

**Region/District:** Lo/We

**Council:** West Lothian

**Location:** West Calder

**Previous owners:** Young's Paraffin Light and Mineral Oil Company Limited

**Types of coal:** Manufacturing

**Sinking/Production commenced:** 1908

**Year closed:** 1962

**Average workforce:** 384

**Peak workforce:** 491

**Peak year:** 1955

**Shaft/Mine details:** surface mines, Main Mine 1 in 3.8 for 44m, South Slope Mine 1 in 4 for 41m (upcast)

**Details in 1948:** Output 420 tons per day, 105,000 tons per annum, stoop and room working. 300 employees. 1 screen, no washer. Canteen (packed meals) and first-aid room. No baths. 100% electricity from Scottish Oils. Report dated 30-07-1948.

**Other details:** Originally opened to supply Pumpherston Oil Works with coal to fire its oil-shale retorts. Closed when the oil works was shut in 1962. Previously linked to new drift mine at Cuthill (West Lothian).

**Figure 5.163**

Figure 5.162: Arniston Colliery, Gore Pit, c.1900. The second of the two Arniston pits, sunk by the Arniston Coal Company in 1878. In the nationalised era, the two pits employed on average 980 miners until closure 1962. Gorebridge Local History Society, SC446574

Figure 5.163: Baads Colliery, West Calder, c.1955. Established in 1908 by Young's Paraffin Light and Mineral Oil Company Limited to produce manufacturing coal to heat its oil-shale retorts. Its surface mines employed on average 384 miners after nationalisation in 1947, production continuing until 1962. SMM:1998.660, SC393140

Figure 5.164: Bilston Glen Colliery, Loanhead, 1989. Developed over an 11-year period by the NCB from 1952, this was one of several prestigious superpit projects in Scotland, and the first of two in Midlothian. It operated from 1963 to 1989 with an average workforce of 2,154 miners. SC376683

Figure 5.165: Biston Glen Colliery, Loanhead, 1989. No. 1 shaft's reinforced concrete headframe, beneath which four 10.7 tonne skips were used to wind coal to the surface using two ground-mounted electric winding engines manufactured by Fullerton, Hodgart & Barclay of Paisley. SC376690

## Bilston Glen

NT 2714 6510 (NT26NE/77)

**Parish:** Lasswade

**Region/District:** Lo/Mi

**Council:** Midlothian

**Location:** Loanhead

**Previous owners:** National Coal Board

**Sinking commenced:** 1952

**Production commenced:** 1963

**Year closed:** 1989

**Year abandoned:** 1989

**Average workforce:** 2,154

**Peak workforce:** 2,367

**Peak year:** 1970

**Shaft/Mine details:** No. 1 shaft (NT 2714 6510) 7.3m diameter, 751m deep, with 2 ground-mounted AC 1082KW friction winders hauling 4 10.7-tonne skips, No. 2 shaft (NT 2725 6507) 6.1m diameter, 733m deep with 746KW AC double-drum ground-mounted winder hauling 2 double-decked cages each carrying 72 men simultaneously. Winding engines by Fullerton, Hodgart & Barclay of Paisley. The coal preparation plant (a Baum-type washer) was completed in 1960, and was designed to treat 640 tons of coal per hour.

**Other details:** One of the NCB's most successful superpit developments, and designed to go much deeper than neighbouring mines into the Midlothian coalfield basin, exploiting the limestone coals, with an intended output of 1 million tons per annum. The scheme, which was one of NCB architect Egon Riss's projects and included baths, canteen and medical facilities, was inaugurated on 19th May 1952, production commencing in 1963. Prior to closure, 45% of output was consumed by electricity generators (SSEB). Bilston Glen witnessed some of the most bitter scenes of unrest in Scotland during the miners' strike of 1984. Demolished shortly after closure in 1988.

**Figures 5.164, 5.165 and 5.166**

## Burghlee

NT 2766 6488 (NT26SE/67)

**Parish:** Lasswade

**Region/District:** Lo/Mi

**Council:** Midlothian

**Location:** Loanhead

**Previous owners:** Shotts Iron Company Limited

**Types of coal:** House and Steam, ironstone and fireclay

**Sinking/Production commenced:** *c.*1860

**Year closed:** 1964

**Year abandoned:** 1965

**Average workforce:** 768

**Peak workforce:** 795

**Peak year:** 1952

**Shaft/Mine details:** 1 shaft, 366m deep, 2 inclines (60 degree incline, to depth of 366m).

**Details in 1948:** Output 883 tons per day, 220,750 tons per annum. 660 employees. No washer, dross to Ramsay Colliery washer. No baths, canteen snack only, first-aid room. Small amount of electricity from SE Scotland Electricity Board, remainder generated at colliery, some supplied to Ramsay Colliery. Report dated 15-07-1948.

**Other details:** Closed shortly after opening of Bilston Glen Colliery nearby.

**Figure 5.167**

## Carberry

NT 3633 7007 (NT37SE/111)

**Parish:** Inveresk

**Region/District:** Lo/Ea

**Council:** East Lothian

**Location:** Musselburgh

**Previous owners:** Deans and Moore, Edinburgh Collieries Company from 1900

**Types of coal:** Gas, House and Steam

**Sinking/Production commenced:** 1866

**Year closed:** 1960

**Year abandoned:** 1960

**Average workforce:** 589

**Peak workforce:** 615

**Peak year:** 1952

**Shaft/Mine details:** 2 shafts, 229m and 241m deep. No. 1 shaft disused after sinking of No. 3 shaft in 1900.

**Details in 1948:** Output 600 tons per day, 150,000 tons per annum. 500 employees. Lurig type washer. Baths, canteen, first-aid room. Electricity supplied by SE Scotland Electricity Board. Report dated 15-07-1948.

Figure 5.166: Biston Glen Colliery, Loanhead, 1989. The surface layout was designed by the NCB Scottish Division's architect, Egon Riss. It was entirely demolished after closure in 1989, and is now the site of an industrial estate. SC376685

Figure 5.167: Burghlee Colliery, Loanhead, 1964. One of several Shotts Iron Company pits in the Lothians coalfield, it was sunk c.1860. After nationalisation, it operated with a workforce on average of 768 miners, closing in 1964, shortly after the opening of neighbouring Bilston Glen. SC871997

Figure 5.168: Carberry Colliery, near Musselburgh. Originally sunk by Deans and Moore in 1868, it was operated by Edinburgh Collieries Limited from 1900. It produced gas, house and steam coal until 1960, employing on average 589 miners in the NCB era. SMM:1996.2588, SC379653

**Other details:** Washery closed in 1953 and coal sent to new washery at Dalkeith 5.
**Figure 5.168**

**Castle Mine**
NT 3847 6617 (NT36NE/90)
**Parish:** Cranston
**Region/District:** Lo/Mi
**Council:** Midlothian
**Previous owners:** National Coal Board
**Sinking commenced:** 1949
**Production commenced:** 1950
**Year closed:** 1958
**Year abandoned:** 1959
**Average workforce:** 102
**Peak workforce:** 150
**Peak year:** 1950
**Other details:** No further data

**Cowdenfoot**
NT 3497 6798 (NT36NW/163)
**Parish:** Dalkeith
**Region/District:** Lo/Mi
**Council:** Midlothian
**Location:** Dalkeith
**Previous owners:** National Coal Board
**Sinking/Production commenced:** 1956
**Year closed:** 1965
**Year abandoned:** 1965
**Other details:** No further data

**Dalkeith 1, 2 & 3**
NT 3637 6858 (NT36NE/91)
**Parish:** Inveresk (Midlothian)
**Region/District:** Lo/Mi
**Council:** Midlothian
**Location:** Smeaton
**Previous owners:** A G Moore & Company
**Types of coal:** House and Manufacturing
**Sinking/Production commenced:** 1903
**Year closed:** 1948
**Year abandoned:** 1948
**Average workforce:** 83
**Peak workforce:** 83
**Peak year:** 1947

**Shaft/Mine details:** 3 surface mines
**Details in 1948:** closed

## Dalkeith 5 & 9

NT 3520 6902 (NT36NE/76)
**Parish:** Inveresk (Midlothian)
**Region/District:** Lo/Mi
**Council:** Midlothian
**Location:** Smeaton
**Previous owners:** A G Moore & Company
**Types of coal:** House and Manufacturing
**Sinking/Production commenced:** 1903
**Year closed:** 1978
**Average workforce:** 617
**Peak workforce:** 898
**Peak year:** 1964
**Shaft/Mine details:** Originally 6 surface mines, eventually only 2 brought into operation. No. 6 was for ventilation and manriding, and both 7 and 8 were for access. Only Nos. 5 and 9 were therefore used for production.
**Details in 1948:** Output 490 tons per day, 122,500 per annum. 401 employees. Dickson and Mann Lurig type washer (replaced in 1953). Canteen (not in use), first-aid room. All electricity bought from SE Scotland Electricity Board. Report dated 15-07-1948.
**Other details:** A large new washery was built adjacent to Mine No. 9 in 1953, and received the output from the Dalkeith and neighbouring collieries (Meadowmill, Winton, Limeylands, Oxenford, Castle, Carberry and Tynemount). Waste was then taken by aerial ropeway to the old bing at Dalkeith 1, 2 and 3. Plans for the development of two new drifts (10 and 11) in the 1950s were never realised.
**Figure 5.169**

Figure 5.169: Dalkeith No. 9 Colliery, Smeaton Junction, 1953. Developed from 1903 by A G Moore and Co, it comprised several surface mines, and included a coal preparation plant completed in 1953, designed also to handle coal from neighbouring pits. After nationalisation, Dalkeith Nos. 5 and 9 operated until 1968 with a combined average workforce of 617 miners. SC446373

## Easthouses

NT 3464 6585 (NT36NW/164)
**Parish:** Newbattle
**Region/District:** Lo/Mi
**Council:** Midlothian
**Location:** Dalkeith
**Previous owners:** Lothian Coal Company
**Types of coal:** House and Steam
**Sinking/Production commenced:** 1909
**Year closed:** 1969
**Year abandoned:** 1971
**Average workforce:** 818
**Peak workforce:** 1,030
**Peak year:** 1965

Figure 5.170: Easthouses Colliery, 1949. Developed by the Lothian Coal Company from 1909, it comprised three surface mines which, in the NCB era, employed on average 818 miners, peaking at 1,030 in 1965. Closure followed four years later in 1969. SMM: 1998.1272.1, SC445860

**Shaft/Mine details:** 3 surface mines, No. 1, No. 2 and Simpson Mine, all 1:2 incline. No. 2 mine was originally main airway, but was enlarged to accommodate a new haulage system. Ventilation/fan at mouth of Simpson Mine.

**Details in 1948:** Output 690 tons per day, 172,500 tons per annum. 633 employees. Output was taken in hutches by endless-rope haulage, and later in mine cars hauled by locomotives, to be washed and screened at Lady Victoria Colliery. Canteen, first-aid room, baths planned. Electricity supplied from Lady Victoria Colliery. Report dated 15-08-48.

**Other details:** Coal sent to Lady Victoria to be screened. New steam-driven balanced-rope haulage installed in 1949. Re-organised in 1955, included the introduction of underground diesel haulage.

**Figure 5.170**

## Edgehead

NT 3718 6494 (NT36NE/73)
**Parish:** Cranston
**Region/District:** Lo/Mi
**Council:** Midlothian
**Location:** Edgehead, Ford
**Previous owners:** Earl of Stair, Fordel Mains (Midlothian) Colliery Company from *c.*1930
**Types of coal:** House
**Sinking/Production commenced:** 1850
**Year closed:** 1959
**Year abandoned:** 1959
**Average workforce:** 39
**Peak workforce:** 47
**Peak year:** 1949
**Shaft/Mine details:** 2 inclines from surface driven by NCB, replacing earlier mine. Surface level is 208m, and foot of inclines 144m.
**Details in 1949:** Output 90 tons per day, 20,611 tons per annum (1948). 47 employees. No washer, but efficient screening plant. Coal dispatched by lorry (no railway). No baths, canteen or medical centre. Electricity from public grid supply. Report dated 21-02-1949.
**Other details:** Transferred to NCB in 1949.

## Emily

(see Arniston)

## Gilmerton

NT 2979 6814 (NT26NE/79)
**Parish:** Edinburgh
**Region/District:** Lo/Ed
**Council:** City of Edinburgh
**Location:** Gilmerton, Edinburgh

**Previous owners:** Fordell Mains (Midlothian) Coal Company, subsequently Gilmerton Colliery Company

**Types of coal:** House and Steam

**Sinking/Production commenced:** 1928

**Year closed:** 1961

**Year abandoned:** 1963

**Average workforce:** 314

**Peak workforce:** 380

**Peak year:** 1950

**Shaft/Mine details:** 2 shafts, each 154m deep, No. 1 4.57m diameter and brick-lined, No. 2 rectangular 5.49m by 1.83m, initially stone-lined, but timber-lined below.

**Details in 1948:** Output 390 tons per day, 97,500 per annum. 326 employees. Feldspar washer built by Campbell Binnie. Baths, canteen (snack), first-aid room. Generated over half electricity on site, remainder supplied by SE Scotland Electricity Board. Report dated 15-07-1948.

**Other details:** Said to be the deepest pit mining the near-vertical edge coals (at least 70 degrees) in 1830 (161m), but mining in the Gilmerton area ceased in 1894. The sinking of a new pit was commenced in 1926, and underground locomotive haulage was introduced in 1931. Problems of underground heating in the seams grew in the 1950s, and closure in 1961 was caused by a fire. In November 2000, serious subsidence in the adjacent Ferniehill area was caused by collapsing limestone mines, and not workings from Gilmerton Colliery.

**Figure 5.171**

Figure 5.171: Gilmerton Colliery, near Edinburgh. Sunk in 1928 immediately to the south of an earlier colliery of the same name, it was owned by the Fordell Mains (Midlothian) Coal Company (subsequently Gilmerton Colliery Co). In the NCB era, it produced coal until 1961 with an average workforce of 314 miners. SMM:1997.853, SC379656

## Gore

(see Arniston)

## Harwood (Adie's Syke)

NS 9996 5879 (NS95NE/4)

**Parish:** West Calder

**Region/District:** Lo/We

**Council:** West Lothian

**Location:** Cobbinshaw Station

**Previous owners:** Glasgow Iron and Steel Company Limited

**Types of coal:** Steam, House and Manufacturing

**Sinking/Production commenced:** 1946

**Year closed:** 1959

**Year abandoned:** 1959

**Average workforce:** 132

**Peak workforce:** 196

**Peak year:** 1958

**Shaft/Mine details:** 2 surface mines (at NT 0000 5880), No. 1 62m long at 1 in 4.5, No. 2 39m long at 1 in 3 (upcast)

**Details in 1952:** Output 90 tons per day, 22,500 per annum, longwall working. 78 employees. 1 jig screen, no washer. No baths or canteen. First-aid room. 100% electricity from public supply. Report dated 30-07-1952.

### Klondyke

(see Newcraighall)

### Lady Victoria (The Lady)

NT 3327 6375 (NT36SW/22)
**Parish:** Cockpen/Newbattle
**Region/District:** Lo/Mi
**Council:** Midlothian
**Location:** Newtongrange
**Previous owners:** Lothian Coal Company
**Types of coal:** Gas, House and Steam
**Sinking/Production commenced:** 1895
**Year closed:** 1981
**Year abandoned:** 1982
**Average workforce:** 1,339
**Peak workforce:** 1,765
**Peak year:** 1953
**Shaft/Mine details:** 1 shaft, 487m in 1948. When originally sunk, it was walled simultaneously, rendering it watertight. Originally 530m deep, it was the deepest in Scotland at the time, and was connected to neighbouring Lingerwood, preventing the need for a second shaft.
**Details in 1948:** Output 1,388 tons per day, 347,000 per annum. 1,080 employees. Campbell Binnie bash type washer, rewasher Baum-type, locally made. Canteen, first-aid room, ambulance. Electricity generated in own power stations, some sent to Easthouses Colliery. Report dated 15-08-1948.
**Other details:** Named in honour of the wife of the Marquis of Lothian. Neighbouring Newtongrange village was built by the Marquis of

Figure 5.172: Lady Victoria Colliery, Newtongrange. The flagship pit of the Lothian Coal Co, whose main offices were situated on the opposite side of the road. When sunk in 1895, it was linked with neighbouring Lingerwood Colliery, preventing the need for a mandatory second shaft. In the nationalised era, it employed on average 1,340 miners, but peaked at 1,765 in 1953. Until the development of the superpits at Bilston Glen and Monktonhall in the 1960s, Lady Victoria was the largest colliery in Midlothian. SMM:1998.0208, SC384741

Figure 5.173: Lady Victoria Colliery, Newtongrange, c.1960. During its working life, it produced a range of house, steam and gas coals. After closure in 1981, almost all the surface buildings of the colliery were preserved, and the site later joined forces with Prestongrange in neighbouring East Lothian to become the Scottish Mining Museum. This view shows the covered walkway spanning the A7 trunk road, connecting the baths and canteen with the pithead. SC446344

Lothian to house miners to work the pit. Became showpiece pit of the Lothian Coal Company, with the largest-diameter shaft in Scotland at the time, served by a large Grant Ritchie of Kilmarnock steam winding engine. A bad fire in the winding engine house led to refurbishment in 1903. The steam winder was later powered by a boilerhouse re-equipped with 12 war-surplus Lancashire boilers dating from c.1920. Prior to the mechanisation of underground haulage, 120 ponies were employed underground. In the NCB era, the colliery also became a major engineering workshop centre for the Lothians area. Baths serving both Lady Victoria and Lingerwood were opened in 1954, and a reinforced concrete heated walkway over the A7 trunk road connected them to the pithead. Following closure in 1981, much of the surface arrangement was rescued from demolition, and with the assistance of Lothian Region, was transformed into the Scottish Mining Museum, operating initially as a dual site with Prestongrange in neighbouring East Lothian.

**Figures 5.172 and 5.173**

**Lingerwood**

NT 3365 6354 (NT36SW/43)

**Parish:** Newbattle

**Region/District:** Lo/Mi

**Council:** Midlothian

**Location:** Newtongrange

**Previous owners:** Lothian Coal Company

**Types of coal:** Steam and House

**Sinking/Production commenced:** 1798 ('Old Engine Pit')

**Year closed:** 1967

**Year abandoned:** 1982

**Average workforce:** 727

**Peak workforce:** 770

**Peak year:** 1951

**Shaft/Mine details:** 2 shafts, and one surface mine. No. 1 (Dickson's Shaft) 253m deep, 5.2m diameter and brick-lined, with electric

Figure 5.174: Lingerwood Colliery, Newtongrange, c.1960. Dating originally from 1898, this pit was later owned by the Lothian Coal Co, No. 1 shaft acting as a second (upcast) shaft for Lady Victoria Colliery. SMM:1996.3411, SC379658

Figure 5.175: Lingerwood Colliery, Newtongrange. After nationalisation in 1947, and reorganisation work in 1955 and in 1961, the workforce averaged 727 miners until closure in 1967. This view shows one of the two headframes, and a cage in the foreground. SMM:1996.2562, SC379657

winder, upcast for Lady Victoria. No. 2 (Engine Pit) downcast, 266m deep, rectangular shaft, with steam winder.

**Details in 1948:** Output 1,000 tons per day, 250,000 tons per annum. 603 employees. Coal washed and screened at Lady Victoria Colliery. First-aid room. Canteen at Lady Victoria. Electricity supplied from SE Scotland Electricity Board. Report dated 15-07-1948.

**Other details:** Pneumatically-operated traversers and tipplers installed at surface for mine-car circuit in 1955. Re-organised in 1961. Baths for 798 men also served Lady Victoria. When Heriot Watt University's mining school was closed, its rescue station was first moved to Lingerwood, and eventually to its current location at Crossgates in Fife.

**Figures 5.174 and 5.175**

## Loanhead

(see Ramsay)

## Loganlea

NS 9768 6192 (NS96SE/20)

**Parish:** West Calder

**Region/District:** Lo/We

**Council:** West Lothian

**Location:** West Calder

**Previous owners:** United Collieries Limited

**Types of coal:** Gas, House and Steam

**Sinking/Production commenced:** *c.*1890

**Year closed:** 1959

**Average workforce:** 510

**Peak workforce:** 675

**Peak year:** 1948

**Shaft/Mine details:** 2 shafts, No. 2 104m (downcast with forcing fan), and No. 3 102m deep (upcast at Foulshiels)

**Details in 1948:** Output 620 tons per day, 155,000 tons per annum, longwall working. 675 employees. 3 screens. Jig and feldspar washer (Dickson and Mann, installed 1912). Baths (1930), canteen (packed meals), and first-aid room. All electricity generated at mine. Report dated 30-07-1948.

**Other details:** New power station built in 1891, resulting in an early example of the introduction of electric haulage, coal cutting and lighting underground in Scotland. Fans at Loganlea forced air into the mine, the upcast shafts being at Foulshiels and Woodmuir nearby.

**Figure 5.176**

## The Moat

(see Roslin)

## Monktonhall

NT 3219 7027 (NT37SW/198)

**Parish:** Newton

**Region/District:** Lo/Mi

**Council:** Midlothian

**Previous owners:** National Coal Board

**Sinking commenced:** 1954

**Production commenced:** 1967

**Year closed:** 1997

**Average workforce:** 1,618

**Peak workforce:** 1,786

**Peak year:** 1971

**Shaft/Mine details:** 2 shafts. No. 1 shaft (upcast) 930m deep, 7.32m diameter and concrete-lined, 2 1,600hp multi-rope tower-mounted friction winders, designed to lift 420 tons of coal per hour in 2 14-ton skips. No. 2 shaft (downcast) also 7.32m diameter, concrete lined, and 920m deep, with 1 multi-rope 1,600hp tower-mounted friction winder with 2-deck cage and counterweight, carrying 2 mine cars (automatic ram loading) or 130 men.

**Other details:** Designed to exploit the deep limestone coals of the Midlothian basin, it was one of the wettest pits of the NCB era, water causing major problems both during its sinking and its operation. Nevertheless, one of the successful NCB new sinkings and superpits, and the last to survive. Surface facilities included a coal preparation plant, a rapid-loading surface bunker and merry-go-round rail facility, workshops, and administration buildings as well as full baths, canteen and medical facilities. Broke productivity records in 1969. Built with the assured market of Cockenzie Power Station nearby, but eventually mothballed in 1987, after which neighbouring Bilston Glen closed in 1989. Re-opened as Britain's first large private mine in 1992, but run by a miners' co-operative. First coalface back in production in 1993, but lacked capital and ran into financial difficulties in August 1994, after which Waverley Mining Finance plc took increasing interest in the mine. Closed finally in 1997. First of towers was demolished in November 1997, the second tower in 1998.

**Figure 5.177, see also 3.36, 3.43 and 3.44**

Figure 5.176: Loganlea Colliery, near West Calder. Originating in the 1890s, and equipped with its own electricity generating station, it was an early example of the application of electric power underground. The colliery produced gas, steam and house coals for United Collieries Limited, and after nationalisation, employed on average 510 miners until closure in 1959. William Aitchison, SC706608

Figure 5.177: Monktonhall Colliery, near Edinburgh, 1970. Developed by the NCB over a 13-year period from 1954, this was the last of the big superpit schemes in Scotland. In 1997, it was also the last to close, having operated with an average workforce of 1,618 miners. The surface buildings have since been demolished. SC446423

Figure 5.178: Newcraighall Colliery, near Edinburgh, also known as the 'Klondyke', 1965. Developed originally by the Benhar Coal Company in 1897, the two surface mines were augmented with the addition of a shaft in 1910. After nationalisation in 1947, the average number of miners employed at the pit was 760, closure eventually occurring in 1968. SMM: 1998.467, SC445887

## Newcraighall (The Klondyke)

NT 3177 7181 (NT37SW/234)

**Parish:** Edinburgh

**Region/District:** Lo/Ed

**Council:** City of Edinburgh

**Location:** Portobello

**Previous owners:** Benhar Coal Company, Niddrie and Benhar Coal Company from 1882

**Types of coal:** House and Manufacturing, and shale

**Sinking/Production commenced:** 1897

**Year closed:** 1968

**Year abandoned:** 1970

**Average workforce:** 760

**Peak workforce:** 810

**Peak year:** 1950

**Shaft/Mine details:** 2 surface mines in 1897. By 1948, 1 shaft (downcast, 4.88m diameter, 249m deep, sunk in 1910), 1 incline (1 in 4, 207m, 14.88m by 7.44m, direct rope haulage).

**Details in 1948:** Output 900 tons per day, 225,000 tons per annum. 686 employees. Simon Carves Baum-type washer. Baths and canteen (opened 1938, for 960 men), first-aid room, ambulance. Electricity bought from SE Scotland Electricity Board. Report dated 15-07-1948.

**Other details:** Also known as the Klondyke. Double shifts introduced in 1959 because of high demand for coal, but in 1962 attempts at improved underground mechanisation failed. In 1967, prior to closure, average daily output through longwall working was 700 tons. Surface buildings demolished and site subsequently occupied by a pub and a hotel.

**Figure 5.178, see also 3.1 and 3.2**

## Oxenford 2

NT 3934 6779 (NT36NE/89)

**Parish:** Cranston

**Region/District:** Lo/Mi

**Council:** Midlothian

**Location:** Ormiston

**Previous owners:** Ormiston Coal Company

**Types of coal:** Gas, House and Coking

**Sinking/Production commenced:** 1926

**Year closed:** 1950

**Average workforce:** 152

**Peak workforce:** 170

**Peak year:** 1948

**Shaft/Mine details:** 1 shaft, 37m deep

**Details in 1948:** Output 163 tons per day, 40,750 tons per annum. 125 employees. Coal screened at Tynemount Colliery. Canteen, first-aid room, ambulance. Electricity supplied from SE Scotland Electricity Board. Report dated 15-07-1948.

## Oxenford 3

NT 3932 6766 (NT36NE/89.01)
**Parish:** Cranston
**Region/District:** Lo/Mi
**Council:** Midlothian
**Location:** Ormiston
**Previous owners:** National Coal Board
**Types of coal:** Gas, House and Coking
**Sinking commenced:** 1949
**Production commenced:** 1952
**Year closed:** 1957
**Year abandoned:** 1962
**Average workforce:** 250
**Peak workforce:** 250
**Peak year:** 1952
**Other details:** none

## Ramsay

NT 2827 6571 (NT26NE/93)
**Parish:** Lasswade
**Region/District:** Lo/Mi
**Council:** Midlothian
**Location:** Loanhead
**Previous owners:** Shotts Iron Company Limited from 1865
**Types of coal:** House and Steam, and originally, blackband ore
**Sinking/Production commenced:** *c.*1850
**Year closed:** 1965
**Year abandoned:** 1966
**Average workforce:** 346
**Peak workforce:** 385
**Peak year:** 1952
**Shaft/Mine details:** 2 shafts, 302m and 174m deep. Shafts located at NT 2827 6571 and NT 2830 6575.
**Details in 1948:** Output 122 tons per day, 30,500 tons per annum. 163 employees. Coppee Baum-type washer. Baths (1932), first-aid room, canteen (snack, added 1939)). Electricity mostly from Burghlee Colliery power station, a little from SE Scotland Electricity Board. Report dated 15-07-1948.
**Other details:** Also known as Loanhead Colliery. New washery (Blantyre Engineering Company) added in 1929, at which time one of the shafts was reconstructed (4.88m diameter and concrete lined) and deepened from 176m to 285m.
**Figure 5.179**

Figure 5.179: Ramsay Colliery, Loanhead. Sunk in the mid-19th century to exploit ironstone, and owned by the Shotts Iron Company from 1865. After nationalisation in 1947, the workforce averaged 346 miners, closure occurring in 1965. The baths and much of the surface arrangement survive, and are now occupied by small businesses, including a scrap yard. SMM:1996.3417, SC379668

Figure 5.180: Roslin Colliery, also known as 'the Moat'. Another Shotts Iron Company colliery, dating from 1901. Pithead baths were built at the colliery in 1930, and an associated brickworks operated between 1937 and the 1970s. During the nationalised period, the colliery employed a workforce of on average 754 miners, closing in 1969. SMM:1997.603, SC379669

## Roslin (known as 'The Moat')

NT 2631 6354 (NT26SE/66)

**Parish:** Lasswade

**Region/District:** Lo/Mi

**Council:** Midlothian

**Location:** Roslin

**Previous owners:** Shotts Iron Company Limited

**Types of coal:** Steam and House

**Sinking/Production commenced:** 1901–3

**Year closed:** 1969

**Year abandoned:** 1969

**Average workforce:** 754

**Peak workforce:** 770

**Peak year:** 1952

**Shaft/Mine details:** 2 shafts located at NT 2631 6354 and NT 2632 6357. No. 1 downcast, 282m deep, 6.1m by 1.83m, timber-lined rectangular shaft, with 600hp 'Ingles' steam winder. No. 2 upcast, 294m deep, 3.35m diameter. Deepened in 1922. Pumping at bottom of No. 2 shaft.

**Details in 1948:** Output 777 tons per day, 194,250 tons per annum. 765 employees. Coal washed at Ramsay Colliery. Baths (1930), first-aid room, canteen (snack). Some electricity generated at colliery, remainder bought from SE Scotland Electricity Board. Report dated 15-07-1948.

**Other details:** Screening extended in 1920s, and baths installed in 1930 – said to be the first in the Lothians. Associated brickworks, established in 1937, operated until the 1970s. Spontaneous combustion underground caused problems, particularly in 1953.

**Figure 5.180**

## Whitehill (Rosewell)

NT 2865 6205 (NT26SE/71)

**Parish:** Lasswade

**Region/District:** Lo/Mi

**Council:** Midlothian

**Location:** Rosewell

**Previous owners:** Lothian Coal Company

**Types of coal:** Steam and House

**Sinking/Production commenced:** 1850

**Year closed:** 1961

**Year abandoned:** 1962

**Average workforce:** 562

**Peak workforce:** 590

**Peak year:** 1950

**Shaft/Mine details:** 3 shafts, each 91m deep, and 1 surface mine

**Details in 1948:** Output 640 tons per day, 160,000 tons per annum. 495 employees. Campbell Binny washer. Baths (1936), first-aid room, canteen. Most electricity DC and generated at colliery, some being bought from SE Scotland Electricity Board. Report date 15-07-1948.

**Other details:** Possibly the first mine in Scotland to use coal cutters powered by compressed air. NCB mechanised underground haulage as part of reconstruction in 1954. Associated brickworks operated from *c.*1870 until 1977. Rosewell village, which contains many brick terraces of miners' houses, is one of the most important surviving mining villages in Scotland.

**Figure 5.181**

Figure 5.181: Whitehill Colliery, Rosewell, 1905. A Lothian Coal Company pit dating from 1850, it produced steam and house coal until its closure in 1961. After nationalisation, it employed on average 562 miners. An adjacent brickworks continued production until 1977. SMM:1999.286, SC445885

## Woodmuir

NS 9721 6073 (NS96SE/10)

**Parish:** West Calder

**Region/District:** Lo/We

**Council:** West Lothian

**Location:** West Calder

**Previous owners:** Woodmuir Coal Company, United Collieries Limited from 1902

**Types of coal:** Coking, House, Manufacturing and Steam

**Sinking/Production commenced:** 1896

**Year closed:** 1963

**Average workforce:** 255

**Peak workforce:** 273

**Peak year:** 1963

**Shaft/Mine details:** 2 shafts, No. 5 90m (downcast) and No. 6 91m deep (upcast)

**Details in 1948:** Output 285 tons per day, 73,500 tons per annum, longwall working. 210 employees. 2 screen tables. Jig and feldspar washer (Dickson and Mann). No baths. Canteen (packed meals), first-aid room. All electricity from Loganlea Colliery. Report dated 30-07-1948.

**Other details:** Range of ruined beehive coke ovens survive nearby.

**Figure 5.182**

Figure 5.182: Woodmuir Colliery, near West Calder. Established by the Woodmuir Coal Company in 1896, it was operated by United Collieries Limited from 1902, and produced house, steam and coking coals, the latter being processed in adjacent coke ovens, the remains of which can still be seen. In the NCB era, it employed on average 255 miners until its closure in 1963. SMM:1996.481, SC379670

## Woolmet

NT 3128 6998 (NT36NW/134)

**Parish:** Newton

**Region/District:** Lo/Mi

**Council:** Midlothian

**Location:** Millerhill

**Previous owners:** Niddrie and Benhar Coal Company

**Types of coal:** House and Steam, and shale

**Sinking/Production commenced:** 1898

**Year closed:** 1966

**Year abandoned:** 1967

**Average workforce:** 863

**Peak workforce:** 960

Figure 5.183: Woolmet Colliery, Millerhill, near Edinburgh, 1925. Sunk by the Niddrie and Benhar Coal Company in 1898, it operated until 1966, at which time its workforce was transferred to neighbouring Monktonhall Colliery, which had just commenced production. SMM:1996.1420, SC379674

**Peak year:** 1963

**Shaft/Mine details:** 1 shaft, 224m deep, 1 surface mine

**Details in 1948:** Output 825 tons per day, 206,250 tons per annum. 708 employees. Simon Carves washer, built in 1920s. Baths (1936), canteen, first-aid room, ambulance. Electricity bought from SE Scotland Electricity Board. Report dated 15-08-1948.

**Other details:** On closure, all men were transferred to newly opened Monktonhall, after which it was retained as a training centre.

**Figure 5.183 and 5.184**

Figure 5.184: Woolmet Colliery, Millerhill, near Edinburgh, c.1960. During the nationalised era, the average workforce employed at the pit was 863, peaking at 960 in 1963. SC446509

## Licensed Mines in Midlothian 1947–97

| Name | Licensee | Grid Ref | Opened | Closed |
|---|---|---|---|---|
| Aitkendean (Gorebridge) | W Walker | NT 319 620 | 1957 | 1973 |
| Baads 42 (West Calder) | Young's Paraffin Light & Mineral Oil Co Ltd | NT 002 609 | pre-1947 | c. 1947 |
| Barleydean (Gorebridge) | Temple Farm Coal Company | NT 295 607 | 1967 | 1973 |
| Blinkbonny (Gorebridge) | D Beattie | NT 354 271 | 1967 | 1995 |
| Cornton (Penicuik) | Cornton Coal Company Limited | NT 205 590 | pre-1947 | c. 1964 |
| Edgehead (Ford) | Fordell Mains Coal Company | NT 371 651 | c. 1850 | NCB |
| Guildyhowers (Gorebridge) | W T Bathgate (Lime Works) Limited | NT 340 584 | pre-1947 | c. 1950 |
| Halkerston (Gorebridge) | Edward L Gray | NT 349 587 | 1958 | 1958 |
| Monktonhall (Edinburgh) | Monktonhall Mineworkers Limited | NT 321 702 | 1992 | 1997 |
| Nunnery (Cousland) | Dalkeith Transport & Storage Limited | NT 379 681 | 1955 | 1955 |
| Temple Braidwood (Gorebridge) | McNeill & Knox | NT 315 585 | 1955 | 1986 |
| Temple Farm/High Temple (Gorebridge) | Temple Farm Coal Company | NT 321 579 | 1958 | 1968 |

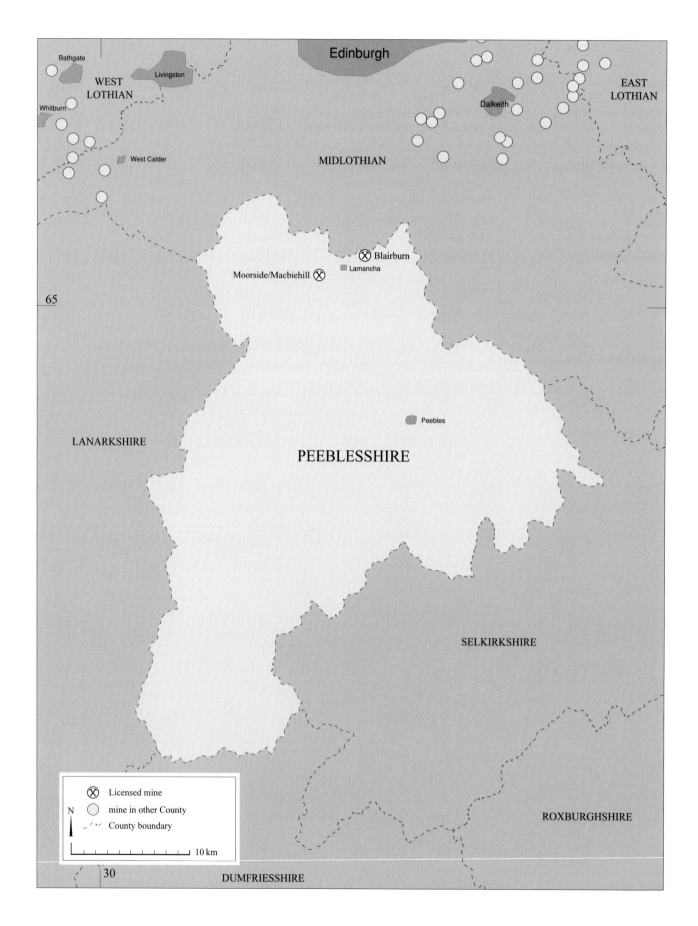

# Peeblesshire

Very little significant deep coal mining occurred within Peeblesshire in the nationalised era, but a coal outcrop close to Lamancha was sufficient to support two small short-lived private mines during the 1950s.

## Licensed Mines in Peeblesshire 1947–97

| Name | Licensee | Grid Ref | Opened | Closed |
|------|----------|----------|--------|--------|
| Blairburn (Lamancha) | Flockhart & Buchan | NT 221 541 | 1959 | 1959 |
| Moorside/Macbiehill (Lamancha) | William Potter | NT 183 525 | 1955 | c. 1959 |

Figure 5.185 left: Map of licensed collieries in Peeblesshire

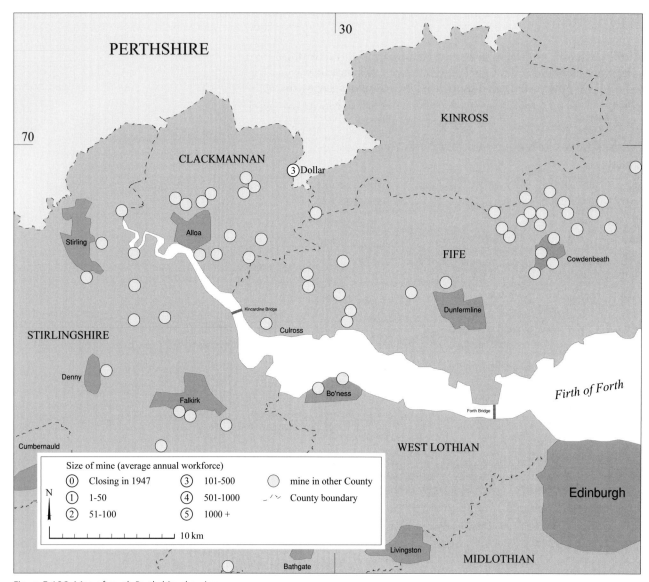

Figure 5.186: Map of south Perthshire showing
Dollar Mine

# Perthshire

Perthshire is not renowned for being a coal-producing county, but a small coalfield in north-east Clackmannanshire overlaps into the county, the entrance to Dollar mine being located a short distance outside Clackmannanshire in the Perthshire portion of Muckhart parish.

## Dollar

NS 9709 9797 (NS99NE/31)

**Parish:** Muckhart

**Region/District:** Ta/Pe

**Council:** Clackmannanshire/Perthshire

**Location:** Dollar

**Previous owners:** Alloa Coal Company

**Types of coal:** Steam and House, latterly Upper Hirst

**Sinking/Production commenced:** 1943

**Year closed:** 1973

**Year abandoned:** no data

**Average workforce:** 135

**Peak workforce:** 454

**Peak year:** 1963

**Shaft/Mine details:** 3 surface mines, gradient 1 in 9. Mines No. 1 and 3 at NS 9706 9979, and Mine No. 2 at NS 9703 9803.

**Details in 1948:** Output 180 tons per day, 45,000 tons per year. 89 employees. Dross sent to Devon colliery for washing. No baths or canteen. First-aid room. AC electricity provided by Devon Colliery. Report dated 25-08-1948.

**Other details:** Mines 4 & 5 were driven by the NCB from 1956, entering production in 1960. A highly productive pit, it was never connected underground to the Longannet complex, but did supply Kincardine Power Station.

**Figure 5.187**

Figure 5.187: Dollar Mine, to the east of Dollar, 1964. The complex was begun by the Alloa Coal Company in 1943, and originally comprised three surface mines. Two new mines were developed by the NCB from 1956, entering production in 1960, producing Upper Hirst coal for Kincardine Power Station. After 1947, the mines employed on average 135 miners, and ceased production in 1973. SMM:1996.02444, SC379599

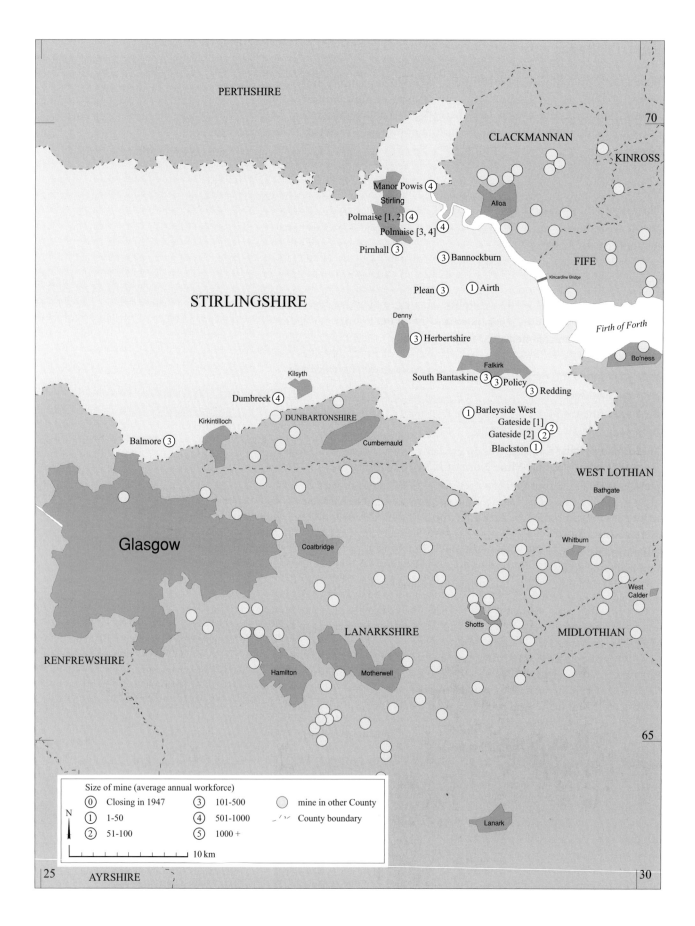

PERTHSHIRE

CLACKMANNAN

KINROSS

Manor Powis ④

Stirling

Polmaise [1, 2] ④

Polmaise [3, 4] ④

Pirnhall ③

Alloa

STIRLINGSHIRE

③ Bannockburn

FIFE

Plean ③    ① Airth

Kincardine Bridge

Denny

Firth of Forth

③ Herbertshire

Falkirk

Bo'ness

Kilsyth

South Bantaskine ③ ③ Policy

③ Redding

Dumbreck ④

① Barleyside West

Gateside [1]

Kirkintilloch    DUNBARTONSHIRE

Gateside [2] ②    ②

WEST LOTHIAN

Cumbernauld

Blackston ①

Balmore ③

Bathgate

Whitburn

Glasgow

West Calder

Coatbridge

Shotts

LANARKSHIRE

MIDLOTHIAN

RENFREWSHIRE

Hamilton    Motherwell

70

65

Size of mine (average annual workforce)

| ⓪ | Closing in 1947 | ③ | 101-500 | ◯ mine in other County |
| ① | 1-50 | ④ | 501-1000 | County boundary |
| ② | 51-100 | ⑤ | 1000 + |

N

10 km

Lanark

25    AYRSHIRE    30

# Stirlingshire

The majority of collieries in Stirlingshire were concentrated in a north–south axis in the far east of the county, from the Stirling area in the north to the Falkirk area in the south. Two outlying pits were also to be found on the southern edge of the county, towards Dunbartonshire. Although by 1947 Stirlingshire was nowhere near the largest coal-producing county in Scotland, it is historically very significant, containing, for example, the Slamannan plateau, which was one of the most important sources of coal and ironstone during the 19th century. Indeed, Stirlingshire was itself the home of the Carron Company, founded in 1760, and the original launching point for the modern iron industry in Scotland. Other local consumers of coal were the many specialist iron foundries for which the Falkirk area was world famous, and like Glenboig (across the border in North Lanarkshire), there was a substantial refractory industry exploiting high-quality local fireclays.

With vulcanism very evident in the north around Stirling, it is not surprising that the pits in this area, notably the two Polmaise collieries, and Manor Powis, were significant producers of anthracite (see Glossary). Elsewhere, collieries produced a mixture of house, manufacturing and steam coals, but coking coal was also produced by some mines, such as Bannockburn, Plean 3 and Dumbreck, the latter two having their own coke ovens. In addition, fine local fireclays were extracted in some cases, and in the later stages of their lives, both Polmaise 3 and 4 and Bannockburn produced Upper Hirst coal for burning in the big power stations in west Fife.

Of the 17 collieries to have been operated by the NCB in Stirlingshire, 16 were inherited from the private sector in 1947. The value of the anthracite extracted in the pits near Stirling is reflected in the fact that both Polmaise 3 and 4 and Manor Powis benefited from major reconstructions in the 1950s. A prestigious new sinking was also begun by the NCB at Airth in 1958, but the embarrassing failure of the big new collieries nearby at Glenochil in Clackmannanshire and Rothes in Fife created sufficient concern for this project to be abandoned a year later. The coals which the Airth project was supposed to exploit were later targeted by an extension under the Forth of the Longannet complex in the late 1990s, but this too came to a sudden end with the flooding of Longannet Mine in 2002.

Prior to nationalisation, no single company had dominated in Stirlingshire, but the biggest players were inevitably connected to the iron industry. The Carron Company owned the two pits at Pirnhall and Bannockburn, and Bairds operated Dumbreck, including its coke ovens. Other important operators were Archibald Russell, who established the two Polmaise complexes, as well as Policy Colliery, all near Stirling. In terms of size of workforce, the largest mines were Manor Powis with an average of over 800 miners, followed by the two Polmaise collieries, each with over 600, and Dumbreck with over 500.

In the decades following nationalisation, mining activity had peaked in 1947 at 16 mines and approximately 5,000 miners; the number of mines declined steadily throughout the 1950s before plunging to six in 1960, and to only two by 1964. With the closure of Manor Powis in 1972, Polmaise 3 and 4 was the only operating colliery in the county, and survived long enough to be a flashpoint in the genesis of the

Figure 5.188 (left): Map of National Coal Board (NCB) collieries in Stirlingshire

Figure 5.189: Airth Colliery. One of the NCB's new sinkings, construction work commenced in the late 1950s. The project was terminated in the aftermath of the failed Rothes and Glenochil schemes, and never produced coal. SC446346

miners' strike in 1984. It never recovered, and was closed for good in 1987, signalling the end of coal mining in Stirlingshire.

The previous importance of the Slamannan area is reflected in the fact that many private licensed mines were opened there in the nationalised era, taking advantage of small coal deposits left behind after the major period of mining activity in 19th and early 20th centuries. However, there were many private mines elsewhere, such as those around Avonbridge, and Stirlingshire as a whole was home to at least 60 such operations between 1947 and 1978.

### Airth 1 & 2
NS 8618 8637 (NS88NE/65)
**Parish:** St Ninians
**Region/District:** Ce/St
**Council:** Stirling
**Location:** Airth
**Previous owners:** National Coal Board
**Sinking commenced:** 1958
**Production commenced:** never produced
**Year closed:** 1959
**Shaft/Mine details:** 2 shafts, No. 1 at NS 8618 8637, and No. 2 at NS 8617 8633
**Other details:** Originally conceived as a major coking-coal colliery, with a planned daily output of 4,000 tons. The project was abandoned, probably in the face of mounting and very costly problems with contemporary sinkings at Rothes and Glenochil. By 2000, the area was being worked by Scottish Coal's new developments, driving under the Forth from Longannet. However, this project was curtailed in 2002 after serious underground flooding.
**Figure 5.189**

### Balmore
NS 6050 7418 (NS67SW/57)
**Parish:** Baldernock
**Region/District:** St/St
**Council:** East Dunbartonshire
**Location:** Balmore, Torrance
**Previous owners:** Hugh Forrester & Company
**Types of coal:** Manufacturing, House and Coking
**Sinking/Production commenced:** 1943
**Year closed:** 1960
**Year abandoned:** 1960
**Average workforce:** 131
**Peak workforce:** 173
**Peak year:** 1947
**Shaft/Mine details:** 2 surface mines, No. 1 165m in length, dipping 1 in 4 (NS 6050 7418), No. 2 73m (NS 6044 7426)

**Details in 1948:** Output 120 tons per day, 32,400 tons per annum. 196 employees. No washer. No canteen or baths. First-aid room. Electricity supplied by Clyde Valley Supply. Report dated 19-08-1948.

## Bannockburn

NS 8395 8890 (NS88NW/45)

**Parish:** St Ninians

**Region/District:** Ce/St

**Council:** Stirling

**Location:** Cowie

**Previous owners:** Carron Company

**Types of coal:** Coking, House and Steam (including navigation), later Upper Hirst

**Sinking/Production commenced:** 1894

**Year closed:** 1953

**Year abandoned:** 1955

**Average workforce:** 344

**Peak workforce:** 563

**Peak year:** 1947

**Shaft/Mine details:** Two rectangular shafts each 293m deep (No. 1 NS 8396 8884, and No. 2 NS 8393 8885, concrete-lined). No. 3 Shaft (NS 8310 8879) added in 1902–3.

**Details in 1948:** Output 400 tons per day, 100,000 tons per annum. 415 employees. Baum box-type washer. Baths (1931), canteen and first-aid room. Electricity supplied from Pirnhall Mine. Report dated 18-08-1948.

**Other details:** Drift mine opened in 1953 and successfully worked the Upper Hirst seam until 1964.

**Figure 5.190**

Figure 5.190: Bannockburn Colliery, Cowie, c.1900. A Carron Company colliery dating from 1894 and developed to produce coking coal. After nationalisation, it yielded house and steam coal, and from 1953, a new drift mine produced Upper Hirst coal until 1964. In the nationalised era, the original colliery employed on average 344 miners, and ceased production in 1953. SMM:1996.02586, SC381826

## Barleyside West

NS 8628 7646 (NS87NE/71)
**Parish:** Falkirk
**Region/District:** Ce/Fa
**Council:** Falkirk
**Location:** Falkirk
**NCB Sub-area:** Stirling/Bo'ness
**Types of coal:** House
**Sinking/Production commenced:** unknown
**Year closed:** 1949
**Average workforce:** 17
**Peak workforce:** 18
**Peak year:** 1947
**Other details:** none

## Blackston Mine

NS 9170 7360 (NS97SW/37)
**Parish:** Muiravonside
**Region/District:** Ce/Fa
**Council:** Falkirk
**Location:** Avonbridge
**Previous owners:** Brownieside Coal Company Limited
**Types of coal:** House
**Sinking/Production commenced:** 1928
**Year closed:** 1948
**Year abandoned:** 1948
**Average workforce:** 14
**Peak workforce:** 22
**Peak year:** 1948
**Shaft/Mine details:** Main mine 24m at 1 in 3 to Coxrod seam. Air mine 24m at 1 in 3 to Coxrod.
**Details in 1948:** Output 25 tons per day, 6,875 tons per annum, longwall working. 22 employees. Bar screen. No washer. No baths. Gateside's canteen available. First-aid services. All electricity from public supply. Report dated 28-07-1948.

## Dumbreck

NS 7015 7759 (NS77NW/59)
**Parish:** Kilsyth
**Region/District:** St/CN
**Council:** North Lanarkshire
**Location:** Kilsyth
**Previous owners:** William Baird & Company, later Bairds & Scottish Steel Limited
**Types of coal:** Manufacturing, Coking, House and Steam
**Sinking/Production commenced:** 1887

Figure 5.191: Dumbreck Colliery, near Kilsyth. Sunk by William Baird and Company in 1887, it is said to have had one of Scotland's first electrically-powered underground haulage and pumping plants in 1893. In addition to coking, house and steam coals, No. 3 shaft also produced ironstone. Dumbreck is also remembered for a disaster in 1938 when eight men perished. SC446566

**Year closed:** 1963

**Average workforce:** 575

**Peak workforce:** 669

**Peak year:** 1947

**Shaft/Mine details:** 3 shafts, No. 1 412m (NS 7015 7759), No. 2 411m (NS 7015 7757), and No. 3 243m deep (emergency shaft, NS 7018 7758, originally ironstone)

**Details in 1948:** Output 680 tons per day, 183,600 tons per annum. 644 employees. Coppee Baum-type washer. Baths under construction (completed 1950), canteen, first-aid room. 90% electricity supply from Clyde Valley. Report dated 19-08-1948.

**Other details:** Said to have had one of Scotland's first electrically-powered underground haulage and pumping plants in 1893, situated off No. 2 shaft. Famous for its coking coal, and had own coke ovens (upgraded in 1938). Retained by NCB after vesting because of its high-quality coal. Also remembered because of the tragedy of 1938 when eight men died after being overcome by smoke and fumes.

**Figures 5.191 and 5.192**

Figure 5.192: Dumbreck Colliery, near Kilsyth, 1941. The colliery produced coking coal which it processed in its own coke ovens. During the NCB era, an average of 575 miners were employed at the pit, peaking at 669 in 1947. Production ceased in 1963. SC458785

**Gateside 1**

NS 9259 7398 (NS97SW/35)

**Parish:** Muiravonside

**Region/District:** Ce/Fa

**Council:** Falkirk

**Location:** Avonbridge
**Previous owners:** John Hunter and Sons
**Types of coal:** House and fireclay
**Sinking/Production commenced:** 1938
**Year closed:** 1949
**Year abandoned:** 1949
**Average workforce:** 78
**Peak workforce:** 78
**Peak year:** 1947
**Shaft/Mine details:** Main mine 24m at 1 in 2.5
**Details in 1948:** Output 90 tons per day, 24,750 tons per annum, longwall working. 78 employees. Bar screen, no washer. No baths, but canteen. First-aid services. All electricity from public supply. Report dated 28-07-1948.

### Gateside 2

NS 9237 7455 (NS97SW/36)
**Parish:** Muiravonside
**Region/District:** Ce/Fa
**Council:** Falkirk
**Location:** Avonbridge
**Previous owners:** John Hunter and Sons
**Types of coal:** House
**Sinking/Production commenced:** 1939
**Year closed:** 1952
**Year abandoned:** 1952
**Average workforce:** 72
**Peak workforce:** 78
**Peak year:** 1950
**Shaft/Mine details:** Main mine 37m at 1 in 1.5, air pit 17m to mid-coal
**Details in 1948:** Output 100 tons per day, 27,500 tons per annum, longwall working. 74 employees. Bar screen, no washer. No baths, but canteen and first-aid services. All electricity from public supply. Report dated 28-07-1948.

### Herbertshire

NS 8170 8234 (NS88SW/58)
**Parish:** Denny
**Region/District:** Ce/Fa
**Council:** Falkirk
**Location:** Denny
**Previous owners:** Robert Addie and Sons Collieries Limited
**Types of coal:** Steam
**Sinking/Production commenced:** 1889
**Year closed:** 1959

Figure 5.193: Herbertshire Colliery, Denny, 1910. Originally an ironstone pit dating from 1889, it was operated by Robert Addie and Sons Collieries Limited. After 1947, it was worked by an average of 272 miners until closure in 1959. An adjacent brickworks continued to operate until 1980. Falkirk Museums P13106, SC706273

Figure 5.194: Manor Powis Colliery, Causewayhead, Stirling, c.1955. Sunk by the Manor Powis Coal Company in search of anthracite, production commenced in 1914. No. 3 shaft, which had a steel-framed winding tower, was added during an NCB reconstruction in the 1950s. The pit employed on average 802 miners in the nationalised era, eventually closing in 1972. SC446511

**Year abandoned:** 1961

**Average workforce:** 272

**Peak workforce:** 394

**Peak year:** 1948

**Shaft/Mine details:** 3 shafts, No. 3 247m, No. 4 241m (NS 8172 8229), and No. 5 263m (NS 8177 8239)

**Details in 1948:** Output 300 tons per day, 80,000 tons per annum. 395 employees. No washer. Baths (1939), canteen, first-aid. Electricity and compressed air generated at colliery. Report dated 23-08-1948.

**Other details:** Originally an ironstone pit. Well known for methane problems. Associated brickworks operated from *c.*1900 until 1980.

**Figure 5.193**

## Manor Powis 1, 2 & 3

NS 8287 9477 (NS89SW/76)

**Parish:** Dunipace

**Region/District:** Ce/St

**Council:** Stirling

**Location:** Causewayhead, Stirling

**Previous owners:** Manor Powis Coal Company

**Types of coal:** Anthracite and Steam, later Upper Hirst

**Sinking/Production commenced:** 1911–14

**Year closed:** 1972

**Year abandoned:** 1973

**Average workforce:** 802

**Peak workforce:** 985

**Peak year:** 1963

**Shaft/Mine details:** 3 shafts, No. 1 373m (NS 8287 9477), No. 2 373m (NS 8289 9477), and No. 3 97m (NS8292 9478)

**Details in 1948:** Output 647 tons per day, 177,900 tons per annum. 600 employees. 3 Baum-type washers. Baths (1932, for 1,216 men, with 64 shower cubicles, the first two-storeyed baths in Scotland), canteen, ambulance room. All electricity supplied by Scottish Central Electricity Board. Report dated 13-08-1948.

**Other details:** Said to produce the finest anthracite in Scotland. Shafts 1 and 2 both had wooden headgear, No. 2 thought to have been salvaged from a pit in Cambuslang. Both replaced with steel structures in 1931. NCB reconstruction in mid 1950s, and included the construction of a tower-mounted winder. A satellite drift mine was opened in 1954 to work the Upper Hirst seam, and did so successfully, supplying Kincardine Power Station for many years.

**Figures 5.194, 5.195 and 5.196**

## Millhall

(see Polmaise 1 & 2)

Figure 5.195: Manor Powis Colliery's surface mine, Causewayhead, Stirling. The NCB added a new drift mine, which opened in 1954, producing Upper Hirst coal for Kincardine Power Station. SC381867

Figure 5.196: Manor Powis Colliery, Causewayhead, Stirling. Drawing of pithead baths, completed in 1932 with the support of the Miners' Welfare Fund to the design of its Architect in Scotland, J A Webster. These were the first two-storeyed pithead baths to be built in Scotland. SC675012

Figure 5.197: Pirnhall Colliery, Whins of Milton, Stirling, c.1938. A Carron Company colliery comprising two surface drift mines, sunk in 1933 to produce coking coal. After 1947, the colliery operated with an average workforce of 174 miners, closing in 1963. The National Archives, COAL80/778/6, SC614081

Figure 5.198: Plean No. 4 Colliery, near Stirling. One of a group of collieries dating from the 1930s, and sunk to extract coking coal. The colliery closed in 1962, having been operated by on average 441 miners during the nationalised era. SMM:1996.3409, SC381869

Figure 5.199: Plean No. 4 Colliery, near Stirling. This view of its surface arrangement was taken in the early 1930s, prior to the construction of the pithead baths. The National Archives, COAL80/779/11, SC614070

## Pirnhall

NS 8021 8953 (NS88NW/46)

**Parish:** St Ninians

**Region/District:** Ce/St

**Council:** Stirling

**Location:** Whins of Milton, Stirling

**Previous owners:** Carron Company

**Types of coal:** Coking, House and Steam

**Sinking/Production commenced:** 1933

**Year closed:** 1963

**Year abandoned:** 1963

**Average workforce:** 174

**Peak workforce:** 241

**Peak year:** 1957

**Shaft/Mine details:** 2 surface mines, No. 1. at NS 8017 8954, No. 2. at NS 8021 8953

**Details in 1948:** Output 170 tons per day, 48,000 tons per annum. 142 employees. Dross washed at Bannockburn Colliery. Baths (1938), canteen and first-aid room. All electricity supplied from Bannockburn Colliery. Report dated 16-08-1948.

**Figure 5.197**

## Plean 3, 4 & 5 (East Plean)

NS 8395 8621 (NS88NW/47)

**Parish:** St Ninians

**Region/District:** Ce/St

**Council:** Stirling

**Location:** Plean, Stirling

**Previous owners:** Plean Colliery Company

**Types of coal:** Coking, Steam and House

**Sinking/Production commenced:** 1932

**Year closed:** 1962

**Year abandoned:** 1963

**Average workforce:** 441

**Peak workforce:** 528

**Peak year:** 1958

**Shaft/Mine details:** 3 shafts, No. 5 402m (NS 8395 8621, sunk in 1931), No. 4 165m deep (used only for ventilation and pumping, NS 8291 8705), No. 3 shaft nearby (NS 8218 8634)

**Details in 1948:** Output 400 tons per day, 114,406 tons per annum. 461 employees. Baum feldspar washer. Baths (1932) and canteen (at No. 5 pit). AC electricity bought from public supply, DC generated at colliery. Report dated 25-08-1948.

**Other details:** An explosion in 1922 killed 12 men. Coke ovens at Plean 3.

**Figures 5.198, 5.199, 5.200 and 5.201**

**Policy**

NS 8843 7882 (NS87NE/59)

**Parish:** Falkirk

**Region/District:** Ce/Fa

**Council:** Falkirk

**Location:** Slamannan Road, Falkirk

**Previous owners:** Callender Coal Company

**Types of coal:** House, Steam, and fireclay

**Sinking/Production commenced:** 1924

**Year closed:** 1959

**Year abandoned:** 1960

**Average workforce:** 155

**Peak workforce:** 185

**Peak year:** 1948

**Shaft/Mine details:** 2 surface mines, dipping at 1 in 4

**Details in 1948:** Output 175 tons per day, 48-50,000 tons per annum. 186 employees. No washer. Canteen, first-aid room. All electricity AC, purchased from Scottish Midland Electric Supply Company Report dated 12-08-1948.

**Polmaise 1 & 2 (Millhall)**

NS 8135 9214 (NS89SW/44)

**Parish:** St Ninians

**Region/District:** Ce/St

**Council:** Stirling

**Location:** Millhall, Stirling

**Previous owners:** Archibald Russell Limited

**Types of coal:** Anthracite, House and Steam

**Sinking/Production commenced:** 1904

**Year closed:** 1958

**Year abandoned:** 1958

**Average workforce:** 659

**Peak workforce:** 723

**Peak year:** 1951

**Shaft/Mine details:** 2 rectangular shafts, each 146m and served by steam winders. No. 1 (downcast) at NS 8135 9214, and No. 2 (upcast) at NS 8130 9213.

**Details in 1948:** Output 650 tons per day, 160,000 tons per annum. 714 employees. Washer manufactured by Campbell Binnie Reid, and a Huntington and Heberlein Gum Washer. Baths (1933), canteen, Stirling Royal Infirmary 1.5km away. Electricity AC, supplied from Polmaise 3 and 4. Report dated 12-08-1948.

**Other details:** Connected to Polmaise 3 & 4 in 1931. Briquetting plant also added. Boilerhouse chimney 38m high.

**Figure 5.202**

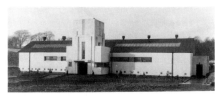

Figure 5.200: Plean No. 4 Colliery, near Stirling. View of new pithead baths, completed with the assistance of the Miners' Welfare Fund in 1932. The National Archives, COAL80/779/1, SC614054

Figure 5.201: Plean No. 3 Colliery, near Stirling, 1942. This aerial view was taken by the RAF during World War II, and clearly shows the colliery's coke ovens in operation. SC705331

Figure 5.202: Polmaise 1 & 2 Colliery, Millhall, Stirling, c.1910. Established by Archibald Russell Limited in 1904, it produced anthracite as well as steam and house coals. After nationalisation, it continued in production with an average workforce of 659 miners until its closure in 1958. SMM:1996.3421, SC381873

Figure 5.203: Polmaise 3 & 4 Colliery, Fallin, c.1910. The second of the two Polmaise collieries, it was also sunk by Archibald Russell Limited in 1904, and produced anthracite as well as house and steam coals. Some of the steam coals were of Navigation quality. SMM:1996.3416, SC381881

## Polmaise 3, 4 & 5 (also known as Fallin)

NS 8386 9138 (NS89SW/43)

**Parish:** St Ninians

**Region/District:** Ce/St

**Council:** Stirling

**Location:** Fallin

**Previous owners:** Archibald Russell Limited

**Types of coal:** House, Steam (including navigation), and Anthracite

**Sinking/Production commenced:** 1904

**Year closed:** 1987

**Year abandoned:** 1987

**Average workforce:** 606

**Peak workforce:** 778

**Peak year:** 1957

**Shaft/Mine details:** 3 shafts in 1948, Nos. 3 and 4 375m deep, but No. 4 wound from 311m. No. 5 shaft 301m, sunk in 1940 (for ventilation, pumping and upcast). No. 3 at NS 8386 9138, rectangular, downcast and wood-lined; No. 4 at NS 8384 9139, rectangular, upcast and wood-lined; No. 5 at NS 8378 9142, circular and concrete-lined, with tower-mounted electric winder.

**Details in 1948:** Output 700 tons per day, 175,000 tons per annum. 574 employees. Coppee Baum washer. Canteen, baths (1934, for 720 men, with 48 shower cubicles), first-aid. Electricity AC, generated at colliery (also supplies Polmaise 1 and 2). Report dated 16-08-1948.

**Other details:** Shafts 4 and 5 had timber headframes until replacement with steel structures in the early 1940s. The NCB reconstruction scheme of 1954 added a tower-mounted winder and mechanised underground transport. The complex included a washery, which also handled coal from Polmaise 1 & 2. By 1960, production had diverted to the Upper Hirst seams, supplying Kincardine Power Station. In 1984, Polmaise 3 & 4 was one of the major flashpoints triggering the miners' strike. A miners' memorial was unveiled on 2nd July 1994.

**Figures 5.203 and 5.204**

Figure 5.204: Polmaise 3 & 4 Colliery, Fallin, c.1970. The tower-mounted winder was added during an NCB reconstruction in 1954, and the colliery later converted to producing Upper Hirst coal for Kincardine Power Station. Throughout the nationalised period, it employed on average over 600 miners, and after being a major flashpoint in the initiation of the 1984 miners' strike, closed for good in 1987. Morris Allan of Dunfermline, SC381877

## Redding

NS 9132 7815 (NS97NW/77)

**Parish:** Grangemouth

**Region/District:** Ce/Fa

**Council:** Falkirk

**Location:** Polmont

**Previous owners:** James Nimmo & Company

**Types of coal:** Steam and House

**Sinking/Production commenced:** *c*.1894

**Year closed:** 1958

**Year abandoned:** 1959

**Average workforce:** 235

**Peak workforce:** 285

**Peak year:** 1948

**Shaft/Mine details:** 2 shafts, 88m and 64m deep, east shaft NS 9132 7816, west NS 9130 7816

**Details in 1948:** Output 300 tons per day, 82,500 tons per annum. 285 employees. No washer. No baths, canteen (packed meals), first-aid, ambulance and doctors all at Falkirk Infirmary. Electricity supply DC, generated at colliery. Report dated 12-08-1948.

**Other details:** Remembered for Scotland's worst flooding disaster on 25th September 1923 during which 66 men were trapped, only 14 being rescued alive.

**Figure 5.205**

Figure 5.205: Redding Colliery, Polmont, 1923. A James Nimmo and Company colliery dating from the 1890s. The pit is perhaps best known for Scotland's worst flooding disaster, which trapped 66 miners, 52 of whom were lost. This view of the surface arrangement was taken at the time of the disaster on 25th September. After nationalisation in 1947, the colliery employed on average 235 miners until its closure in 1958. SMM:1996.2353, SC381918

## South Bantaskine

NS 8756 7917 (NS87NE/60)

**Parish:** Falkirk

**Region/District:** Ce/Fa

**Council:** Falkirk

**Location:** Slamannan Rd, Falkirk

**Previous owners:** Callender Coal Company

**Types of coal:** House and Steam

**Sinking/Production commenced:** 1946

**Year closed:** 1959

**Year abandoned:** 1959

**Average workforce:** 107

**Peak workforce:** 141

**Peak year:** 1954

**Shaft/Mine details:** 3 surface mines, one to Mid-Drumgray, two to Armadale Main Coal

**Details in 1948:** Output 85 tons per day, 25,000 tons per annum. 84 employees. No washer, no baths, no canteen. All electricity supplied by Scottish Midland Electric Supply Company. Report dated 25-08-1948.

## Licensed Mines in Stirlingshire 1947–97

| Name | Licensee | Grid Ref | Opened | Closed |
|------|----------|----------|--------|--------|
| Arnloss 2 (Slamannan) | P & J Horne | NS 871 724 | 1954 | c. 1969 |
| Banknock 3 (Longcroft) | Joseph Bergin | NS 795 800 | 1937 | c. 1962 |
| Barnsmuir (Slamannan) | P & J Horne | NS 870 704 | 1949 | 1955 |
| Blackbraes 8 (California) | R Pringle & Company | NS 905 752 | 1952 | 1952 |
| Blackbraes 9 (California) | R Pringle & Company | NS 790 753 | 1952 | 1958 |
| Boagstown (Avonbridge) | James Bell | NS 903 749 | 1951 | 1960 |
| Bonnyburn (Bonnybridge) | Bonnyburn Fireclay Company | NS 834 803 | 1950 | 1953 |
| Bonnyside (Bonnybridge) | J Dougall & Sons Limited | NS 834 794 | pre-1947 | c. 1960 |
| Carse (Airth) | Joseph Jack | NS 893 877 | 1953 | 1954 |
| Craigmad (California) | James Penman | NS 905 757 | 1955 | 1955 |
| Drum (Bonnybridge) | Bonnybridge Silica & Fireclay Co Ltd | NS 201 680 | c. 1947 | c. 1957 |
| Easter Jaw (Falkirk) | James Broadby/James Drysdale | NS 873 748 | 1938 | 1953 |
| Easter Whin 1 & 2 (Harthill) | David Graham | NS 868 691 | 1953 | 1957 |
| Elrig (Avonbridge) | James Drysdale | NS 883 747 | 1941 | 1948 |
| Foggermountain (Avonbridge) | L Moore, later Caledon Coal Company | NS 912 736 | 1955 | 1958 |
| Gateside (Stanburn) | Morton & Smith | NS 925 743 | 1970 | 1972 |
| Glenellrigg (Slamannan) | McNeill & Taylor | NS 880 739 | 1949 | 1952 |
| Glenend 1 (Maddiston) | Thomas Heeps & Sons | NS 928 766 | 1948 | 1957 |
| Glenend 2 (Maddiston) | Thomas Heeps & Sons | NS 925 764 | 1956 | c. 1973 |
| Glenend 3/4 (Maddiston) | Thomas Heeps & Sons | NS 929 763 | 1958 | c. 1973 |
| Grayrigg (Avonbridge) | Alex Mc Neill | NS 913 750 | 1954 | 1957 |
| Greencraig 1 (Avonbridge) | Greencraig Coal Company | NS 912 741 | 1938 | 1950 |
| Greencraig 2 (Avonbridge) | Greencraig Coal Company | NS 910 743 | 1950 | 1966 |
| Hairstanes (Avonbridge) | A H Adams | NS 933 735 | 1959 | 1968 |
| Hairstanes 2 (Avonbridge) | A H Adams | NS 931 731 | 1958 | 1961 |
| Hillfarm 1/2 (Avonbridge) | Hillfarm Coal Company Limited | NS 939 733 | pre-1947 | 1961 |
| Hillfarm 3 (Avonbridge) | Hillfarm Coal Company Limited | NS 939 735 | 1954 | 1961 |
| Hillfarm 4 (Avonbridge) | Hillfarm Coal Company | NS 938 733 | 1960 | 1962 |
| Hillfarm 5 (Avonbridge) | Hillfarm Coal Company | NS 931 727 | 1962 | 1968 |
| Hillfoot (Avonbridge) | Hillfoot Coal Company | NS 942 737 | pre-1947 | 1956 |
| Holehousemuir (Slamannan) | P & J Horne | NS 866 703 | pre-1947 | 1950 |
| Howierigg (Falkirk) | Rumford Coal Company | NS 847 787 | pre-1947 | 1950 |
| Jawcraig Mid (Falkirk) | John McKechnie | NS 848 751 | 1955 | 1957 |
| Kendieshill (Avonbridge) | J Drysdale | NS 937 756 | 1950 | 1968 |
| Knowehead (Longcroft) | J Bergin | NS 794 802 | pre-1947 | 1947 |
| Lime Road (Camelon) | J McCaig & Sons | NS 859 793 | 1948 | 1960 |
| Lippie (Boagstown, Avonbridge) | J Walker | NS 891 743 | 1957 | c. 1959 |
| Livingstone (Banknock) | Cloybank Minerals | NS 783 798 | c. 1933 | 1952 |
| Lochend (by Falkirk) | Alec McNeill | NS 882 748 | 1959 | c. 1965 |
| Longriggend 1 (Caldercruix) | John Jenkins | NS 827 702 | 1953 | 1953 |
| Maddiston (Polmont) | Maddiston Coal Company | NS 937 767 | pre-1947 | 1949 |
| Mavisbank / Pirleyhill | Callendar Brick & Fireclay Company Limited | NS 888 778 | 1962 | 1967 |

| Name | Licensee | Grid Ref | Opened | Closed |
|---|---|---|---|---|
| Milnquarter (Bonnybridge) | John G Stein & Company Limited | NS 829 791 | pre-1947 | c. 1957 |
| North Dyke (Slamannan) | James Penman | NS 857 747 | 1953 | 1954 |
| Oakerdykes 2 (Slamannan) | Mrs A White | NS 846 743 | pre-1947 | 1950 |
| Oakerdykes 3 (Slamannan) | Mrs A White | NS 845 747 | pre-1947 | 1948 |
| Oakerdykes 4 (Slamannan) | Mrs A White, White & Caine | NS 843 745 | 1950 | 1978 |
| Pirleyhill 1 (Falkirk) | Callendar Brick & Fireclay Company | NS 889 776 | 1947 | 1955 |
| Pirleyhill 2 (Falkirk) | Callendar Brick & Fireclay Company | NS 888 778 | 1951 | c. 1967 |
| Pirleyhill 3 (Falkirk) | Callendar Brick & Fireclay Company | NS 888 778 | 1954 | c. 1967 |
| Reddingmuir (Polmont) | James Penman | NS 909 769 | 1948 | 1949 |
| Redhall (Slamannan) | Hutchison Caine | NS 880 717 | 1957 | 1961 |
| Redhall 2 (Slamannan) | Hutchison Caine | NS 881 713 | 1959 | c. 1970 |
| Righead (Avonbridge) | Morton Bros | NS 901 743 | 1948 | c. 1967 |
| Slamannan (Slamannan) | James Drysdale | NS 857 738 | 1954 | c. 1957 |
| Snabhead (Avonbridge) | A McNeil | NS 920 753 | 1953 | 1954 |
| Stanley (Banknock) | Cloybank Minerals | NS 775 800 | pre-1947 | 1949 |
| Summerhouse (Falkirk) | Maddiston Coal Company | NS 884 766 | pre-1947 | 1947 |
| Threaprig (Slamannan) | J Frew | NS 835 747 | 1958 | 1970 |
| Wester Burnhead 2 (Avonbridge) | William Walker & Sons | NS 882 695 | 1947 | 1952 |

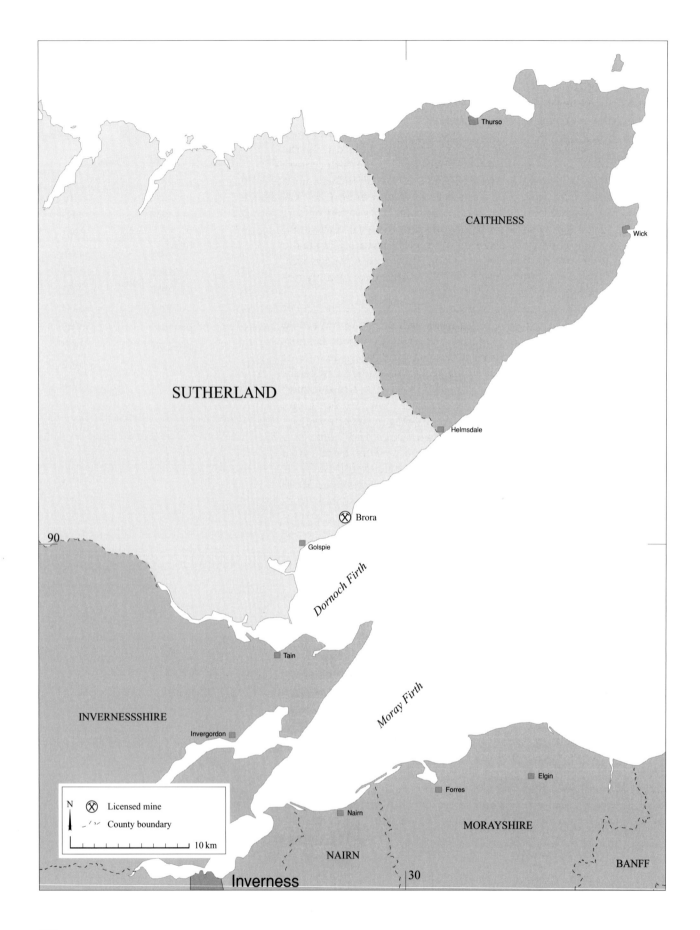

CAITHNESS

Thurso

Wick

SUTHERLAND

Helmsdale

⊗ Brora

90

Golspie

*Dornoch Firth*

Tain

*Moray Firth*

INVERNESSSHIRE

Invergordon

Elgin

Forres

N ⊗ Licensed mine
- · · - County boundary

Nairn

MORAYSHIRE

10 km

NAIRN

30

BANFF

Inverness

# Sutherland

The most northerly coalfield in Scotland, Brora is of the Jurassic period, and produced coal in relatively thin seams which was of indifferent quality compared with the older Carboniferous coals of central Scotland. The presence of pyrites was also a source of heating and spontaneous combustion problems which eventually proved to be the mine's downfall. Coal was probably mined as early as the 16th century in Brora, but was formally established by the Duke of Sutherland in 1811, around which period local salt pans were also built. Coal was exported from the town's harbour, but was also used by a variety of local industries, including an adjacent brickworks and the local textile mill, which were also eventually owned by T M Hunter Limited. Despite the occasional underground fire, the mine was considered to be gas free, and carbide lamps were used until the 1950s. Miners were even permitted to smoke tobacco underground.

At the time of coal nationalisation in 1947, the NCB considered Brora Colliery to be unprofitable and not worth taking into the state sector, despite the fact that it employed more than 30 miners, the normal maximum for private licensed mines. In the ensuing decades, it struggled to survive, benefiting from the establishment of a new brickworks and a briquetting plant in 1953–4. However, it had difficulty enticing the local distilleries to use the coal, the complaint being that it did not raise steam quickly enough. In 1960, an underground fire caused extensive damage, but the mine stayed open after going into financial liquidation, having been rescued by a workers' co-operative with assistance from the 'Highland Fund'. It was renamed 'Ross Pit' in 1961, subsequent modernisation involving the truncation of one of the wooden headgear and the replacement of the other with a steel structure. In 1964, a visitor recorded the fact that the miners were diversifying by growing mushrooms in boxes in the return airway, having previously experimented with forced rhubarb.

After another fire in 1967, and with reserves in the main part of the mine dwindling, a new mine, Ross No. 2, was commenced in 1969. This was assisted with a grant of £100,000 from the Highlands and Islands Development Board. Ross No. 1 was closed shortly afterwards, and production from the new mine commenced in September. By 1971, mining conditions were deteriorating, and miners were being lured away by alternative employment in the North Sea oil and gas industries, and by oil platform construction yards at Nigg and Ardersier. Mining ceased on 22nd June 1973, but the colliery was retained on a care and maintenance basis for some months before a small workforce of six miners recommenced production following its purchase by Mr E E Pritchard of Stockport. He subsequently died of a heart attack when inspecting the underground workings in 1974, and the mine closed permanently in 1975 after attempts failed to sell it to a new owner. The surface remains were rapidly cleared away and the shafts capped by the NCB. Records of the mine were rescued by Dr J K Almond, and a book on its history was later published by John S Owen in 1995.

Figure 5.206 left: Map of Sutherland showing the position of Brora colliery

**Brora**

NC 8991 0403 (NC80SE/44)

**Parish:** Clyne

**Region/District:** Hi/Su

**Council:** Highland

**Owners:** Marquess of Stafford/Dukes of Sutherland, later T M Hunter Limited, a miners' co-operative, and from 1973, Mr E E Pritchard

**Sinking/Production commenced:** 1811, new mine (Ross No. 2) in 1969

**Year closed:** 1975

**Abandoned:** 1975

**Shaft/Mine details:** 2 shafts, No. 1 NC 899 040 and No. 2 895 040. New surface mine, Ross No. 2 (1969) at NC 8980 0400

**Other details:** The only Jurassic coal mine in Scotland. Never taken into state ownership by the NCB. Re-named Ross Pit, the original shafts becoming Ross No. 1, the new drift mine, Ross No. 2, being driven in 1969. Ross No. 1 was closed after a fire in 1969 forced the fire brigade to flood the mine. Ross No. 2 closed six years later. Pithead baths were built by the Miners' Welfare Fund in 1936. An associated brickworks (re-established in 1954) produced bricks and tiles using local clays and shale from the mine, and operated from c.1907 to c.1971, supplying common building brick to customers all over the Highlands.

## Licensed Mines in Sutherland 1947–97

| Name | Licensee | Grid Ref | Opened | Closed |
| --- | --- | --- | --- | --- |
| Ross 1 (Brora) | Brora Coal Company, E E Pritchard | NC 899 040 | 1811 | 1969 |
| Ross 2 (Brora) | Brora Coal Company, E E Pritchard | NC 895 040 | 1969 | 1975 |

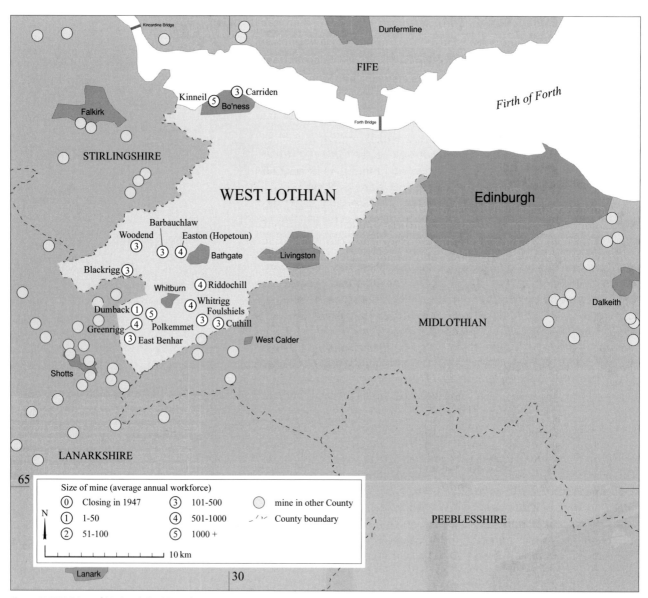

Figure 5.207: Map of National Coal Board
(NCB) collieries in West Lothian.

# West Lothian

West Lothian was the fifth most important coal producer in Scotland, after Fife, Lanarkshire, Ayrshire and Midlothian, employing over 7,000 miners at its peak in 1956. Formerly known as Linlithgowshire, and more generally associated with oil-shale mining, its coal mines were concentrated in the south-west of the county along the border with Midlothian and Lanarkshire, and as far north as Armadale and Bathgate. The second area of mining was in the far north-west on the Firth of Forth at Bo'ness, an area which witnessed the earliest coal-mining, in association with medieval monastic leases, and later the operation of salt pans.

Of the 16 collieries to have operated in West Lothian in the nationalised period, 14 were inherited from the private sector in 1947 as working entities. A further two, Dumback and Cuthill, were added by the NCB in the mid-1950s, when the number of mines in operation at any one time peaked at 14. As was the case elsewhere in Scotland, mine closures accelerated in the late 1950s, as did the decline in the number of working miners. For much of the 1960s, the number of operating collieries remained at four, dropping in 1973 to two, leaving only Polkemmet from 1978. Polkemmet itself finally closed in 1986 as a result of the 1984-5 miners' strike, but at 40 years, was one of the longest-serving collieries in Scotland in the nationalised era.

The output of West Lothian mines included steam, manufacturing, and gas coals, along with a small amount of anthracite from Woodend Colliery. As with the other coalfields, blaes and fireclays were also produced for use by brick and refractory works which were often associated with or close to the collieries themselves. However, perhaps the most important output was coking coal, especially from Kinneil and Polkemmet, the latter being one of Ravenscraig Steelworks' (in Motherwell, Lanarkshire) most important domestic suppliers. The importance of the iron and steel industry as a market is reflected in the names of the owners of the mines prior to nationalisation, which included William Dixon (Polkemmet), Bairds & Scottish Steels (Easton, Carriden and Riddochhill), and the Coltness Iron Company (Woodend). Other companies with interests in West Lothian included R Forrester, and United Collieries Limited.

Although no major new sinkings occurred in West Lothian, one of the biggest reconstruction schemes was at Kinneil Colliery in Bo'ness, and was again designed by NCB architect, Egon Riss. The project commenced in 1951, and took five years to complete. Kinneil was later connected to Valleyfield Colliery under the Firth of Forth in 1965. The other major reconstruction took place at Polkemmet, and was completed in 1958. Inevitably, these two collieries were the largest in the county, Polkemmet averaging a workforce of almost 1,500 throughout the post-war period, and Kinneil almost 1,200. Other large employers included Whitrigg with an average workforce of over 950, and Easton (formerly Hopetoun) with 715.

Sixteen small licensed private mines are known to have operated in the county since nationalisation, some originating well before 1947. The largest concentrations were around Whitburn, Linlithgow and Armadale, but all had closed by 1972. Twelve years later, the closure of Polkemmet in the aftermath of the miners' strike signalled the end of deep mining in West Lothian.

**Barbauchlaw**

NS 9430 6880 (NS96NW/23)

**Parish:** Bathgate

**Region/District:** Lo/We

**Council:** West Lothian

**Location:** Armadale

**Previous owners:** Robert Muir & Company

**Types of coal:** Steam, and fireclay

**Sinking/Production commenced:** 1900

**Year closed:** 1952

**Year abandoned:** 1953

**Average workforce:** 104

**Peak workforce:** 115

**Peak year:** 1947

**Shaft/Mine details:** Main mine 122m at 1 in 3.5. Air mine 46m, 1 in 3 to clay seam.

**Details in 1948:** Output 90 tons per day, 24,750 tons per annum, longwall working. 91 employees. Bar screen only. No washer. No baths, no canteen. Fist aid services. All electricity from public supply. Report dated 28-07-1948.

**Other details:** Associated brickworks operated from *c.*1890 until 1971.

**Blackrigg 1, 2 & 3**

NS 9140 6753 (NS96NW/24.00)

**Parish:** Bathgate

**Region/District:** Lo/We

**Council:** West Lothian

**Location:** Blackridge

**Previous owners:** United Collieries Limited

**Types of coal:** Steam, and blaes

**Sinking/Production commenced:** *c.*1860

**Year closed:** 1955

**Year abandoned:** 1955

**Average workforce:** 291

**Peak workforce:** 339

**Peak year:** 1951

**Shaft/Mine details:** 2 shafts, both 99m deep (Shaft No. 3 at NS 9035 6770, site No. NS96NW/24.01)

**Details in 1948:** Output 340 tons per day, 93,500 tons per annum, longwall working. 279 employees. 3 screen tables. Baum-type washer (Blantyre Engineering Co). No baths, small canteen (packed meals only). First-aid service. All electricity from Westrigg Power Station, none from public supply. Report dated 28-07-1948.

**Carriden 1 & 2**

NT 0054 8200 (NT08SW/118)

**Parish:** Bo'ness and Carriden

**Region/District:** Lo/Fa

**Council:** West Lothian

**Location:** Bo'ness

**Previous owners:** Carriden Coal Company, Bairds & Scottish Steel Limited in 1930s

**Types of coal:** Gas, House, Steam and Manufacturing

**Sinking/Production commenced:** 1914

**Year closed:** 1953

**Year abandoned:** 1954

**Average workforce:** 321

**Peak workforce:** 331

**Peak year:** 1948

**Shaft/Mine details:** 2 shafts, each 137m

**Details in 1948:** Output 270 tons per day, 70,000 tons per annum. 316 employees. Dickson and Mann bash tank washer. Canteen, no baths, first-aid. Electricity from public supply. Report dated 16-08-1948.

**Figure 5.208**

Figure 5.208: Carriden Colliery, Bo'ness, c.1930. Sunk originally in 1914 by the Carriden Coal Co, and taken over by Bairds and Scottish Steels Limited. After nationalisation, the pit employed on average 321 miners until its closure in 1953. SMM:1997.0522, SC382411

## Cuthill

NS 9905 6323 (NS96SE/22)

**Parish:** Whitburn

**Region/District:** Lo/We

**Council:** West Lothian

**Previous owners:** National Coal Board

**Sinking commenced:** 1955

**Production commenced:** 1958

**Year closed:** 1960

**Average workforce:** 207

**Peak workforce:** 303

**Peak year:** 1959

**Shaft/Mine details:** 4 surface mines

**Other details:** A short-term NCB drift intended to satisfy demand until output from the new sinkings kicked in. Linked to Baads.

**Figure 5.209**

Figure 5.209: Cuthill Colliery, near Whitburn, 1956. A new development comprising four surface mines, begun by the NCB in 1955, production commencing in 1958. Linked to neighbouring Baads Colliery, it operated until 1960 with an average workforce of 207 miners. SMM:1996.2424, SC382412

## The Dardanelles

(see Polkemmet)

## Dumback 1

NS 9223 6430 (NS96SW/30)

**Parish:** Whitburn

**Region/District:** Lo/We

**Council:** West Lothian

**Location:** near Whitburn
**Previous owners:** National Coal Board
**Sinking/Production commenced:** 1954
**Year closed:** 1959
**Year abandoned:** 1959
**Average workforce:** 26
**Peak workforce:** 26
**Peak year:** 1958
**Other details:** none

## East Benhar Mine

NS 9162 6202 (NS96SW/20)
**Parish:** Whitburn
**Region/District:** Lo/We
**Council:** West Lothian
**Location:** Fauldhouse
**Previous owners:** Barr and Thornton
**Types of coal:** House
**Sinking/Production commenced:** 1940
**Year closed:** 1957
**Year abandoned:** 1957
**Average workforce:** 139
**Peak workforce:** 159
**Peak year:** 1951
**Shaft/Mine details:** Surface mine, to depth of 46m
**Details in 1948:** Output 120 tons per day, 36,200 tons per annum, longwall working. 11 employees. 2 bar screens. No washer. No baths, snacks prepared and sent underground. First-aid services. All electricity DC, and generated at mine. Report dated 28-07-1948.

## Easton (Hopetoun)

NS 9903 6900 (NS96NE/44)
**Parish:** Bathgate
**Region/District:** Lo/We
**Council:** West Lothian
**Location:** Bathgate
**Previous owners:** Balbardie Colliery Company, Bairds & Scottish Steel Limited from 1939
**Types of coal:** Coking and Steam
**Sinking/Production commenced:** 1898
**Year closed:** 1973
**Year abandoned:** 1974
**Average workforce:** 715
**Peak workforce:** 863
**Peak year:** 1958

**Shaft/Mine details:** 2 shafts, No. 1 379m, and No. 2 99m deep

**Details in 1948:** Output 820 tons per day, 205,000 tons per annum, longwall working. 537 employees. 2 screens. Hoyois washer, being substituted with Coppee Washer (under construction). Baths (1939, for 400 men), canteen (packed meals) and first-aid room. All electricity generated at mine. Report dated 30-07-1948.

**Other details:** Winding electrified in 1950s.

**Figure 5.210**

## Foulshiels

NS 9772 6343 (NS96SE/19)

**Parish:** Whitburn

**Region/District:** Lo/We

**Council:** West Lothian

**Location:** Stoneyburn

**Previous owners:** United Collieries Limited

**Types of coal:** House, Steam and Gas, and blaes

**Sinking/Production commenced:** *c.*1900

**Year closed:** 1957

**Year abandoned:** 1958

**Average workforce:** 448

**Peak workforce:** 463

**Peak year:** 1951

**Shaft/Mine details:** 1 shaft, No. 2 pit 112m deep (upcast for Loganlea Colliery)

**Details in 1952:** Output 580 tons per day, 145,000 tons per annum, longwall working. 428 employees. 4 screens, Baum-type washer (Blantyre Engineering Company). Baths (1935, for 450 men, with 37 shower cubicles), canteen (packed meals), first-aid room. All electricity supplied from Loganlea Colliery. Report dated 30-07-1952.

**Figure 5.211**

Figure 5.210: Easton Colliery, near Bathgate. Formerly known as Hopetoun Colliery, the pit was originally sunk in 1898 by the Balbardie Colliery Co, and was later taken over by Bairds and Scottish Steels Limited. A significant producer of coking and steam coals, it employed on average 715 miners during the NCB era, closing in 1973. The baths were built with the assistance of the Miners' Welfare Fund in 1939. SC675069

Figure 5.211: Foulshiels Colliery, Stoneyburn. Sunk in *c.*1900 by United Collieries Limited, its single shaft acted as an upcast shaft for Loganlea Colliery across the border in Midlothian. The colliery worked for ten years after nationalisation, employing on average 448 miners, closing in 1957. The pithead baths were completed in 1935, with assistance from the Miners' Welfare Fund. The National Archives, COAL80/417/2, SC614113

### Greenrigg

NS 9205 6313 (NS96SW/23)

**Parish:** Whitburn

**Region/District:** Lo/We

**Council:** West Lothian

**Location:** Fauldhouse

**Previous owners:** Loganlea Coal Company, United Collieries Limited from 1902

**Types of coal:** Coking, Gas and House, and fireclay

**Sinking/Production commenced:** 1905

**Year closed:** 1960

**Average workforce:** 558

**Peak workforce:** 634

**Peak year:** 1952

**Shaft/Mine details:** 2 shafts, both 143m deep

**Details in 1948:** Output 600 tons per day, 142,000 tons per annum, longwall working. 559 employees. 3 plate tables, Baum-type washer (Norton). Baths (1930), canteen (pieces), first-aid services. All electricity (AC and DC) from Central Power Station (NCB) at Westrigg. Report dated 28-07-1948.

**Other details:** Pithead fire in 1925 destroyed all surface buildings. Clay supplied to Etna Brickworks in Armadale.

### Hopetoun

(see Easton)

### Kinneil

NS 9866 8121 (NS98SE/81)

**Parish:** Bo'ness and Carriden

**Region/District:** Ce/Fa

**Council:** Falkirk

**Location:** Bo'ness

**Previous owners:** Kinneil Cannel & Coking Coal Company

**Types of coal:** House, Coking, Gas and Manufacturing

**Sinking/Production commenced:** 1880–90

**Year closed:** 1982

**Year abandoned:** 1983

**Average workforce:** 1,184

**Peak workforce:** 1,268

**Peak year:** 1960

**Shaft/Mine details:** 4 older shafts, Furnaceyard Pit 311m, No. 2 Snab 351m, No. 4 Snab 128m, and Lothians 196m deep. New project begun in 1951, and completed in 1956. The 2 new shafts were 6.71m diameter and concrete-lined. No. 1 (NS 9866 8121) was downcast and equipped with 2 Koepe single-rope winders mounted in a tower, and raised the coal in 2 sets of two cages, each cage having 3 decks capable of carrying a 2-ton capacity mine car. No. 2 (NS 9879 8128) had a single ground-mounted Koepe winder and steel headframe for winding men and materials, and was upcast. In 1951, the new shafts were expected to exceed 900m in depth. Kinneil was merged with Valleyfield (under the Forth) in 1965.

**Details in 1948:** Output 650 tons per day, 194,584 per annum. 601 employees. Baum type washer with froth flotation tank. Baths (1933), no canteen, first-aid. Electricity AC, none bought from outside. Report dated 19-08-1948.

**Other details:** NCB reconstruction commenced in 1951, and was described in the new colliery brochure as marking '...the beginning of a major project in the great reconstruction programmes for the coal mines of Scotland envisaged in the Board's national plan'. The new design, including the car hall, had emerged following a visit to Bismark Colliery in the Ruhr, West Germany, and was designed to handle 3,000 tons of coal a day. A new Coppee coal washery was added as part of the scheme, which also included new offices, baths, canteen and medical facilities, the new layout of the colliery being overseen by the NCB's Scottish region architect, Egon Riss. Closed on 14th December 1982 because of severe geological conditions.

**Figures 5.212 and 5.213**

Figure 5.212: Kinneil Colliery, Bo'ness, c.1952. Sunk originally by the Kinneil Cannel and Coking Coal Company in the 1880s, the four older shafts were replaced by two new shafts in one of the NCB's most ambitious reconstructions during the 1950s. The new shafts were deepened to over 900 metres. SMM:1998.0399, SC381924

Figure 5.213: Kinneil Colliery, Bo'ness. The new surface arrangement was completed in 1956 to the design of the NCB Scottish Division's architect, Egon Riss. Nine years later, it was linked under the Firth of Forth with Valleyfield Colliery in Fife. During the nationalised period, the colliery operated with an average workforce of 1,184 miners until its closure in 1982. Despite attempts to save some of the surface buildings, they were eventually demolished. SMM:1998.0325, SC381930

Figure 5.214: Polkemmet Colliery, Whitburn, c.1960. One of Scotland's most important producers of coking coal in the late 20th century, Polkemmet was established by William Dixon Limited, production commencing in 1921. A major NCB reconstruction in the 1950s replaced its steam winders with electric motors, and also involved the provision of a new fan house and coal preparation plant. SC446340

Figure 5.215: Polkemmet Colliery, Whitburn, 1961. In the nationalised era, Polkemmet employed more miners than any other Scottish pit, averaging at 1,496 over 39 years of operation. Latterly, it was also an important supplier of coking coal for Ravenscraig, Scotland's last surviving integrated steel works. Its abandonment in 1986 was deeply controversial and was a direct result of the 1984–5 miners' strike, during which a decision was made to switch off the pumps. SMM:1996.0776, SC381932

Figure 5.216: Polkemmet Colliery, Whitburn, 1964. This underground view shows the Jewel seam's cable belt. The reconstruction in the 1950s led in particular to major improvements underground. Continental Conveyors Ltd, SC382418

## Lady of The Dales
(see Whitrigg 2, 4 & 5)

## Polkemmet (The Dardanelles)
NS 9341 6399 (NS96SW/24)
**Parish:** Whitburn
**Region/District:** Lo/We
**Council:** West Lothian
**Location:** Whitburn
**Previous owners:** William Dixon Limited
**Types of coal:** Coking and Gas
**Sinking/Production commenced:** 1913–6, production from 1921
**Year closed:** 1986
**Year abandoned:** 1986
**Average workforce:** 1,496
**Peak workforce:** 1,959
**Peak year:** 1960
**Shaft/Mine details:** 2 shafts, both 6.1m diameter. No. 1 477m (upcast) sunk in 1913; No. 2 479m (downcast), sunk in 1916. Reconstruction completed in 1958 resulted in the replacement of steam with electric winders (No. 1 in 1955 and No. 2 in 1957), the introduction of skip winding in No. 2 shaft (with 8-ton skips), and the construction of a new fan house. Other improvements included a Simon Carves preparation plant, new compressor plant, locomotive shed, electrical workshop, and administration block.
**Details in 1948:** Output 1,320 tons per day, 330,000 tons per annum, longwall working. 981 employees. 4 screens. Baum-type washer (Simon Carves). Baths and canteen (1937), first-aid room. All electricity generated at mine. Report dated 30-07-1948.
**Other details:** In terms of longevity and the number of people employed, Scotland's most important pit. Its baths were the second largest in Scotland. Latterly, a major supplier of coking coal for

Ravenscraig steelworks at Motherwell. The boilerhouse chimney was demolished in 1970. The mine closed in controversial circumstances as a result of flooding which occurred during the 1984–5 miners' strike. After closure, persistent spontaneous combustion problems in the bing generated noxious fumes which, when propelled by a prevailing wind, were instantly recognisable to regular travellers on the M8 motorway.

**Figures 5.214, 5.215 and 5.216**

## Riddochhill

NS 9751 6632 (NS96NE/45)

**Parish:** Livingston

**Region/District:** Lo/We

**Council:** West Lothian

**Location:** Bathgate

**Previous owners:** Gavin Paul and Sons, William Baird & Company from 1922, later Baird & Scottish Steels Limited

**Types of coal:** Gas, House, Manufacturing and Steam

**Sinking/Production commenced:** 1890

**Year closed:** 1968

**Year abandoned:** 1969

**Average workforce:** 541

**Peak workforce:** 687

**Peak year:** 1958

**Shaft/Mine details:** 1 shaft, 289m (downcast), return mine 1 in 3 for 69m (upcast)

**Details in 1948:** Output 420 tons per day, 105,000 tons per annum, longwall working. 404 employees. 2 screens. Baum-type washer (Blantyre Engineering Company). Baths and canteen (1938), and first-aid room. All electricity supplied from Easton Colliery. Report dated 30-07-1948.

**Other details:** Skip winding replaced hutches in the 1950s. The M8 motorway was built over the site after closure.

**Figures 5.217 and 5.218**

Figure 5.217: Riddochhill Colliery, near Bathgate, 1957. Sunk originally by Gavin Paul and Sons in 1890, the pit was taken over by William Baird and Company in 1922. NCB reconstruction work in the 1950s replaced hutches with a skip-winding system. SMM:1996.02447, SC382416

Figure 5.218: Riddochhill Colliery, near Bathgate. During the nationalised era, the pit operated with an average workforce of 541 miners until its closure in 1968. SMM:1996.0472, SC382415

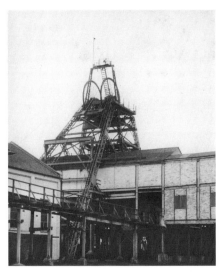

Figure 5.219: Whitrigg Colliery, near Whitburn, also known as 'Lady of the Dales'. Established in c.1900 by R Forrester and Company Limited, the colliery comprised three shafts and two surface mines, a third mine being added in 1952. SMM:1996.0426, SC380995

## Whitrigg 2, 4 & 5 (Lady of the Dales)

NS 9670 6457 (NS96SE/21)

**Parish:** Whitburn

**Region/District:** Lo/We

**Council:** West Lothian

**Location:** Whitburn

**Previous owners:** R Forrester & Company Limited

**Types of coal:** Gas, House, Manufacturing and Steam

**Sinking/Production commenced:** c.1900

**Year closed:** 1972

**Year abandoned:** 1972

**Average workforce:** 953

**Peak workforce:** 1,208

**Peak year:** 1952

**Shaft/Mine details:** 3 shafts, 2 surface mines: No. 5 pit 323m deep (downcast), No. 4 mine 1 in 6 for 34m, No. 1 pit 77m deep, No. 2 pit 106m deep. No. 6 mine was added later, beginning production in 1952, ceasing in 1958, average workforce 109, peaking in 1952 with 111 employees.

**Details in 1948:** Output 1,260 tons per day, 315,000 tons per annum, longwall working. 910 employees. 4 screens. Baum-type washer (Norton Engineering Company). Baths (1932), canteen and first-aid room. All electricity generated at mine. Report dated 30-07-1948.

**Other details:** Also known as Lady of the Dales.

**Figures 5.219 and 5.220**

## Woodend

NS 9220 6950 (NS96NW/07)

**Parish:** Torphichen

**Region/District:** Lo/We

**Council:** West Lothian

**Location:** Armadale

**Previous owners:** Coltness Iron Company Limited

Figure 5.220: Whitrigg Colliery, near Whitburn, c.1960. After nationalisation in 1947, the colliery operated with an average workforce of 953 miners, closing in 1972. SC446541

**Types of coal:** Anthracite, House and Manufacturing

**Sinking/Production commenced:** *c.*1870

**Year closed:** 1965

**Average workforce:** 397

**Peak workforce:** 488

**Peak year:** 1954

**Shaft/Mine details:** 2 shafts, No. 5 101m deep (to Ironstone), Air Shaft 49m (to Mill Coal)

**Details in 1948:** Output 300 tons per day, 82,500 tons per annum, longwall working. 325 employees. 3 tables, but only 2 shakers. Baum-type washer (Blantyre Engineering Company). Baths, canteen, first-aid services. All electricity AC and generated at colliery. Report dated 28-07-1948.

**Other details:** The washery was built in the 1930s, but was retained after closure in 1965 to treat coal from other local mines.

**Figure 5.221**

Figure 5.221: Woodend Colliery near Armadale. Sunk in c.1870 by the Coltness Iron Company to extract anthracite, the colliery closed in 1965, having employed on average 397 miners in the nationalised era. However, its Baum-type coal washer, which was built in the 1930s by the Blantyre Engineering Co, was retained after closure to treat coal from neighbouring mines. SMM:1996.03418, SC380994

## Licensed Mines in West Lothian 1947–97

| Name | Licensee | Grid Ref | Opened | Closed |
|------|----------|----------|--------|--------|
| Airngarth 1 (Linlithgow) | Dugald McKechnie | NS 998 794 | 1958 | 1961 |
| Airngarth 2 (Linlithgow) | Dugald McKechnie | NS 996 795 | 1961 | c. 1967 |
| Blackburnhall (Bathgate) | David Allan & Company | NS 990 649 | pre-1947 | 1947 |
| Brunton (Westfield) | William McCracken | NS 295 672 | 1955 | 1966 |
| Couston (Armadale) | Couston Mining Company | NS 949 699 | 1952 | 1972 |
| Dumback 1 (Whitburn) | Whitburn Coal Company | NS 722 644 | 1942 | 1954 |
| Dumback 2 (Whitburn) | J & D McWilliam | NS 927 644 | pre-1947 | 1966 |
| Dumback 3 (Whitburn) | J & D McWilliam | NS 922 644 | 1962 | 1968 |
| East Bonhard (Bridgeness) | Woodhead Coal Company | NT 023 798 | 1968 | 1971 |
| Heads (Whitburn) | Thomas Green & Sons | NS 934 635 | 1964 | 1965 |
| Muckraw (Avonbridge) | A Allan & Sons Limited | NS 913 714 | pre-1947 | 1947 |
| Murraysgate (Whitburn) | Thomas Green & Company | NS 930 639 | pre-1947 | 1965 |
| Northrigg 1 (Bathville) | United Fireclay Company | NS 953 665 | unknown | 1963 |
| Stonehead (Fauldhouse) | Kerr & Adams Limited | NS 942 624 | 1951 | 1960 |
| Woodbank (Armadale) | Woodbank Mining Company, William Park | NS 937 700 | 1950 | 1965 |
| Woodhead Linlithgow | Savage & Moyser | NS 982 697 | 1957 | 1967 |

# Glossary

**Abandonment Plans**

The plans required by law to be deposited with the 'District Inspector' upon the closure of a coalmine, or seam in a mine. They show the workings in relation to surface features and record information with respect to orientation, contours, scales, boundaries, and faults. The plans also show the last recorded workings for the safety of any future operations, and can be of relevance when assessing the causes and risks of subsidence. Since privatisation, coal records have been co-ordinated on a national (UK) basis by the Coal Authority at Mansfield, and information is available online at www.coal.gov.uk/Services/miningrecords.htm

**Aberfan Disaster**

One of the most tragic disasters in the history of the British coal-mining industry occurred on 21st of October 1966 when, after a period of heavy rainfall, a spoil heap above the mining village of Aberfan in South Wales slipped down the hillside, engulfing a farm and the local primary school, killing 144, of whom most were children, and injuring a further 28 people. The disaster prompted a reappraisal of tipping practices in the coal industry, and in subsequent decades resulted in the clearance and rehabilitation of coal-mining landscapes throughout the UK, and the disappearance of the surface remains of many coal mines in Scotland.

**Acid Battery**

A battery suitable in size and weight for use with hand-held and miner's cap lamps. They are routinely recharged in the Lamp Room of the mine at the surface.

**Adit**

A tunnel driven into a hill or valley side in connection with mineral working, and used for transport, ventilation or drainage (or all three).

**Admiralty List of Mines**

Several coal mines, mostly in the western section of the South Wales coalfield, produced steam coal especially suited for use in high-performance steamships, and were placed on the Royal Navy's Admiralty List. A small number of Scottish mines, such as Valleyfield in Fife, were noted for their Navigation Coal, supplying Navy vessels nearby at Rosyth.

**Aerial Ropeway**

An overhead system for transporting coal or colliery waste from the pithead, usually comprising buckets suspended from overhead cables supported on towers. Aerial ropeways were often used to transport coal from one colliery to utilise screening plant elsewhere, but were also used to lift and dispose of colliery waste onto spoil heaps, as was the case at Comrie and Lindsay Collieries in Fife.

**Afterdamp**

A residue of non-inflammable gases left after a Methane explosion in a coal mine, with high carbon dioxide and carbon monoxide levels, and very little oxygen. Afterdamp was often responsible for more deaths than the explosions themselves.

**Agent**

In the pre-nationalisation era, a senior official controlling a group of mines, or single large colliery.

### Air Compressor

A machine for pressurising air which can then be transmitted and used as a safe power source for driving other machinery, especially underground. Compressed air, which is stored in a sealed tank or air receiver until it is required for use, was widely used in coal mines, but was mostly superseded by the introduction of flameproof and intrinsically-safe electrical machinery. However, in many mines such as Cardowan in Lanarkshire, air-compressor houses were a significant element in the surface arrangement around the pithead. Major Scottish manufacturers of air compressors included Alley & McLellan of Glasgow.

### Airlock

A double door in a tunnel controlling the flow of air within the mineworkings, ensuring that the fresh air in the intake airway is not short-circuited to the return airway – a situation which would deprive parts of the mine workings of fresh air. Also used to refer to a casing at the top of the upcast shaft designed to minimise the leakage of surface air into the ventilation fan.

### Air Pit

A term often seen on maps, and used to describe a shaft used specifically for ventilation.

### Airway

A passage specifically used for carrying air, ventilating the underground workings of a mine. An Intake Airway draws air from the Downcast Shaft in the mine, whilst the Return Airway carries stale air out of the mine through the Upcast Shaft. Most airways are formed by the mineworkings themselves, and the flow of air is controlled through a system of Airlocks, which act as regulator doors ensuring that all working areas are adequately ventilated.

### Ambulance Room

A room at the pithead containing first-aid equipment to be used in the event of an accident. With welfare improvements following the formation of the Miners' Welfare Fund in the 1920s, and privatisation in 1947, most pits were eventually equipped with more comprehensive medical facilities.

### Anderson Boyes ('AB')

A highly successful Scottish company founded in the late 19th century and based at the Flemington Electrical Works near Motherwell, specialising in the production of flameproof and intrinsically-safe electrical machinery and switchgear for use underground, particularly in coal mines. 'AB' coal-cutting machinery was exported throughout the world, and the company eventually merged with its Glasgow rival, Mavor & Coulson to become Anderson Mavor. After many decades of production, the company was closed down in the late 1990s after being taken over by an American company, Long Airdox.

### Anthracite

A valuable hard, black coal containing a high proportion of fixed carbon and low levels of volatile matter, resulting in smokeless properties and a high heat value. Often associated with areas where volcanic intrusions of magma near coal measures have acted as a natural coke oven, driving off impurities. However, mining in such disturbed geological zones can be problematic, given the associated faulting, and the possibility that the coal measures may have been burned out through being too close to the magma. It was, nevertheless, often considered to be worth the risk, and many small Scottish collieries produced anthracite for much of the 20th century, such as the Gartshore pits in the Kilsyth area of Dunbartonshire, and a cluster in the Stirling area such as the Polmaise pits, and Manor Powis.

### Ash Content

The proportion of incombustible material in coal which is left behind as a residue after it is burned. Ash content varies with different types of coal. Until 2002, much of the coal burned in Scotland's biggest power station, Longannet, was 'Upper Hirst', which has a relatively high ash content. Waste ash from Longannet was deposited in the Firth of Forth near Culross, creating lagoons and connecting Preston Island to the mainland. Similarly, ash from Cockenzie Power Station in East Lothian is disposed of on the foreshore to the west of the site.

### Balance Rope

A rope suspended from beneath a cage or skip in a shaft to counterbalance the winding rope in a balanced-rope winding system, the original form of which was the Koepe Winder.

### Bank

Colliery surface (i.e. the top of the pit/shaft or mine), comprising a raised working area around the shaft top where hutches or mine cars were handled. Hutches and/or mine cars and/or tubs full of coal arriving at the surface were pushed across the bank and tipped through the screens (usually using tipplers) into waiting railway wagons at ground level.

### Banksman or Bankman

The person in charge of the shaft and cage or Skip at the surface or pit Bank of the colliery, where Hutches or Mine Cars were handled.

## Bar Screen

A mechanical device for separating different sizes of coal comprising a number of parallel inclined bars spaced at regular distances apart, the coal sliding down along the bars under gravity. Many of the smaller Scottish collieries were fitted with bar screens rather than full-scale coal-preparation plants.

## Bash Tank

A device used for coal preparation in which water is forced through a perforated plate by a plunger into the tank, causing coal and dirt to separate. Lady Victoria Colliery in Midlothian retains an example which has been restored by the Scottish Mining Museum.

## Baths

A building usually containing showers, a hot water system (calorifier), changing rooms and lockers, provided by the mine company to permit miners to wash after the end of their shift. Provision to provide pithead bathing facilities became increasingly widespread after the formation of the Miners' Welfare Fund in 1920, but those at Welleseley (Fife) and Douglas Castle (Lanarkshire) were built in 1915 and 1919, respectively.

## Battery Locomotive

An approved flameproof railway locomotive powered by electricity, suitable for use underground in areas where steam, diesel or conventional electric locomotives would risk causing a fire or explosion. In mines, they were often used for hauling tubs, hutches and mine cars.

## Baum Washer or Baum Jig

A system for cleaning coal dating from 1892. Coal was separated from dirt by immersing it in water where, with the assistance of compressed air, the lighter coal rose and heavier shale and stone sank. Many Scottish pits were equipped with Baum washers at the pithead, such as Frances Colliery (Fife) in 1923. Baum washers were manufactured by a number of companies, such as Campbell Binnie Reid, the Blantyre Engineering Co., and Coppee.

## Beam Engine

An atmospheric or steam engine in which the piston rod of the vertical cylinder is attached to one end of a large pivoted beam, the other end of which is connected to a plunger rod for operating a pump, or in a rotative beam engine, the crank of a flywheel. In the early 19th century, it was often used for both pumping and winding in mines. The best-preserved survivor in Scotland is at Prestongrange Industrial Museum (formerly Prestongrange Colliery) in East Lothian.

## Bell Pits

One of the earliest methods for working coal lying very near the surface. They consisted of shallow shafts which were belled out at the bottom as far as was thought safe at the time. After it was considered too dangerous to work, the shaft would be abandoned and another sunk nearby. The remains of bell pits can be found in many parts of the Scottish coal fields such as at Muirkirk (Ayrshire) and Wilsontown (Lanarkshire), and often have associated remains of often circular horse-engine platforms, as well as small spoil heaps.

## Belt Conveyor

A moving continuous belt on which coal, spoil and other materials are carried to and from the working face. Often, a series of conveyors would be used to carry coal from the face to the shaft, or bunkers at the foot of the shaft. In incline mines, they often carried coal from the Drift to the surface. Belt conveyors could vary in length from a few metres to several kilometres, as was the case at the Longannet complex, where coal was fed directly into the power station.

## Bevin Boys

During World War II, a severe shortage of manpower in coal mining resulted in a Ministry of Labour and National Service scheme, promoted by the then Minister of Labour, Ernest Bevin. From December 1943 until the end of the war, 48,000 Bevin Boys were directed to work in the coal mines rather than serve in the armed forces, representing ten per cent of male conscripts aged between 18 and 25. The severe coal shortage in the post-war years resulted in the scheme being retained after 1945, and men were able to avoid National Service by working in the mines until 1960. The scheme resulted in the building of a number of special hostels in Scotland, such as that at Dungavel.

## Bing

A Scottish term describing a spoil heap of waste separated from the coals following extraction from a mine. Bings form a very distinctive part of mining landscapes, but since the Aberfan disaster, and also as a result of spontaneous combustion problems as well as for aesthetic reasons, many have been cleared away in environmental improvement programmes.

## Bituminous Coal

The class of coal which covers most British coals, which rank from high-volatile low rank types with poor coking qualities to medium low-volatile prime coking coals, suitable for use in the iron and steel industry, and in gas works. Bituminous coals, which are black, are intermediate between brown coal (lignite) and anthracite (the purest form of coal). They smoke when burned, and with the introduction of clean-air legislation in the UK in the 1950s, the urban market for coal suffered badly.

## Blackband

Carbonaceous ironstone in clay beds within geological strata, mingled amongst coal measures, and containing sufficient combustible matter to allow it to be calcined (see calcining) without extra fuel. Many pits in central Scotland yielded both coal and ironstone, and these ores were immensely important in the development of Scotland's iron industry, notably in the Monklands to the east of Glasgow, where David Mushet first discovered blackband in 1801.

## Blackdamp

Blackdamp (or Chokedamp) comprises a dense mixture of nitrogen and carbon dioxide which is often found in mines, and which is extremely dangerous because of its insufficient oxygen content. Although unspectacular compared with explosive gases, it was one of the most serious hazards, particularly in early coalmines.

## Black Powder

Similar to gunpowder, a form of low explosive manufactured by milling a mixture of charcoal, sulphur and saltpetre, and frequently used in mining and civil engineering projects before the introduction of much more powerful nitro-glycerine based high-explosives, pioneered by Alfred Nobel.

## Blaes

A form of shale or mudstone found in coal measures, and therefore frequently a waste product of coal mining that was dumped in spoil heaps known as bings. Also often used as a source of stiff clay suitable for the manufacture of common bricks at colliery brickworks, saving fuel in the kilns because of the combustible matter naturally present within the shale.

## Blowing Fan

see Forcing Fan

## Brakesman

The operator of the winding-engine at a colliery.

## Braking

The action of operating a winding engine at a colliery. As well as traditional drum and disc braking on winding engines, dynamic braking (achieved by switching from AC to DC) was also used.

## Brickworks

With nationalisation, the NCB acquired in 1947 85 brickworks throughout the UK, and a number of pipeworks and specialist ceramics works producing fireclay products. Many of these were old, small, and not very efficient, and some were elements integrated within the surface arrangements of collieries. By 1951 the number had been reduced to 75 brickworks and five pipeworks producing 473 million bricks about a quarter of which were used within the NCB both at mines, and for housing projects. In 1956 the production rose to 522 million (about 12% of the national output). Most production was of common bricks, using blaes from adjacent colliery bings. In Scotland, most NCB brickworks were sold to the Scottish Brick Corporation.

## Briquette

Inflammable material such as sawdust, peat or coal dross re-constituted and compressed (with a binder, such as bitumen) into fuel briquettes for both industrial and domestic use. The most common types encountered in the UK are smokeless, and can therefore be used legally within smokeless zones defined by clean-air legislation. Examples include 'Coalite' and 'Phurnacite'. In Scotland, a number of briquette plants were built, an example being the Niddrie Nuggeting plant in Edinburgh. Earlier plants, such as that at Lugar in Ayrshire, manufactured briquettes from a mixture of pitch and coal dust.

## British Coal Corporation

The new name of the National Coal Board, brought about by the Coal Industry Act of 1987. The intention was to engineer a new start following the debilitating miners' strike of 1984–5, the corporation subsequently being known as 'British Coal' until privatisation in 1994.

## Brora Coalfield

Situated in Sutherland in the far north-east of Scotland, this is a Jurassic (as opposed to Carboniferous) coalfield, and in geological terms is therefore the youngest coalfield in Britain. It was never operated by the NCB, but a private mine produced coal until a fire closed the workings in the early 1970s.

## Bunker

A reservoir of coal, usually fed by a belt conveyor and located at the bottom of a shaft to enable the continuous loading of skips, which are then raised to the surface by a winding engine.

## Bunkering

The storage of coal, normally occurring at any transfer point in the coal clearance system where congestion may occur, and used to smooth out peaks and troughs in the coal-loading cycle. Many mines had bunkering facilities both at the pithead and underground, especially where skips were used in preference to mine cars. Monktonhall Colliery near Edinburgh was one of the last Scottish collieries to retain a large surface bunkering system feeding an associated complex of railway sidings.

## Cage

A platform with sides and a roof in which men, materials and minerals are hauled up and down a mine shaft. Some cages had two or three decks.

## Caking Coal

A coal which, when distilled, softens to a semi-liquid before burning to coke as more volatiles are driven off. It is used in the making of metallurgical coke.

## Calcining

The roasting of ores or rocks to drive off unwanted impurities, and in some cases, induce the disintegration of the material being calcined.

## Canaries

Birds taken underground in coal mines to detect the presence of a hazardous atmosphere, such as carbon monoxide. At the first sign of the gas, canaries become distressed and in some cases lose consciousness, falling from their perch. A small gas bottle of oxygen was often carried to resuscitate the bird. Although precise electronic means of detecting gas have been used for many years, canaries were still required by law to be kept at rescue stations, and were retained in bird houses at many collieries.

## Canteen

A facility usually forming part of the surface arrangement of a colliery, providing drinks, hot meals and other consumables (such as chewing tobacco) for miners. At smaller mines, canteens sometimes provided only drinks and snacks, whilst the new generation of superpits such as Rothes (Fife) and Killoch (Ayrshire) had large canteens that were fully integrated into the surface facilities. Many canteens were built rapidly after 1943 in order to ensure that miners directly received extra war-time rations designed to enhance their productivity.

## Carbide Lamp

A type of miner's lamp which uses calcium carbide as a fuel. Water is added to the calcium carbide, generating acetylene gas which, when it burns, produces a bright flame, thereby creating illumination which was better than that from oil lamps. However, they were not considered safe enough to be used in gassy mines. Carbide lamps were used in many Scottish mines well into the 20th century, but were eventually replaced by battery-powered electric cap lamps.

## Carbon Dioxide

A colourless, non-toxic gas with a sometimes pungent acid smell which, with nitrogen, forms Blackdamp and can deprive miners of sufficient oxygen. Dangerously high levels of carbon dioxide (and by implication, low levels of oxygen) are detected by the flame of a Safety Lamp being extinguished (typically in air containing less than 17% oxygen).

## Carboniferous

A geological period from 354 to 290 million years ago in which coal-bearing rocks were formed. Almost all Scotland's coalfields are carboniferous, some having upper and lower series of strata in which coal has been exploited. The only significant exception to this pattern is the Sutherland coalfield in the north-east Highlands around Brora, which, like the North Sea oil and gas fields, is Jurassic in origin.

## Carbon Monoxide

Also known as Whitedamp, a combustible, tasteless and almost odourless gas which is the product of incomplete combustion. It is usually associated with underground fires and explosions in coal mines. Extremely poisonous even when mixed with other gases such as carbon dioxide (when it is known as Afterdamp), it was traditionally detected using Canaries until the development of sophisticated electronic detection equipment. It poisons by irreversibly combining with haemoglobin in the blood.

## Central Rescue Station

A service designed to provide emergency expertise in a mining area to deal with underground emergencies. Dating originally from 1902 with the establishment of Tankersley rescue station in Yorkshire, England, the first central station, which served the Lancashire and Cheshire coalfield, was established in 1908 at Howe Bridge. After the passing of the Mines Rescue Act in 1911, it became mandatory for colliery owners to make adequate provision for mines rescue work. Several central rescue stations were established subsequently in Scotland, the only remaining example being at Crossgates in Fife, which survived privatisation and the disappearance of deep mining, to become a health and safety training organisation.

## Central Workshops

The National Coal Board established a number of purpose-built central workshops in Scotland to serve the different coalfield areas. Examples included those at Newbattle (adjacent to Lady Victoria Colliery), Cowdenbeath, Glenburn, Lugar and Alloa.

## Check Board

A board on which checks (also known as tokens) are hung when not carried by miners, and used to monitor and identify the miners working underground at any given time.

## Check Office

An office, usually close to or on the way to the pithead, from which each miner is issued with a numbered brass tag (or Token) for identification, the purpose being to ensure that every miner going to work underground has returned to the surface at the end of the shift (see also Time Office).

## Chewing Tobacco

Except in Naked Light Pits, smoking was strictly forbidden underground for fear that a naked light might ignite gas and cause an explosion. Smoking tobacco, cigarettes, lighters and matches were therefore classed as Contraband. Many miners therefore chewed tobacco, which, after 1947, was often sold in colliery canteens in packs specially supplied to the NCB.

## Chokedamp

A sometimes lethal mixture of carbon dioxide and Nitrogen (see Blackdamp)

## CISWO

See Coal Industry Social and Welfare Organisation

## Clean Air Act

The burning of coal in cities both in households and by industry generated increasing quantities of smoke, posing serious respiratory problems, and damaging the facades of buildings. This culminated in the Great London Smog from 4th to 9th December 1952 which was estimated to have killed 4,000 people. Subsequent legislation, commencing with the Clean Air Act of 5th July 1956, seriously affected the NCB's marketing of household fuels, and marked the beginning of the decline in the domestic market for coal. Alternative smokeless fuels such as briquettes and nuggets were produced by the coal industry, but cleaner forms of domestic warmth such as electricity and oil-fired central heating gained ground rapidly, and were ultimately overtaken by natural gas from the 1970s.

## Clean Coal

Coal that has waste and dirt removed by a coal preparation process.

## Coal

A combustible normally black rock formed by the decomposition of vegetation, and in Scotland, found in geological strata dating from the Carboniferous and Jurassic periods. Although exploited by man for many hundreds of years, the mining of coal in Scotland grew to prominence during the 18th century, fuelling salt, iron, glass, and chemical industries, eventually becoming the largest single employer in the Scottish economy, and providing the energy for a huge range of industries. Coal also became the principal supplier of fuel for electricity power stations, and for town gas works. Coal extraction peaked in 1913, but declined rapidly in the final decades of the 20th century, Scotland's last deep mine closing in 2002.

## Coal Cutter

A machine used to undercut coal at the coal face. Examples of different types of coal cutters include bar machines, disc coal cutters, and chain machines.

## Coal Drops

A railway facility, sometimes known as staithes, in which the bottoms of loaded wagons are opened, allowing the coal to fall into hoppers or bunkers below. These were often associated both with coal merchants in cities, and with port facilities where coal was loaded onto ships at adjacent jetties, such as at Methil in Fife.

## Coal Dross

A residue comprising small particles of coal previously regarded as waste, but latterly used to produce fuel briquettes, and increasingly valuable with the introduction of pulverised fuel systems for boilers. The power stations situated adjacent to Barony Colliery in Ayrshire and at Methil in Fife were built to consume coal slurry, which comprised dross with a high moisture content.

## Coalfield

A geologically defined area within which the underlying rocks contain coal that can be extracted either by deep mining or open-cast operations. In Scotland, the most important coalfields are of Carboniferous origin, and stretch from Mull of Kyntyre in the west, through the highly-populated central Lowlands, to Fife and the Lothians in the east. A coalfield yet to be exploited exists in southern Dumfriesshire around Canonbie, and a small Jurassic coalfield was exploited until the 1970s in the Brora area of Sutherland in the north-east Highlands. Very small quantities of poor-quality coal were also occasionally extracted from pockets of sedimentary rocks elsewhere in the Highlands, such as on the Isle of Skye.

## Coal Gas

A fuel gas produced in local gas works (and therefore also known as Town Gas) generated by the distilling (roasting) of coal in closed retorts. Important by-products of this process include coke and coal tar, from which a huge range of chemicals was produced. The coal gas comprised on average 50% hydrogen, 30% methane, and 8% carbon monoxide, the latter rendering it poisonous. Used initially to fuel street and factory lighting, it became more widely distributed through growing networks of gas mains for domestic and industrial consumers. The network was gradually expanded through the laying of new pipelines, eventually to become a gas grid which was later taken over by 'natural' gas from the North Sea and Morecambe Bay. Newer coal gas technologies include the Lurgi Process, and an experimental plant was built at Cardenden in Fife in the 1950s.

## Coal Industry Social and Welfare Organisation (CISWO)

The Coal Industry Social and Welfare Organisation, which was formed in 1952 to take over the duties of the Miners' Welfare Fund following the nationalisation of the coal industry in 1947. CISWO's headquarters in Scotland is at 50 Hopetoun Street, Bathgate, West Lothian. As the coal industry has shrunk, so the focus of CISWO has shifted from the provision of facilities towards welfare issues, and the support of the miners' welfare institutions which survive in many of the former coal-mining communities in Scotland, and the UK as a whole.

## Coal Preparation Plant

A complex where coal is separated into different sizes mechanically, often also incorporating a washery where coal is separated from dirt using water-based processes. Early types included the Baum (dating from 1892) and Lurig washers, the technology and equipment having since evolved to become more specialised and sophisticated, although still relying on differing specific-gravities to separate dirt from the coal. In Scotland, many larger collieries such as Seafield (Fife) had their own coal preparation plant,

sometimes also serving neighbouring pits. In addition, central facilities were occasionally built not immediately adjacent to a mine, such as that constructed by the Coppee Company (Great Britain) Ltd in 1941 for the Dalmellington Iron Company at Waterside in Ayrshire. In 2003, the coal preparation plant at Killoch Colliery in Ayrshire was still in operation, processing coal from local open-cast mines. Specialist manufacturers included Simon Carves, Coppee, Campbell Binnie and Reid, Dickson and Mann, Norton-Harty, and the Blantyre Engineering Company.

## Coal Reserves

Estimated quantities of coal that are calculated to be present in coal measures within a specific area, and often used to determine the potential longevity and viability of a colliery. Not always as accurate as might have been expected, as was demonstrated when some of the anticipated coal reserves beneath the showpiece Glenochil mine (Clackmannanshire) in the 1950s proved to have already been worked.

## Cobbles

A large official size of coal, pieces of coal ranging from five to ten centimetres across.

## Coke

A hard, dry fuel, predominantly made up of carbon, produced by heating bituminous coal to a very high temperature in the absence of air in a process which drives off bitumen, sulphur and other volatile impurities. The latter often comprised potentially valuable by-products such as gas and coal tar. Coke was used in a variety of processes where the impurities generated by burning untreated coal would have damaged the product. The most important consumer was the iron industry.

## Coke Ovens

Chambers or retorts, the early brick-built examples being referred to as 'Beehive Ovens', in which coal was roasted at a high temperature to drive off impurities. The remains of banks of coke ovens can be found at a number of locations in Scotland, adjacent to former collieries such as that at Woodmuir near West Calder in what was Midlothian. Coke was used predominantly by the iron industry, and large modern iron and steel works tended to build their own massive battery coke ovens, as was the case at Ravenscraig Works in Craigneuk, near Motherwell, Lanarkshire.

## Coking Coals

Coals with caking properties rendering them suitable for the manufacture of coke (see caking coals), notable concentrations of which were found in North Lanarkshire, East Dunbartonshire and Stirlingshire.

## Colliery

A place where coal is extracted by means either of vertical shafts sunk into the rocks below, or by inclined Surface Mines driven at an angle into the ground, or a combination of both shafts and surface mines. Collieries therefore tend to comprise not only holes in the ground, but an assemblage of buildings and structures designed to facilitate the extraction and subsequent processing of the coal, and the disposal of any associated waste, often onto an adjacent bing. In Scotland, some collieries were relatively modest operations, comprising little more than a shallow drift mine and a few sheet-metal sheds. In contrast, others, especially the showpiece superpits designed in the 1950s immediately after nationalisation, were large integrated complexes intended to extract coal for many decades. As well as having integrated welfare and office facilities, they were usually connected to mainline railways, and were equipped with coal-preparation plants.

## Compressed Air

Air that is under pressure greater than that of the atmosphere which, when generated by an air-compressor and subsequently stored under pressure in air receivers, can be used to power a range of equipment. In coal mining, compressed air was used to power a range of air-powered or pneumatic equipment such as drills for boring holes for explosive charges. Where mines were especially gassy and there was a risk of explosion because of the presence of methane from the coal measure, there was often a preference for pneumatic power tools. Compressed air was also used on the surface for processes including coal washing.

## Compressor

See Air Compressor

## Contraband

Materials that are forbidden underground in mine workings, mostly referring to items such as cigarettes and matches which would be dangerous underground. Those found with contraband were normally sacked. Searches for contraband were carried out on a regular basis, both at the surface, and underground.

## Conveyor

Most often a continuous moving belt or chain that transports large volumes of material, and which were extensively used in Scotland's coal mines. One of the longest conveyors in the world was built in the Longannet complex in Fife and Clackmannanshire, serving the power station at Longannet. Other types of conveyor include scraper and shaking varieties.

## Cornish Engine

Originating in Cornwall where it was widely used in the tin-mining industry, a steam-powered Beam Engine developed from Newcomen's atmospheric engine by Richard Trevithick, and later adopted in mines throughout the UK and overseas in countries such as Mexico. In Scotland, the finest survivor is at Prestongrange Colliery in East Lothian, now the Prestongrange Industrial Museum.

## Creeper

A slowly-moving endless chain with projecting bars designed to catch hutches, tubs or mine cars and move them up an inclined stretch of track, usually around a narrow-gauge network adjacent to the shaft at the pithead of a coal mine. Hutches normally left the cage at the top of the shaft, travelling down a gentle gradient to the screens or conveyors where they were emptied by being turned upside-down in a tippler, subsequently completing a circuit by being hauled up an incline by a creeper, returning them to the cage decking level, after which they were returned to the pit bottom for re-loading. A good example of a hutch circuit incorporating a creeper survives at Lady Victoria Colliery in Midlothian (now the Scottish Mining Museum).

## Davy Safety Lamp

A pioneering miner's lamp invented by Sir Humphry Davy in 1815. The lamp incorporated a cylinder of fine wire gauze which surrounded the naked flame, protecting it from direct contact with explosive gases in the mine. By adjusting the flame of the lamp, it was possible to detect the presence of dangerous gases, and its introduction in coal mines subsequently saved many lives. Variations and improvements on the Davy Lamp were adopted throughout the world.

## Deck

The area around the shaft collar where men and materials enter the cage to be lowered underground. Deck is also a term used to describe the floor or floors of the cage itself. Cages were originally single-deck, but double- and treble-deck cages became common during the 20th century.

## Deep Mine

A term used to describe incline or surface mines and shafts created to access coal buried deep underground, as opposed to open-cast mines which extract coal by clearing away overburden and over-lying strata to gain direct access to the coal.

## Dense Medium Plant

A type of coal preparation plant, often built at larger collieries, and sometimes at port facilities where coal is imported. It is designed to wash and grade coal, separating it from waste. The process involves adding an inert substance such as finely ground magnetite to water to make it more dense, allowing the coal to be floated off and separated from other material. Dense medium plants proved to be more efficient than Baum washers, and were introduced to a number of larger collieries, including Lady Victoria in Midlothian, and most recently, Killoch in Ayrshire.

## Deputy

An official appointed by a colliery manager with responsibilities for underground operations and safety within his prescribed district of the mine.

## Detonator

A device which, in combination with a fuse or electricity, is designed to provide sufficient impulse to detonate high explosives. Detonators were widely used in coal mines following the introduction of specialised mining explosives, and were manufactured in Scotland at Westquarter near Falkirk in Stirlingshire by the Nobels Explosives Company.

## Dip

The angle at which a structure or rock bed, such as a coal seam, is inclined from the horizontal. Coal seams are almost always inclined. Working downhill is said to be the Dip, and working up hill to be the Rise. Near-vertical seams are known as Edge Coals, as was the case at Gilmerton Colliery on the south side of Edinburgh.

## Dook

An inclined underground roadway, usually driven downhill and following inclined strata.

## Downcast Shaft

The shaft, or division of a shaft, down which fresh air descends into a mine as part of a the mine ventilation system.

## Dragline

A large excavation machine used in open-cast mining to remove overburden (layers of rock and soil) covering a coal seam. Most commonly manufactured by Bucyrus, a company founded in the early 1880s and based in Wisconsin, USA, but most famous for building excavators used in the building of the North American railroads, and the Panama Canal.

## Drift

A horizontal or near-horizontal underground drivage from the surface that follows along the length of a seam, vein or rock formation. It serves the same purpose as a shaft, but is usually considerably cheaper (see also Fan Drift).

## Drift Mine

See Drift and also Surface Mine

## Dross

See Coal Dross

## Dyke

Normally a long and thin vertical body of igneous rock that, when in its molten state and under high pressure, exploited fissures in older rocks. It often constitutes a natural interruption of a coal seam, causing it either to be thrown down or up, creating difficult mining conditions. Although dykes sometimes caused the burning out of coal measures when too close, they were also a factor in the purification of coal, driving off impurities and producing pockets of anthracite in parts of the Scottish coalfields, especially the Kilsyth area of Dunbartonshire, and to the east of Stirling at Polmaise and Manor Powis (see also Sill).

## Edge Coal

Highly inclined seams (with an inclination of more than 1 in 1), often set at an almost vertical angle, as was the case at Gilmerton Colliery, on the south side of Edinburgh.

## Emergency Winder

Portable equipment which was used to raise a cage in the event of the failure of the main winding equipment. Emergency facilities were required at every colliery to operate in the event of a power failure.

## Endless Rope Haulage

Haulage system comprising a continuously moving rope running between two pulleys, one at each end of the system, tubs or hutches being attached and detached as required. Used in Scotland both on the surface and underground, a good example was constructed at Kingshill Colliery in Lanarkshire in the 1950s, connecting the coal preparation plant with the new No. 3 shaft on the other side of the moor. 'Under-rope' haulage involved the use of 'Smallman clips' with a rope passing under the hutches, whilst 'over-rope' haulage used 'bulldog clips' or 'lashing chains', with the rope on top of the hutches.

## Explosives

Chemicals widely used in the mining industry to reduce the burden of physical work by blasting roadways and advancing coal faces. Initially low explosives such as blackpowder were employed, but were not always effective and were dangerous to use in gassy mines. These were subsequently replaced by a new generation of high explosives, most based on nitro-glycerine following the invention of Dynamite by Alfred Nobel in 1866. Latterly, the use of explosives became heavily regulated, especially in mines where Permitted explosives such as Penobel became standard. Huge quantities of these explosives were manufactured by Nobel's Explosives at their Ardeer factory near Irvine, and many of Nobel's new products were tested in Scottish mines.

## Face

The exposed area of a coal seam from which coal is extracted in a mine. Larger collieries had several faces operating simultaneously.

## Fall

A body of rock or coal which has fallen from the roof in a mine. Geological conditions in some mines were more prone to falls than others, and falls sometimes proved fatal, killing or injuring miners, in some cases also cutting off escape routes and air supply.

## Fan

A mechanical device used to force movement of air through the workings of a colliery, thereby achieving ventilation by ensuring a supply of fresh air, and reducing the accumulation of dangerous gases. Collieries used several types of fan, but the main fan was usually part of the surface arrangement and situated at the end of a short tunnel or Fan Drift connected to the Upcast shaft, and was designed to ventilate the mine continuously. In modern mines, a standby fan was always available in case there was a breakdown, or if maintenance was required. Older mines with extended workings often required booster fans to enhance the flow of air, and portable auxiliary fans were sometime used where necessary.

## Fan, Auxiliary

A small, portable fan used to supplement the ventilation of a localised working place within a mine, particularly narrow coal drivages which are not ventilated by the normal air current.

## Fan, Booster

A large supplementary fan installed in the main air current, working in tandem with the main fan, and normally required where extended workings had rendered the existing ventilation inadequate.

## Fan Drift

An air-tight tunnel or incline mine connecting a fan to a shaft.

## Fan House

Normally a building within the surface arrangement of a colliery containing the main fan responsible for ensuring the continuous ventilation of the mine-workings below. Not all collieries had a fan house of their own, some relying on the fans of a neighbouring colliery.

## Fault

A fracture in geological strata such as coal measures, and caused by earth movement, often leading to substantial displacement, and major disruption to mining activities. On occasion, faults can also expose strata, making pockets of minerals more easily accessible from the surface.

## Feldspar Washer

A type of coal preparation plant which operates on a similar basis to the Baum Washer, but which has beds of feldspar designed to collect smaller particles (usually less than one-quarter of an inch diameter).

## Fines

Tiny coal or shale particles usually less than 1.5mm in size, rarely more than 3mm. Often a by-product of coal preparation. The power stations at Barony in Ayrshire and Methil in Fife were designed to burn fines slurry from the local collieries.

## Fireclay

A form of clay with refractory qualities caused by naturally high levels of silica or alumina. Often were suitable for the manufacture of a wide range of heavy ceramics goods such as firebricks and other refractories, fireclays were found in coal measures, and were often extracted at coal mines in Scotland. Many mines had adjacent brickworks, and fireclays were frequently added to mixes of clay to produce bricks of differing quality and appearance. Some collieries, such as Prestongrange in East Lothian, had ceramics works of their own specialising in fireclay products, especially salt-glazed sewer pipes.

## Firedamp

An explosive mixture of gas formed mostly from methane and air, and caused by the decomposition of coal or other carbonaceous matter. One of the most serious hazards in coal mines, it was detected using safety lamps, and latterly by modern electronic equipment.

**Flame Safety Lamp**

A contemporary version of the original Davy Lamp, and used by colliery officials to check for the presence of gas underground. Also sometimes referred to as a 'Gauze Lamp', or in Scotland, a Glennie (a corruption of the name 'Clanny', a lamp designer).

**Flue Gas De-sulphurisation**

A process which removes sulphur compounds formed during coal combustion from flue gases prior to their emission into the atmosphere. Many coal-burning power stations in the UK were required to install flue-gas de-sulphurisation plants because of the high sulphur content of the coal that they burned, but in Scotland, the 'Upper Hirst' coal mined in the Longannet complex was so low in sulphur that no such facility was required.

**Fly Ash**

Waste produced by the combustion of coal which can sometimes be used as a lightweight construction fill material, or as a cementitious binder in concrete. Until the closure of the Longannet colliery complex, Scotland's principal coal-burning power stations burned 'Upper Hirst' coal, the main advantages of which was that it was relatively cheap to mine, and had very low sulphur levels. However, its high ash content led to the generation of large quantities of fly ash, much of which has been used in the Firth of Forth in a land reclamation process comprising a complex of lagoons. Fly ash is also dumped on the foreshore adjacent to Cockenzie Power Station in East Lothian.

**Forcing Fan**

A fan which ventilates a Mine or 'heading' by blowing the air, as opposed to extracting it by suction, an example being Kames Colliery in Ayrshire. In England and Wales, this is often referred to as a Blowing Fan. This was relatively unusual, most pits using extractor fans as the main form of ventilation.

**Gas (in mines)**

In the context of coal mines, several gases were hazardous, sometimes leading to injury and death. Methane (see Firedamp) was a common cause of explosions and fires, and Carbon Monoxide (a component of Afterdamp) was poisonous and often fatal. Mines with a tendency to suffer from high levels of Methane were often referred to as 'gassy pits'. Carbon Dioxide (a component of Afterdamp and Blackdamp) was also a hazard when present in so high a proportion that it deprived miners of Oxygen.

**Gas (fuel)**

See Coal Gas

**Glennie**

See Flame Safety Lamp

**Graded Coal**

Coal was graded according to size, the largest being Cobbles (large cobbles were 200–100mm, and ordinary cobbles 100–50mm), and 'Trebles' (75–50mm). 'Doubles' were an intermediate size (50–25mm), and 'Singles' half the size again (25–12.5mm). A further category was 'Smalls', which could vary from a maximum 50mm to a more normal 12.5mm. The coal had often been processed by a coal-preparation plant, and was therefore referred to as 'Washed' (e.g. 'Washed Singles').

**Haulage**

Methods of transporting men, materials and coal, both on the surface and underground. Early sources of motive power in mines included women, children, horses and ponies, the latter being gradually replaced in the 20th century by battery-electric and diesel locomotives. Fixed haulage using ropes, cables and chains was sometimes powered by horse engines in early mines, later to be replaced by engines driven by steam, compressed air, hydraulics and electricity. Fixed-winch systems included direct haulage, where the free end of the rope is attached to the hutches which are pulled towards the winch. Endless rope haulage was also widely employed in Scotland, and comprised a continuously moving rope running between two pulleys, one at each end of the haulage system, tubs being attached and detached as required. The introduction of belt conveyors for transporting coal and materials as well as for manriding also led to great improvements in productivity and working conditions.

**Headframe**

The structure on top of a shaft which supports the winding rope pulleys or sheaves, and therefore the cages or skips in the shaft below. Early headframes were of relatively small timber or cast-iron construction, later evolving into substantial steel or reinforced concrete structures in the 20th century. Headframes have acquired an iconic power, becoming a symbol of mining throughout the world, although many collieries have incline surface mines which do not require headframes.

**Headgear**

The Headframe located above a mine shaft (see above), but sometimes also used to describe the entire structure around a shaft, including the winding pulleys or sheaves.

## Headstocks

See Headgear

## Heapstead

The collective term used to describe the entire Surface Arrangement around a colliery shaft, including the Headgear, loading and screening plant, tub circuits, winding and surface pumping engines, fan and compressor houses etc. and their respective buildings.

## Heating

Heat generated in coal seams due to the oxidation of coal, and often an early sign of spontaneous combustion.

## Horse Engine or Horse Gin

A winding mechanism powered by a horse or horses, often associated with Bell Pits, where they were used to lift coal out of the ground. The engine was usually situated on a flat area adjacent to the shaft, and the remains of such platforms can often be found next to Bell Pits in Scotland.

## House Coal

A type of coal deemed to be suitable for domestic use. Many Scottish collieries produced house coal, and their market was therefore hit hard by the introduction of Clean Air legislation in the 1950s.

## Hutch

A mine car used to carry coal and other materials within mine workings, cages (which carried them to the surface), and around rail circuits both at the pithead and the pit bottom. Often referred to as tubs elsewhere in the UK.

## Hydraulic Prop

A steel support used in underground workings, and comprising two telescoping steel cylinders extended by hydraulic pressure, and adjusted to variable height, usually by a hand-operated pump built into the prop.

## Incline

An inclined roadway either at the surface or underground (or both), which can be used for transporting men and materials, sometimes by a self-acting mechanism.

## Incline Mine

See Surface Mine

## Ingaun e'e

An early term for a drift mine.

## Inrush

An uncontrolled flood of water and saturated debris into mine workings. One of the most famous in Scotland occurred at Knockshinnoch Castle Colliery in Ayrshire where, in 1950, workings encroached too close to the surface, breaching the strata and allowing a large area of bog on the moor above to drain into the mine. The ensuing inrush trapped 129 miners, of whom 116 were eventually rescued in heroic circumstances through neighbouring disused workings.

## Intake Airway

See Airway

## Ironstone

A rock containing sufficiently large quantities of iron compounds to merit commercial exploitation as iron ore. In Scotland, ironstones were sometimes found within the Carboniferous coal measures, and Blackband ironstone was especially important in the development of the iron industry, particularly in the Monklands area to the east of Glasgow.

## Jig

A form of 'shaker conveyor' which was often used to sort coal. When submerged under water, it was referred to as a Jig-washer, and was used at the pithead to sort and prepare coal prior to sale.

## Jig-washer

See Jig

## Joiners' Shop

A building in which a wide range of joinery work was completed for use both on the surface and underground. Considerable quantities of wood were used, despite the increasing use of steel underground, and the Joiners' Shop was therefore an important constituent of the surface arrangement of larger collieries.

## Jurassic

The geological period from 206 to 144 million years ago, associated with dinosaurs, and vegetation the decay of which produced the oil and gas fields of the North Sea. The Brora coalfield in Sutherland also originates from the Jurassic period, but Scotland's other productive coalfields are from the older Carboniferous period.

## Kilns

Many Scottish collieries had brickworks adjacent to or combined with their surface arrangement, taking advantage of the ready supply of fuel, and the Blaes (shales and clays) of the associated 'Bing'. The kilns took the form of small intermittent round or rectangular structures, but also large continuous kilns which were responsible for providing the bricks from which many miners' houses were built during the 20th century.

## Koepe Sheave

The wheel used in a Koepe Winder in place of a conventional winding drum, usually fitted with replaceable friction linings.

## Koepe Winder

A friction winding system designed for deep shafts, named after its German inventor, Friederich Koepe (1835–1922), who introduced it to Hannover Colliery in Westphalia in 1877. The system involves the replacement of the conventional winding drum (onto which the rope was wound and fixed) with a large wheel or sheave over which the rope was hung, the sheave also being lined with grooved friction material permitting it to act as a drive wheel. The system is kept in balance by the use of a balance rope, and is consequently more efficient, using less power and having less inertia than is the case with a conventional winding drum. The Koepe sheave could be mounted in a headgear, or alongside the winding engine in a tower. Modern Multi-rope Friction Winders are based on the Koepe system.

## Lamp Cabin

See Lamp Room

## Lamp Room

The surface building at a mine where the electric and Flame Safety Lamps were stored, charged and maintained. Also known as the Lamp Cabin, this was the place from which miner's collected their lamps at the start of a shift, and to which they returned the lamps at the end of the shift.

## Landsale

Coal sold direct from a colliery and transported by road (as distinct from sea-sale, canal-sale or rail sale). At the time of nationalisation in 1947, many of the smaller Scottish collieries were dependent on landsale.

## Level

A nearly horizontal passageway or tunnel in a mine, often acting as a drainage adit.

## Licensed Mines

The Coal Act of 1938 vested ownership of royalties for unworked coal reserves with the newly-formed Coal Commission, from which licences for the mining of coal had subsequently to be acquired. After nationalisation in 1947, the National Coal Board took up this responsibility, as well as taking direct control and ownership of all large coal mines in the UK. Smaller mines, however, were licensed to individuals and companies if the number of men employed underground was at no time likely to exceed, or greatly exceed 30. In early 1947, over 430 out of the 481 existing small mines in the UK had sought and been granted licences, and the number grew in subsequent years. In Scotland, there were over 100 licensed mines in 1945, and many new operations were commenced well into the 1960s. However, numbers then steadily declined, and by 2003, all had ceased to operate. It is also important to note that mining activity did not always occur where a licence was granted. The licence merely grants permission to mine, and it is up to the licensee to commence mining if circumstances permit.

## Locomotive Haulage

A term used in the context of the underground transport of coal, which was greatly enhanced by the introduction of locomotives. As coal mines grew larger and the distances between the coalfaces and the shafts grew longer, so the issue of haulage became more important. The development of flameproof diesel and electric engines enabled the safe introduction of locomotives underground, and was a common feature of reconstructed pits and new sinkings in the NCB era. Locomotives tended to work along the main roadways of mine complexes, bringing coal to the pit bottom where the mine cars could either be emptied into bunkers for skip winding, or loaded directly into the cage in hutches and taken to the pithead.

## Longwall Mining

A method of extracting coal by cutting progressively along a long face or wall between two perpendicular parallel side roads, the space from which the coal is extracted either being filled with waste, or being allowed to collapse. The technique is said to have originated in Shropshire in the late 17th century, and in modern mines, the longwall face can be several hundred metres long. Where geological conditions permitted, many Scottish pits converted to longwall mining following reconstruction and modernisation after nationalisation in 1947.

## Lurgi Process

A high-pressure gasification process which converts coal into a fuel gas. British Coal operated an experimental Lurgi gasification plant next to its Westfield opencast mine near Cardenden in Fife.

## Magazine

A building within the surface arrangement of a colliery, or an underground store where explosives are stored prior to use in the mine.

## Manager

The person vested with overall responsibility for the operation of a mine. Most managers were appointed after many years of mining experience, and as the industry evolved, were expected to have acquired professional mining qualifications. Mine Managers were a distinct professional group within the coal industry.

## Manriding

A term used to describe several ways in which men are transported to and from the coal face. This includes being wound up and down a shaft in a cage, riding in mine cars, or in a purpose-built train, or riding on a conveyor belt. The illegal riding of conveyor belts and mine cars was a common cause of injury in mines.

## Man Winding

The process of transporting miners and other personnel up and down the shaft of a colliery. Many collieries had a shaft dedicated to winding men and materials.

## Mavor & Coulson

One of Scotland's most important coal-mining machinery manufacturers, based in Bridgeton in Glasgow, and pioneering the use of electrical equipment underground. It was eventually merged with its principal rival, Anderson Boyes of Motherwell, to become Anderson Mavor.

## Medical Centre

An important part of the surface arrangement of modern coal mines in the NCB era. In 1947, many older Scottish collieries had only first-aid rooms and morphia administration schemes, so the provision of adequate medical facilities was a major priority in the nationalised era.

## Methane

A potentially explosive gas formed naturally from the decay of vegetative matter, occurring in varying quantities within coal seams. Methane is the explosive component of Firedamp, and is kept within safe limits through the use of mine ventilation systems. In some pits where there were high levels of methane, it was possible to extract it for use on the surface. This was the case at Cardowan Colliery near Glasgow, where it was used to fuel the boilers of the colliery itself, and of the adjacent whisky bottling plant.

## Mice

A common pest in coal mines, especially where pit ponies were used, but unlike rats, not considered to be a serious problem.

## Mine

A general term that is used in a variety of ways. Frequently used in the context of a Colliery or coal Pit, but it is often used in Scotland to describe an incline or drift roadway driven into the strata from the surface, as compared with a vertical shaft. This is sometimes referred to as a Surface Mine.

## Mine Car

A modern version of the traditional coal hutch, but with a greater capacity, and designed to be suitable for locomotive haulage. Mine cars carried both coal and materials, but were also adapted for Manriding.

## Miners' Welfare Fund (MWF)

An organisation founded by the Mining Industry Act of 1920 with the purpose of enhancing the well-being, recreation and living conditions of workers in or around coal mines. It raised funds by obliging all coal mine owners to pay the equivalent of one penny per ton of coal output. A second levy was added by an amendment to the Act in 1926, specifically with the aim of funding the construction of pithead baths. Other pithead facilities to be supported by the Fund included first-aid and ambulance stations and associated medical training schemes, and less glamorous projects such as bicycle sheds. Away from the mines, the MWF helped to fund a wide range of community facilities and activities, often relating to music, drama, and literature, and also awarded scholarships to the members of mining families. In addition, often in partnership with local authorities, it helped to establish technical colleges, and maintained Miners' Institutes, local libraries and reading rooms, sports and other recreational facilities. It also maintained a number of convalescent homes for miners suffering from physical injuries and diseases associated with mining. After the nationalisation of coal in 1947, the National Coal Board briefly took over the MWF's responsibilities, eventually transferring them in 1952 to CISWO, the Coal Industry Social and Welfare Organisation

## Morphia Administration Scheme

A facility at some collieries through which morphia could be administered to employees who had been injured in the mine. At the time of nationalisation in 1947, many smaller Scottish collieries did not have proper medical facilities, and such schemes were designed to alleviate the discomfort of injured miners whilst medical assistance was sought either from a neighbouring town or village, or from an adjacent colliery.

## Multi-rope Friction Winder

A balanced-rope winding system based on the principles of the Koepe Winder which utilises friction created by the weight of the ropes, skips and/or cages to assist the grip of ropes on the driving sheaves of the winder. The use of several ropes over parallel sheaves increased capacity and safety, and was especially suited for use in deep shafts, and where high output of coal was anticipated. The new generation of Scottish superpits was therefore equipped with these systems in the 1950s and 1960s, either with ground-mounted winding engines as at Bilston Glen, or with tower mounted winders as at Rothes, Seafield, Killoch and Monktonhall Collieries.

## NACODS

The National Association of Colliery Overmen, Deputies and Shotfirers, a Trade Union representing specialist and professional groups within the British coal-mining industry.

## Naked Light Pit

A pit where conditions were considered to be sufficiently safe to permit naked lights underground.

## National Coal Board (NCB)

The organisation established by The Coal Industry Nationalisation Act (1946), and which took control of the British coal-mining industry on behalf of the State following nationalisation on 1st January 1947. In addition to the mines themselves, the NCB took control of a wide range of businesses associated with the former coal companies, and the estate that it acquired included a huge stock of miners' housing. The NCB maintained direct control of the larger coal mines, but also licensed private individuals and companies to operate small mines. In the aftermath of the miners' strike in the early 1980s, the NCB was renamed the British Coal Corporation in 1987, and was subsequently privatised in 1994.

## Nationalisation Act

The act of parliament, known as The Coal Industry Nationalisation Act (1946) which created the National Coal Board on 1st January 1947.

## National Union of Mineworkers (NUM)

A British trade union representing mineworkers, both underground and at the surface, which was created at the same time as the National Coal Board on the nationalisation of the coal industry on 1st January 1947. It was formed through the merger of the Miners' Federation of Great Britain (MFGB), and the South Wales Miners' Federation.

## Navigation Coal

Coal especially suited for use in high-performance steamships. Mines producing Navigation Coal, such as Valleyfield in Fife, were placed on the Royal Navy's Admiralty List.

## New Sinkings

A term used to describe the new coal mines sunk by the NCB following nationalisation. The flagship new sinkings were sometimes referred to as Superpits, and their surface structures were mostly designed by the NCB's Scottish Area architect, Egon Riss. Although the first to be completed, Rothes (Fife) and Glenochil (Clackmannanshire) were spectacular failures, others such as Killoch (Ayrshire), Seafield (Fife), and Monktonhall and Bilston Glen (Midlothian) were successful.

## Nugget

A type of smokeless coal marketed by the National Coal Board in the aftermath of clean-air legislation which prohibited the burning of ordinary coal in many urban areas. Nuggets, which were manufactured in Niddrie on the south side of Edinburgh, were promoted as being a clean fuel in the hope that they would slow the advance of oil-fired central heating in the 1960s.

## Oil Shale

A sedimentary brown or black rock which yields oil when roasted or 'distilled' in retorts. In Scotland, major deposits of oil shale were found in Midlothian and West Lothian to the west of Edinburgh, and supported an oil industry pioneered by James 'Paraffin' Young in the second half of the 19th century. The industry, which sat alongside adjacent coal measures, has left a lasting legacy of pink shale-oil bings in the Broxburn and West Calder areas, Young's businesses also providing the foundation for the petrochemical multinational company, BP.

## Onsetter

The person responsible for loading hutches/mine cars into cages at the pit bottom.

## Open-cast Mining

A system of mineral exploitation which involves extraction directly from the surface by means of quarrying. With advancing technology and ever bigger excavators, it has been possible to dig bigger holes, in many cases extracting coal from areas that had previously been mined conventionally. For many years, one of the largest open-cast sites in the UK was at Westfield in Fife, near Cardenden. Since 2002, open-cast mining has been the only form of coal extraction in Scotland.

### Outcrop

A place where rock or a specific mineral deposit is naturally exposed, and can therefore be seen at the surface. In some situations, this provides an opportunity for successful mining activity.

### Overburden

Layers of soil and rock overlying coal seams or minerals which must be removed to permit open-cast mining to commence. In most open-cast coal mines in Scotland, the overburden is carefully replaced after the coal has been removed.

### Penobel

A form of high explosive particularly suitable for use in coal mines, manufactured by the Nobel's Explosives Company, whose largest factory was at Ardeer in Ayrshire (see also Permitted Explosives).

### Permitted Explosives

A type of industrial explosive which has been specially developed and tested for use in coal mines where Firedamp and coal dust occur. The explosives contain salts designed to suppress flame and noxious gas, thereby reducing the risk of a secondary uncontrolled explosion caused by gas and coal dust. An example of a permitted explosive was Penobel which was produced by Nobel's Explosives in their Ardeer factory in Ayrshire. Many of Nobel's permitted explosives were tested in local Ayrshire mines. The first list of permitted explosives in Britain was published in 1899.

### Permitted Lights

Underground illumination that was permitted in a colliery under the mining legislation, depending on the nature of the mine in question. Some pits were perceived to be safe enough to permit the use of naked-flame lamps, but by the latter half of the 20th century, safety lamps and flameproof approved equipment became mandatory. Some collieries, such as Kames in Ayrshire, were thought to be gas free, but explosions proved this to be untrue.

### Picker

A person employed to pick out foreign matter from coal at a Picking Table or Picking Belt. In general, underground mining was considered to be a predominantly male activity, but coal picking was

thought to be suitable for women, and for men who had been injured underground and for whom arduous activity was no longer possible.

### Picking Belt

A moving conveyor belt of steel or rubber on which raw coal from the mine is spread out, from which large stones or other debris were manually picked out from the coal by coal Pickers. Picking belts were often an important element within a colliery's surface arrangement.

### Picking Table

A flat or slightly sloping table (either revolving or in bench form) onto which raw coal from the mine was tipped to be sorted by the Pickers.

### Pillar and Stall

A mining technique, also referred to as stoop and room in Scotland, where the coal is excavated leaving behind pillars of coal to support the roof, reducing the risk both of collapse during mining, and of subsequent subsidence at the surface. This contrasts with Longwall mining where all the coal is removed. The residual pillars are often sufficient to justify open-cast mining in formerly worked areas where the coal is relatively close to the surface, and in Scotland several good examples have been exposed, as was recently the case at Benhar in east Lanarkshire.

### Pit

A collective term widely used to describe a coal mine, but often associated with the shaft of a colliery.

### Pit Bottom

The bottom of the shaft of a colliery, sometimes used also to describe all the adjacent roadways and equipment.

### Pit Head

The surface of a colliery surrounding a shaft or mine, incorporating buildings, decking and associated equipment.

### Pithead Baths

A building containing showers supplied by hot water, usually at or close to the pithead, within which miners could wash after a shift working in the colliery. Pithead baths were rare in Scotland in the early 20th century, the first examples being introduced at the Wellesley in Fife (by the Wemyss Coal Company 1915), and at Douglas Castle in Lanarkshire (by Wilsons and Clyde Coal Company in 1919). The establishment of the Miners' Welfare Fund in 1920 provided means by which the provision of baths could be funded at more collieries, and many were built in the 1930s to the design of J A Webster. However, even in the late 1940s, after nationalisation, many collieries were not equipped with baths, and one of the priorities of the new National Coal Board was to improve such facilities. At a time when most miner's dwellings were not equipped with plumbed bathrooms, the provision of baths at the pithead constituted a huge improvement in living conditions in most coal communities.

## Pit Ponies

Small horses or ponies (such as Shetland ponies) used for haulage underground within the workings of a mine. During the 20th century, pit ponies were gradually replaced by mechanical haulage systems, and with the onset of reconstruction programmes, by locomotives. Where pit ponies were retained, spilled feed sometimes caused vermin infestation problems, as was the case at Glencraig, Nellie and Mary Collieries in Fife, where rats and mice became a serious problem.

## Pit Prop

Traditionally a timber post used as roof support. Modern manifestations include telescopic steel props which may be mechanically or hydraulically lengthened to suit the space being mined. The stores of most active collieries tend to contain large stocks of pit props.

## Plough or Plow

A coal cutting machine first developed in Germany in the 1940s to work Longwall faces. It comprises a cutter fixed with blades which is pulled along the Longwall face by a chain, the extracted coal being loaded onto and taken away by a conveyor. It was not particularly suited to Scotland's coal mines, where geological disturbance had often created disruption to coal seams. However, it was used at Bedlay in north Lanarkshire to work thin seams of coking coal.

## Pneumoconiosis

A chronic and debilitating disease of the lung arising from the regular breathing in of coal dust, and one of the most serious hazards associated with coal mining.

## Ponies

See Pit Ponies

## Portal

The surface entrance to a drift, surface or incline mine, tunnel or adit, or the masonry, concrete or timber structure surrounding and reinforcing the immediate entrance to a mine.

## Power Loader

Power-operated machines used for loading coal or other materials into mine cars, or onto conveyors, into bins or bunkers, or into road vehicles.

## Preparation of Coal

The process of cleaning and grading run-of-the-mine coal, rendering it suitable for a specific use.

## Prop

See Pit Prop

## Pulleys

The wheels or Sheaves on top of the Headframe above the shaft over which the winding rope passed from the winding engine to the top of the cages.

## Pulverised Coal

Finely ground coal reduced to sufficiently small particles to allow it to be blown through a nozzle by compressed air, permitting immediate and efficient combustion. Many modern coal-fired boilers are equipped with pulverised fuel systems, as opposed to automatic chain-grate stokers.

## Pumphouse

A place where pumps are situated, usually at the pit bottom, their purpose being to prevent the flooding of the mine workings with accumulations of water.

## Rats

A frequent vermin problem in mines, especially where pit ponies remained in operation, and where spilt animal feed was unavoidable. Rats posed health problems such as Weil's Disease, and extermination programmes in mines were therefore common, especially in the NCB era.

## Reclamation

The process of restoring an Opencast mine after coal extraction has been completed to something approaching the original landscape. Inevitably, physical evidence of earlier coal-mining activity is erased by Opencast Mining, and is not reinstated by the reclamation process.

## Reconstruction

Following nationalisation in 1947, many unproductive coal mines were closed, but others were earmarked for modernisation in sometimes substantial reconstruction programmes. These frequently involved the major reorganisation of surface facilities, as was the case at Barony Colliery in Ayrshire and Kinneil Colliery in West Lothian. At the same time, there was often major investment underground, including electrification, the mechanisation of haulage, and the introduction of mechanised Longwall mining in place of Stoop and Room working.

### Redd

A term used to describe residual waste and stone generated by the coal cleaning process. Redd was normally dumped on adjacent colliery spoil heaps known as bings, but in some instances was subsequently used as a constituent of stiff clay in brickworks.

### Reserves

Coal that can be economically mined, and that has been identified within defined boundaries relating to a colliery. The perceived extent of reserves determines the longevity of a colliery, and where reserves are thought to be limited, closure normally ensues. In some cases, such as Glenochil in Clackmannanshire, reserves were seriously overestimated, and substantial investment by the NCB proved to be wasted.

### Respirators

Devices issued to miners to assist breathing when confronted by dust, smoke and toxic gases underground. Self Rescuers were respirators issued as standard to all miners following the disaster at Michael Colliery in Fife in 1967.

### Respiratory Diseases

Lung diseases, often caused by the regular inhalation of dust in the workplace. The most common examples within the coal industry were Pneumoconiosis and Silicosis, which were caused by the inhalation of coal and stone dust in the mineworkings when driving roadways, at the pithead, and in coal preparation and briquetting plants.

### Retreating Longwall System

A coal-mining system involving first driving two roads to the boundary of the area of coal to be extracted, and then mining in a single face without pillars back towards the shaft.

### Return Airway

See Airway

### Road or Roadway

A tunnel driven to the coalface for access, transport and carrying services and ventilation air.

### Roadheader

A powerful self-propelled electrical machine with a moveable cutting head, capable of driving a tunnel continuously through coal or rock and loading out spoil for removal. Roadheaders were manufactured by Anderson Boyes of Motherwell (latterly Long Airdox) up until the closure of the factory in the late 1990s.

### Rope

In modern coal mining, a steel cable attached to the cage in a shaft, or steel cable or chain used in haulage both at the surface and underground.

### Rope Guides

Steel rope suspended in a vertical shaft preventing the swinging of cages or skips. Modern systems often involved eight rope guides, four for each cage or skip in a balanced-rope arrangement, and sometimes an additional pair of rubbing ropes to prevent a potential collision between skips or cages.

### Rope Haulage

A means of propelling loaded and empty mine cars using wire ropes along roadways and up and down inclines, usually where the slope was too steep for locomotives.

### Run-of-the-mine Coal

Un-prepared coal taken straight from the mine.

### Safety Fuse

A fuse comprising blackpowder encased in waterproofing compounds, used to fire detonators and in turn explosive charges within a mine. The fuse burns at a known rate, thereby permitting the safe use of explosives. Widely used before the introduction of electric detonators, safety fuse was made at a number of factories in the UK, including Ardeer in Ayrshire.

### Safety Lamp

An approved lamp designed for use in gassy coal mines, used latterly to test for the presence of inflammable gas, and the absence of oxygen. See also Davy Lamp.

### Salt Pans

Places where coal was used to boil sea water to produce salt. Salt pans were found along the shores of the Firth of Clyde and particularly the Firth of Forth, where there were plentiful supplies of local coal. Surviving 18th-century remains of salt pans can still be seen at Preston Island near Culross in Fife (next to the ruins of a coal mine developed by Sir Robert Preston), and further west at St Monans.

### Scottish Coalfields

The principal Scottish coalfields are Carboniferous, occurring in a series of basins along the central lowlands from the Firths of Tay and Forth in the east to the Firth of Clyde in the west. Two isolated extensions of the Ayrshire coalfield in the west can be found at Machrihanish in Argyll on the Mull of Kintyre, and around Sanquhar and Kirkconnel in northern Dumfriesshire. The Canonbie coalfield in southern Dumfriesshire has never been exploited, and in the Highlands, a smaller Jurassic coalfield at Brora in Sutherland was worked until the early 1970s.

## Screens

A mechanically-driven arrangement of fixed or vibrating bars or perforated steel sheets for grading coal by size. Screens were to be found at most collieries, and were a component of modern coal preparation plants. Some were fixed installations, and others portable, depending on the scale and arrangement of the colliery.

## Seam

The term normally used to describe the layer or stratum of coal within the rock. Unfortunately, local geological conditions determine that most seams do not lie horizontally, and in some cases are fractured and thrown by faults, complicating the mining process. Seams were often given descriptive names, such as Jewel, Parrot etc.

## Self-advancing Roof Supports

A fully-mechanised assembly of hydraulically-operated steel supports used in longwall mining which can be moved forward as the face advances.

## Self-rescuer

A type of respirator comprising a small filtering device carried by a coal miner underground, either on a belt or in a pocket, to provide immediate protection against carbon monoxide and smoke in case of a mine fire or explosion, thereby assisting survival and escape. Self-rescuers became mandatory for miners working underground following the disaster at Michael Colliery in Fife in 1967.

## Shaft

A vertical hole or Pit sunk through the strata for the purpose of finding or mining coal or ore, and through which coal, men and materials were raised and lowered. Shafts were also used to provide ventilation to underground workings, and to pump away water. Most shafts were lined with wood, stone, brick or concrete, particularly as technology permitted the mining at greater depths, and as large winding engines and associated pit-head buildings were constructed to handle greater quantities of coal.

## Shale

Consolidated beds of rock comprising large proportions of mud or clay, but also containing varying quantities of combustible matter. Frequently found within coal measures, shale makes up a substantial proportion of spoil heaps or Bings. Such shales were sometimes used to form a stiff plastic clay mixture suitable for brick manufacture, their combustible contents rendering the bricks self-firing, saving fuel in the kilns.

## Shale oil

See Oil Shale

## Shearer

An electrically-driven mining machine for Longwall faces that uses a rotating action to 'shear' coal as it progresses along the face. Shearers and other pieces of electrical mining equipment were produced in great quantities in Scotland, particularly by Anderson Boyes of Motherwell and Mavor & Coulson of Glasgow.

## Shearer Loaders

Machines which shear and load the coal in a single operation.

## Sheave

A term used in mining for a pulley either on the headframe above a shaft, or in underground rope haulage.

## Sheave Wheel

A large grooved wheel in the top of a headframe over which the hoisting rope passes above a mine shaft.

## Silicosis

A Respiratory Disease caused by the inhalation of silica dust, commonly associated with underground mining activity.

## Sill

A horizontal intrusion of usually hard igneous rock within strata which can disrupt mining activity and which, in some cases, has affected adjacent coal seams (see also Dyke).

## Sinking

The process of mining, the purpose of which is to create a shaft or incline mine from which minerals can be extracted.

## Skip

A container in which coal was carried up the shaft from the pit bottom, where it was loaded from underground bunkers. Skips were often used to replace conventional cages, hutches and mine cars in some reconstructed Scottish collieries, and in more modern pits such as Bilston Glen, Kames, Killoch, Monktonhall, Polkemmet, and Seafield.

## Slag

Waste material from a smelter, such as a blast furnace in an iron works. Slag can be re-used for a variety of purposes, such as in the production of cement. The term is also sometimes loosely used to describe colliery waste.

## Slag Heap

A spoil heap adjacent to a metallurgical works (such as an iron and steel works) containing slag from smelters. Bings or colliery spoil heaps are often mistakenly referred to as slag heaps.

## Smithy Coal

A low-sulphur non-caking coal used in blacksmith's hearths.

## Smokeless Fuel

Following the first clean-air legislation in 1956, many domestic users of solid fuel in British cities were forbidden from using coal. Alternative smokeless products were produced by the coal industry in new factories such as that at Lugar in Ayrshire, and the Niddrie Nuggeting Plant near Edinburgh.

## Splint Coal

A non-caking low-sulphur coal used from c.1830 to 1940 for iron smelting without the need for prior coking.

## Spoil Heap

A tip adjacent to a mine containing waste from the workings. Colliery and shale-oil waste tips are also known as a Bings in Scotland. Many have been cleared away in the UK because of safety fears following the collapse of a spoil heap in Aberfan in Wales in 1966, which killed over 100 children at the local school, and also for aesthetic reasons.

## Spontaneous Combustion (bings)

The bings (spoil heaps) of some collieries in Scotland suffer from spontaneous combustion, some fires burning below the surface for many years. The associated fumes are often unpleasant, and the fires themselves can be very hazardous, especially where unaccompanied children are nearby. For this reason, the clearance of bings in environmental improvement schemes has been a priority in some parts of the Scottish coalfields.

## Spontaneous Combustion (underground)

A problem associated with some coal measures resulting in the outbreak of sometimes fatal underground fires. Some coal seams, especially the Dysart Main and Lower Dysart at Frances and Seafield in East Fife, were more susceptible than others. In 1967, it was also responsible for the closure of Michael Colliery not far away. Seafield Colliery in Fife pioneered a system which automatically added nitrogen to air in the mine to suppress spontaneous combustion.

## Staithes

Quayside facilities with provision for the unloading of coal from railway wagons (see Coal Drops).

## Steam Winding

The use of steam to drive a winding engine above a shaft. The introduction of steam power revolutionised mining in the mid-18th century, and from about 1790, most larger mines were equipped with steam engines to drive their winders. Steam was widely used in Scotland until the second half of the 20th century, after which it was rapidly replaced by electricity. Nevertheless, some steam winders survived, and were a source of great interest within the industry. Scotland's last steam winder (built by Murray & Paterson of Coatbridge) was at Cardowan Colliery, which closed in 1983. The winding engine was subsequently dismantled and can now be seen at Summerlee Industrial Museum in Coatbridge. A steam winding engine can also be seen at the Scottish Mining Museum's Lady Victoria Colliery in Newtongrange, Midlothian.

## Stell

A form of prop often used to support a face or a roof, preventing collapse.

## Stinkdamp

A term used to describe hydrogen sulphide, an extremely poisonous gas often found in old mine workings.

## Stone dust

Inert dust placed in mine roadways to suppress coal-dust explosions, should gas be accidentally ignited within a mine.

## Stoop and Room Working

A term used in Scotland to describe the Pillar and Stall method of working.

## Stores

Mining required a great volume of equipment and materials, and this had to be stored when not being used underground. Valuable items tended to be protected from the elements inside buildings, but more robust materials (such as pit props) were often stored outside in sometimes substantial fenced compounds.

## Subsidence

The gradual or sudden sinking of the surface as a result of the collapse of underground mine-workings. Buildings, structures and surface features above the subsidence area can be affected, and this is one of the main reasons for the maintenance of Abandonment Plans which record the extent of mine-workings underground.

## Superpit

See New Sinkings

## Surface Arrangement

A general term used to describe the collection of buildings and plant at the surface around shafts and surface mines leading to underground workings below (see also Heapstead).

## Surface Mine

In Scotland, a term often used to describe an Incline or Drift Mine, where a tunnel is sunk into the ground following the incline of the strata. In some parts of the world, the term Surface Mine is used to describe Open-cast workings.

## Tally

Animal (usually sheep) fat in primitive Tally Lamps. The use of animal fat gave off a strong-smelling smoke, and an odourless substitute made from paraffin wax was preferred by some miners. Also, an object placed in a hutch to show which group of miners had extracted the coal.

## Tally Lamp

A small pot-like container made of tin, fixed to the miners' helmets and fuelled by tallow (sheep fat) or wax with a protruding wick. In the early 20th century, it was replaced by the Carbide Lamp. The lamp was used in mines where gas was not perceived to be a problem.

## Time Office

A building or room within a building at the pithead from which every miner is issued with a numbered token before going underground. The tokens are kept on boards so that it is easy to gauge how many people are underground at any one time (see also Check Office).

## Tipper or Tipplers

A rotary apparatus for emptying tubs or mine cars, onto Screens, or down a chute to a conveyor, railway wagon, boat or lorry.

## Token

Numbered brass discs, sometimes called 'Checks' issued to each miner at the Time Office (or Check Office) prior to going underground. The tokens must be returned on returning to the surface, thereby ensuring that every miner is accounted for.

## Town Gas

See Coal Gas

## Transformers

Electrical devices which transform the voltage of electricity from the level at which it is generated or supplied to one suitable for consumption at a specific site. Modern coalmines used electricity extensively both at the surface and underground, and were often therefore equipped with a range of transformers, depending on the equipment in use.

## Trepanner

A coal-cutting machine for Longwall mining, first introduced in the 1950s by Anderson Boyes of Motherwell, and widely used in Scotland. Used particularly successfully at Highhouse Colliery in Ayrshire.

## Tub

A wagon, normally referred to as a Hutch in Scotland, designed to travel on rails, used both to transport coal and other materials underground, and at the pithead where it was unloaded, and subsequently returned to the mine-workings below. Originally an open-topped container assembled from wood or iron, tubs evolved into steel-built Mine Cars, with a capacity of several tons.

## Upcast Shaft

The shaft, or division of it, up which the return air is vented to the atmosphere.

## Ventilation

The supply of a mechanically-assisted flow of fresh and return air along all underground roadways and workings within a mine, usually achieved with the assistance of fans.

## Washed Coal

Coal which has been cleaned after passing through a washery (see Coal Preparation Plant), and which is ready for use.

## Washer

Mechanical equipment for the wet cleaning of coal (see also Coal Preparation Plant and Washery). Specialist manufacturers included Simon Carves, Coppee, Campbell Binnie and Reid, Dickson and

Mann, Norton-Harty, and the Blantyre Engineering Company.

## Washery

A coal preparation plant in which cleaning and grading of coal is carried out by wet methods (see also Coal Preparation Plant). Specialist manufacturers included Simon Carves, Coppee, Campbell Binnie and Reid, Dickson and Mann, Norton-Harty, and the Blantyre Engineering Company.

## Waste

The material that must, for practical reasons, be removed from a mine, but which is of no value.

## Whin

An intrusion of igneous rock into strata which can disrupt mining activities (see Dyke and Sill).

## Whitedamp

A term used to describe carbon monoxide, a potentially lethal odourless gas which is sometimes present in the Afterdamp following an explosion, or as a result of an underground fire. It is rapidly absorbed by the haemoglobin in blood, progressively excluding oxygen. A presence of 0.1% in air can be fatal within ten minutes. Because of their enhanced sensitivity to the gas, Canaries were taken into the mines to provide an early warning if the gas was present.

## Winder

See Winding Engine

## Winding Engine

A steam or electric engine situated at the pithead adjacent to or in a tower above a shaft, the purpose of which is to rotate a drum, thereby raising or lowering cages or skips within the shaft by means of winding ropes.

## Winding-engine Man

The man operating the winding engine at a colliery. In England and Wales, he is often referred to as the Brakesman.

## Wire Rope

A form of rope made from twisted strands of steel wire, the application of which revolutionised coal mining in the mid-19th century, leading to the development of bigger, more capital-intensive collieries. It was used extensively for underground and surface haulage, but its use for winding in shafts was most significant, allowing much heavier loads to be raised to the surface, often from deeper coal measures. Wire ropes come in many forms, with varying lay and numbers of strands and wires, depending on their intended use. The production of wire ropes for coal mines was a specialist business, and Bruntons of Musselburgh were one of the most important manufacturers in the UK.

## Workings

The often interconnected underground tunnels and openings from which coal and other minerals are extracted.

# Bibliography

Agricola, G (1556), *De Re Metallica*, translated from the first Latin edition by H C and L H Hoover, 1950, Dover Publications Inc: New York

Aitken, R L (1998), *Not Many Noble: the Story of the Lanarkshire Coalfield*, The Old Museum Press: Bramber, West Sussex

Anderson, J (1943), *Coal: A History of the Coal-Mining Industry in Scotland with Special Reference to the Cambuslang District of Lanarkshire*, R E Robertson Ltd: Glasgow

Ashworth, W (1986), *The History of the British Coal Industry, Volume 5: 1946–1982, The Nationalised Industry*, Clarenden Press: Oxford

Benson J, Neville R G and Thompson C H (1981), *Bibliography Of The British Coal Industry. Secondary Literature, Parliamentary and Departmental Papers, Mineral Maps and Plans and a Guide to Sources*, published for the National Coal Board by Oxford University Press: Oxford

Blyth, A (1994), *From Rosewell to the Rhondda: The Story of Archibald Hood, A Great Scots Mining Engineer*, Midlothian District Library Service: Loanhead

Bowman, A I (1970), 'Culross Colliery: A Sixteenth Century Mine', *Industrial Archaeology*, Vol. 7, No. 4, pp. 353-372

Bowman, I (1984), 'Coalmining at Culross: 16–17th centuries', *Forth Naturalist and Historian*, Vol. 7, 1982–3, pp. 84–125

Campbell, A B (1979), *The Lanarkshire Miners: A Social History of their Trade Unions*, John Donald: Edinburgh

Campbell, A B (2000), *The Scottish Miners, 1874–1939, Volume I: Industry, Work and Community*, Ashgate Publishing Limited: Aldershot

Campbell, A B (2000), *The Scottish Miners, 1874–1939, Volume II: Trade Unions and Politics*, Ashgate Publishing Limited: Aldershot

Campbell, R H (1961), *Carron Company*, Oliver and Boyd: Edinburgh

Carron Company (*c.*1959), *The story of Carron Company 1759–1959: Two Hundred Years of Service*, Carron Company: Falkirk

Carvel, J L (1944), *One Hundred Years in Coal: The History of the Alloa Coal Company*, T & A Constable: Edinburgh

Carvel, J L (1946), *The New Cumnock Coal-field: a Record of its Development and Activities*, T & A Constable: Edinburgh

Carvel, J L (1948), *The Coltness Iron Company: A Study in Private Enterprise*, T & A Constable: Edinburgh

Carvel, J L (1949), *Fifty Years of Machine Mining Progress*, Anderson Boyes & Co: Motherwell

Church, R (1986), *The History of the British Coal Industry, Volume 3: 1830–1913, Victorian Pre-eminence*, Clarenden Press: Oxford

Clark, T, ed. (1980), *Comrie Colliery, 1940–1980*, pamphlet produced by Comrie Colliery

Colliery Guardian (1948 to 1988), *Guide to the Coal Fields*, Colliery Guardian: Redhill

Corrins, R D (1994), 'The Scottish Business Elite in the Nineteenth Century: The Case of William Baird and Company,' in Cummings, A J G and Devine T (eds), John Donald: *Industry, Business and Society in Scotland since 1750*, Edinburgh, pp. 58–83.

Cotterill, M S (1983), *Investment and Management in the Lothian Coal Company 1890–1955*, Scottish Mining Museum: Edinburgh (Newtongrange)

Court, W H B (1951), *Coal (History of the Second World War: UK Civil Series)*, London

Dick, D, ed. (2000), *A Scottish Electrical Enlightenment: Celebrating 100 Years of the Institution of Electrical Engineers in Scotland, 1899–1999*, IEE: Glasgow

Douglas G and Oglethorpe M (1993), *Brick, Tile and Fireclay Industries in Scotland*, RCAHMS: Edinburgh

Dron, R W (1902), *The Coalfields of Scotland*, Blackie & Son Ltd: London

Duckham, B F (1970), *A History of the Scottish Coal Industry, Vol. 1, 1700–1815*, David & Charles: Newton Abbot

Flinn, M (1984), *The History of the British Coal Industry, Volume 2: 1700–1830, The Industrial Revolution*, Clarenden Press: Oxford

Grice, C S W (1951), *Rat Extermination in Mines*, Ministry of Fuel and Power, Safety in Mines Research Establishment: Sheffield

Griffin, A R (1977), *The British Coalmining Industry: Retrospect and Prospect*, Moorland Publishing: Buxton

Grimshaw, P N (1992), *Sunshine Miners: opencast coalmining in Britain 1942–1992*, British Coal Opencast: Mansfield

Halliday, R S (1990), *The Disappearing Scottish Colliery: A Personal View of Some aspects of Scotland's Coal Industry since Nationalisation*, Scottish Academic Press: Edinburgh

Hassan, J A (1977), 'The Gas Market and the Coal Industry in the Lothians in the Nineteenth Century,' *Industrial Archaeology*, 1977, pp. 49–73

Hatcher, J (1993), *The History of the British Coal Industry, Volume 1: Before 1700*, Clarenden Press: Oxford

Hayes, G (2000), *Coal Mining*, Shire Publications: Market Risborough

Haynes, W W (1953), *Nationalisation in Practice: The British Coal Industry*, Bailey Bros. & Swinfen: London

Heinemann, M (1944), *Britain's Coal: Wages and Hours, Coal and the Monopolists, Saving Man Power, Control of the Mines, Productivity, The Future of Coal*, Victor Gollancz: London

Hill, A (1991), *Coal Mining: a technical chronology 1700–1950*, The Northern Mine Research Society (NMRS), British Mining Supplement: Sheffield

Hudson, R (2002), 'The changing geography of the British Coal industry: nationalization, privatisation and the political economy of energy supply, 1947–97', *Transactions of the Institute of Mining and Metallurgy, Section A*, Vol. 111, No. 3, pp. 180-187

Hughes, S, Mallaws, B, Parry, M and Wakelin, P (1995), *Collieries in Wales: Engineering & Architecture*, RCAHMW: Aberystwyth

Hughson, I (1996), *The Auchenharvie Colliery: an early history*, The Three Towns Local History Group, Stenlake Publishing: Ochiltree

Hume, J R (1976), *The Industrial Archaeology of Scotland, volume 1. Lowlands and Borders*, Batsford: London

Hutton, G (1996), *Mining: Ayrshire's Lost Industry*, Stenlake Publishing: Ochiltree

Hutton, G (1997), *Lanarkshire's Mining Legacy*, Stenlake Publishing: Ochiltree

Hutton, G (1998), *Mining the Lothians*, Stenlake Publishing: Ochiltree

Hutton, G (1999), *Fife – The Mining Kingdom*, Stenlake Publishing: Ochiltree

Hutton, G (2000), *Mining from Kirkintilloch to Clackmannan & Stirling to Slamannan*, Stenlake Publishing: Ochiltree

Hutton, G (2001), *Scotland's Black Diamonds*, Stenlake Publishing: Ochiltree

Hyde, E D (1987), *Coal Mining in Scotland*, Scottish Mining Museum: Edinburgh (Newtongrange)

Kerr, R D, ed. (1980), *A Glossary of Mining Terms used in Fife*, Fife Colleges: Kirkcaldy

Lochside Coal and Fireclay Co. Ltd. (1955), *Lochside Coal and Fireclay Co. Ltd., Manufacturers of Composition Bricks and Every Description of Fireclay Goods*, Company catalogue: Dunfermline

Lothian Coal Company (1930), *A Pioneer Machine-mining Colliery: The Newbattle Collieries of the Lothian Coal Company Limited* (pamphlet published by the company)

Loynes, E (1984), *The Road to Progress with Safety: The Story of the Institution of Mining Electrical and Mining Mechanical Engineers*, IMEMME

Martin, M, Martin, C, and Sparling, C, The Fife Miners at: www.users.zetnet.co.uk/mmartin/fifepits/

McDougall, I, ed. (1981), *Militant Miners: Recollections of John McArthur, Buckhaven; and letters, 1924–26, of David Proudfoot, Methil, to G Allen Hutt*, Polygon Books: Edinburgh

Meccano Magazine (1955), '*Of General Interest: Light on Dirt*', October, p. 552

Ministry of Fuel and Power (1945), *Coal Mining: Report of the Technical Advisory Committee (Reid Report)*, (Cmd. 6610), HMSO: London

Moffat, A (1965), *My Life With the Miners*, Lawrence & Wishart: London (see chapter 179 in particular, on pit closures 1957–60)

Muir, A (*c.*1945), *The Fife Coal Company Limited: a Short History*, Fife Coal Company: Leven

Muir, A (c.1952), *The Story of Shotts: a Short History of the Shotts Iron Company Limited*, Shotts Iron Company Limited: Edinburgh

NCB (1948), Colliery Questionnaire (data gathered from colliery managers after vesting day), Scottish Division. Unpublished

NCB (1947–95), *Shaft Register*. Unpublished

NCB (1948 and subsequent years until 1982), *Annual Report and Statement of Accounts for the year ended 31st December 1947*, HMSO: London

NCB (1950), *Plan for Coal*, HMSO: London

NCB (c.1953), *Dalkeith Central Coal Preparation Plant*, NCB Press Office (Scottish Division): Edinburgh,

NCB (1955), *Scotland's Coal Plan*, NCB Scottish Division, Edinburgh

NCB (1956), *Investing in Coal: Progress and Prospects under the Plan for Coal*, Ministry of Fuel and Power

NCB (1957), *British Coal: Rebirth of an Industry*, NCB: London

NCB (1958), *A Short History of the Scottish Coal-mining Industry*, NCB Scottish Division: Edinburgh

NCB (1958), *Visit of Her Majesty the Queen and HRH The Duke of Edinburgh to Rothes Colliery on Monday, 30 June 1958*, NCB Scottish Division: Edinburgh

NCB (c.1960), *The Case for Coal*, NCB: London

NCB (1972), *National Coal Board: 25 Years, 1947–72*, NCB: London

NCB (1974), *Coal Preparation Certificate Course: Glossary*, 2nd edition, NCB: London

NCB (1978), *Accidents and Disasters in Coal Mines*, (compiled by the) National Coal Board, Headquarters Library, NCB: London

NCB (1984), *Facts and Figures: Britain's Coal Industry*, NCB: London

NCB (*1984), The Design of Headframes and Winder Towers,* NCB: London

NCB (various dates), colliery brochures for: Barony (c.1950), Benarty (1947), Bilston Glen (1952, 1961), Bowhill (1952 and 1953), Cardowan (c.1958), Castlebridge (1983), Easthouses (1955), Frances (1947), Garscube (c.1950), Glenochil (1952, 1955), Killoch (1952), Kingshill No. 3 (c.1952), Kinneil (1951), Monktonhall (1953), Polkemmet (1958), Rothes (c.1955), and Seafield (1954)

Nott-Bower, G and Walkerdine, R H, eds. (1958), *National Coal Board, The First Ten Years: A Review of the First Decade of the Nationalised Coal Mining Industry in Great Britain*, The Colliery Guardian: London

Owen, J S (1995), *Coal Mining at Brora*, 1529–1974, Inverness Highland Libraries

Page Arnot, R (1955), *A History of The Scottish Miners from Earliest Times*, George Allen & Unwin: London

Parker, J C and Sleight, G E (1957), 'Littlemill Colliery – A Reorganisation in the Ayrshire Coalfield', *Transactions of the Institution of Mining Engineers*, Vol. 116 (1956–7), pp. 1045–62

Payne, P (1961), 'The Govan Collieries, 1804–1805', *Business History*, Vol. III, pp. 75–96

RCAHMS (1998), *Forts, Farms and Furnaces, Archaeology in the Central Scottish Forest*, RCAHMS: Edinburgh

RCAHMS (1995), *Muirkirk, Ayrshire: An Industrial Landscape* (broadsheet), RCAHMS: Edinburgh

Reid, W, Crawford , R and McNeill, K H (1939), *The Layout and Equipment of Comrie Colliery in Fifeshire*, Institute of Mining Engineers and Western Mail and Echo: Cardiff

Scottish Home Department (1944), *Scottish Coalfields: Report of the Scottish Coalfields Committee*, cmd. 6575, HMSO: Edinburgh , 23 and 54–5 (stats on Lanarkshire)

Scottish Mining Museum (1999), Conservation Plan. Unpublished

Slaven, A (1967), 'Coalmining in the West of Scotland in the nineteenth century: the Dixon Enterprises', unpublished BLitt thesis, University of Glasgow

Smith, D L (1967), *The Dalmellington Iron Company: its engines and men*, David and Charles: Newton Abbot

Stephenson, H S (1968), *Fire at Michael Colliery: Report on the causes of, and circumstances attending, the fire which occurred at Michael Colliery, Fife, on 9th September, 1967*, Ministry of Power, HMSO (Cmnd.3657): London

Supple, B (1987), *The History of the British Coal Industry, Volume 4: 1913–1946, The Political Economy of Decline*, Clarenden Press: Oxford

Thornes, R (1994), *Images of Industry: Coal*, RCHME: Swindon

The Times (1952), 'Removing Scenes of Squalor from Scottish Pit-heads: The Architect's part in modernizing mining', *The Times*, 12 November 1952, London

Tweedie, J W (1953), 'Mines in Ayrshire, 1745–1950', in Shaw, J E, ed., *Ayrshire, 1745–1950 – A social and industrial history of the county*, compiled for the Ayrshire Archaeological and Natural History Society, Oliver & Boyd: Edinburgh, pp. 233–41

Wallace, W (1983), *Some notes on the coal industry in Hamilton*, Bell College of Technology: Hamilton

Whatley, C A (1977), 'The Introduction of the Newcomen Engine in Ayrshire, *Industrial Archaeology Review*, Vol. II, No. 1, pp. 69–77

Whatley, C A (1982), 'Scottish Salt Making in the 18th century: a Regional Survey', *Scottish Industrial History*, Vol. 5, Part 2, pp. 2–26

Whatley, C A (1983), *The Finest Place for a Lasting Colliery: Coal Mining Enterprise in Ayrshire c.1600–1840*, Ayrshire Natural History Society: Ayr

Whatley, C A (1987), *The Scottish Salt Industry 1570–1850: an economic and social history*, Aberdeen University Press: Aberdeen

Yeoman, P et al (1999), *The Salt and Coal Industries at St Monans, Fife in the 18th and 19th centuries*, Tayside and Fife Archaeological Committee: Glenrothes

# Appendices

## Appendix A
## Scottish NCB Collieries, ranked by mean annual workforce, 1947–90

| Rank | Mine/Colliery | Mean workforce | Total workforce | Years open | Peak workforce | Peak year | County |
|---|---|---|---|---|---|---|---|
| 1 | Michael | 2,598 | 51,962 | 20 | 3,353 | 1957 | Fife |
| 2 | Seafield | 2,178 | 47,927 | 22 | 2,466 | 1970 | Fife |
| 3 | Bilston Glen | 2,154 | 56,004 | 26 | 2,367 | 1970 | Midlothian |
| 4 | Killoch | 2,014 | 54,357 | 27 | 2,305 | 1965 | Ayrshire |
| 5 | Wellesley | 1,908 | 38,160 | 20 | 2,603 | 1957 | Fife |
| 6 | Monktonhall | 1,618 | 35,588 | 22 | 1,786 | 1971 | Midlothian |
| 7 | Polkemmet | 1,496 | 59,836 | 40 | 1,896 | 1960 | West Lothian |
| 8 | Cardowan | 1,493 | 55,240 | 37 | 1,870 | 1959 | Lanarkshire |
| 9 | Lady Victoria | 1,339 | 46,856 | 35 | 1,765 | 1953 | Midlothian |
| 10 | Bowhill | 1,331 | 23,954 | 18 | 1,490 | 1961 | Fife |
| 11 | Aitken | 1,249 | 19,975 | 16 | 1,431 | 1956 | Fife |
| 12 | Comrie | 1,245 | 49,778 | 40 | 1,498 | 1963 | Fife |
| 13 | Kinneil | 1,184 | 33,150 | 28 | 1,268 | 1960 | West Lothian |
| 14 | Glencraig | 1,105 | 18,786 | 17 | 1,316 | 1950 | Fife |
| 15 | Frances | 1,104 | 45,241 | 41 | 1,482 | 1957 | Fife |
| 16 | Barony 1, 2 & 3 | 1,078 | 45,257 | 42 | 1,695 | 1958 | Ayrshire |
| 17 | Fauldhead 1, 2, 3 & 4 | 1,018 | 21,361 | 21 | 1,155 | 1949 | Dumfriesshire |
| 18 | Castlebridge | 985 | no data | 6 | 985 | 1988 | Clackmannanshire |
| 19 | Blairhall | 982 | 21,600 | 22 | 1,053 | 1947 | Fife |
| 20 | Valleyfield | 982 | 16,690 | 17 | 1,052 | 1959 | Fife |
| 21 | Arniston (Emily & Gore) | 980 | 14,693 | 15 | 990 | 1951 | Midlothian |
| 22 | Rothes | 960 | 4,800 | 5 | 1,235 | 1960 | Fife |
| 23 | Whitrigg 2, 4 & 5 | 953 | 23,824 | 25 | 1,208 | 1952 | West Lothian |
| 24 | Solsgirth (Longannet Complex) | 917 | 18,329 | 20 | 1,007 | 1975 | Clackmannanshire |
| 25 | Kingshill 1 | 876 | 18,381 | 21 | 1,381 | 1951 | Lanarkshire |
| 26 | Auchencruive 1, 2 & 3 | 875 | 12,242 | 14 | 1,081 | 1947 | Ayrshire |
| 27 | Woolmet | 863 | 16,397 | 19 | 960 | 1963 | Midlothian |
| 28 | Douglas | 842 | 16,837 | 20 | 1,031 | 1952 | Lanarkshire |
| 29 | Lindsay | 818 | 14,714 | 18 | 970 | 1957 | Fife |
| 30 | Easthouses | 818 | 17,980 | 22 | 1,030 | 1963 | Midlothian |
| 31 | Manor Powis | 802 | 20,036 | 25 | 985 | 1963 | Stirlingshire |
| 32 | Prestonlinks | 794 | 13,492 | 17 | 820 | 1950 | East Lothian |
| 33 | Bedlay | 792 | 26,909 | 34 | 870 | 1959 | Lanarkshire |
| 34 | Mauchline 1, 2 & 3 | 783 | 14,879 | 19 | 876 | 1958 | Ayrshire |
| 35 | Burghlee | 768 | 13,049 | 17 | 795 | 1952 | Midlothian |
| 36 | Newcraighall | 760 | 15,952 | 21 | 810 | 1950 | Midlothian |

| Rank | Mine/Colliery | Mean workforce | Total workforce | Years open | Peak workforce | Peak year | County |
|---|---|---|---|---|---|---|---|
| 37 | Roslin | 754 | 15,822 | 21 | 770 | 1952 | Midlothian |
| 38 | Castlehill | 749 | 13,480 | 18 | 770 | 1972 | Fife |
| 39 | Lingerwood | 727 | 14,537 | 20 | 770 | 1951 | Midlothian |
| 40 | Easton | 715 | 15,012 | 21 | 863 | 1958 | West Lothian |
| 41 | Prestongrange | 686 | 10,290 | 15 | 700 | 1952 | East Lothian |
| 42 | Auchencruive 4 & 5 | 659 | 17,112 | 26 | 958 | 1957 | Ayrshire |
| 43 | Polmaise 1 & 2 | 659 | 7,249 | 11 | 723 | 1951 | Stirlingshire |
| 44 | Kinglassie | 652 | 12,398 | 19 | 764 | 1956 | Fife |
| 45 | Auchengeich | 650 | 11,701 | 18 | 860 | 1956 | Lanarkshire |
| 46 | Minto | 640 | 12,784 | 20 | 730 | 1957 | Fife |
| 47 | Thankerton 1, 3 & 6 | 636 | 3,816 | 6 | 766 | 1947 | Lanarkshire |
| 48 | Kingshill 2 | 624 | 10,604 | 17 | 686 | 1954 | Lanarkshire |
| 49 | Bothwell Castle 1 & 2 | 623 | 3,111 | 5 | 751 | 1947 | Lanarkshire |
| 50 | Bogside | 618 | 18,561 | 30 | 875 | 1971 | Fife |
| 51 | Dalkieth | 617 | 18,506 | 30 | 898 | 1964 | Midlothian |
| 52 | Gartshore 9 & 11 | 616 | 12,919 | 21 | 740 | 1952 | Dunbartonshire |
| 53 | Devon | 615 | 7,984 | 13 | 813 | 1954 | Clackmannanshire |
| 54 | Mary (Lochore) | 614 | 11,051 | 18 | 780 | 1957 | Fife |
| 55 | Polmaise 3 & 4 | 606 | 24,812 | 41 | 778 | 1957 | Stirlingshire |
| 56 | Wester Auchengeich | 601 | 12,611 | 21 | 850 | 1959 | Lanarkshire |
| 57 | Kingshill 3 | 600 | 13,795 | 23 | 769 | 1958 | Lanarkshire |
| 58 | Lumphinnans 11 12 | 596 | 11,909 | 20 | 764 | 1947 | Fife |
| 59 | Carberry | 589 | 7,652 | 13 | 615 | 1952 | Midlothian |
| 60 | Glenochil | 587 | 3,521 | 6 | 908 | 1960 | Clackmannanshire |
| 61 | Pennyvenie 2, 3 & 7 | 581 | 18,578 | 32 | 725 | 1961 | Ayrshire |
| 62 | Lochhead (including surface mine) | 580 | 13,324 | 23 | 841 | 1957 | Fife |
| 63 | Knockshinnoch Castle | 578 | 12,135 | 21 | 755 | 1956 | Ayrshire |
| 64 | Dumbreck | 575 | 9,188 | 16 | 669 | 1947 | Stirlingshire |
| 65 | Fleets | 570 | 6,834 | 12 | 854 | 1947 | East Lothian |
| 66 | Kames 1 & 2 | 569 | 11,936 | 21 | 634 | 1957 | Ayrshire |
| 67 | Whitehill | 562 | 7,858 | 14 | 590 | 1950 | Midlothian |
| 68 | Greenrigg | 558 | 7,250 | 13 | 634 | 1952 | West Lothian |
| 69 | Hamilton Palace | 552 | 6,621 | 12 | 600 | 1948 | Lanarkshire |
| 70 | Gartshore 3 & 12 | 547 | 5,465 | 10 | 611 | 1948 | Dunbartonshire |
| 71 | Benhar | 545 | 8,173 | 15 | 772 | 1947 | Lanarkshire |
| 72 | Riddochill | 541 | 11,358 | 21 | 687 | 1958 | West Lothian |
| 73 | Hopetoun | 538 | 2,686 | 5 | 622 | 1951 | West Lothian |
| 74 | Auchlochan 2, 6, 7, 9 & 10 | 534 | 11,250 | 21 | 687 | 1952 | Lanarkshire |
| 75 | Loganlea | 510 | 6,118 | 12 | 674 | 1947 | West Lothian |
| 76 | Fordell | 509 | 9,672 | 19 | 622 | 1959 | Fife |
| 77 | Blantyreferme 1 & 2 | 507 | 7,594 | 15 | 656 | 1950 | Lanarkshire |
| 78 | Littlemill 2 & 3 | 495 | 13,853 | 28 | 810 | 1960 | Ayrshire |
| 79 | Southfield | 492 | 5,406 | 11 | 556 | 1947 | Lanarkshire |
| 80 | Bardykes | 479 | 7,176 | 15 | 775 | 1947 | Lanarkshire |

| Rank | Mine/Colliery | Mean workforce | Total workforce | Years open | Peak workforce | Peak year | County |
|------|---------------|----------------|-----------------|------------|----------------|-----------|--------|
| 81 | Jenny Gray | 467 | 5,603 | 12 | 492 | 1950 | Fife |
| 82 | Blantyre | 460 | 4,601 | 10 | 533 | 1947 | Lanarkshire |
| 83 | Balgonie | 448 | 5,815 | 13 | 490 | 1952 | Fife |
| 84 | Foulshiels | 448 | 4,477 | 10 | 463 | 1951 | West Lothian |
| 85 | Baton | 445 | 1,777 | 4 | 772 | 1950 | Lanarkshire |
| 86 | Northfield & Hall | 443 | 6,196 | 14 | 484 | 1958 | Lanarkshire |
| 87 | Newbattle (surface) | 441 | 441 | 1 | 441 | 1947 | Midlothian |
| 88 | Plean | 441 | 7,049 | 16 | 528 | 1958 | Stirlingshire |
| 89 | Calderehead 3 & 4 | 437 | 4,807 | 11 | 617 | 1947 | Lanarkshire |
| 90 | Cowdenbeath 7 & 10 | 435 | 5,652 | 13 | 454 | 1950 | Fife |
| 91 | Ayr 9 & 10 (Enterkine) | 434 | 5,209 | 12 | 665 | 1948 | Ayrshire |
| 92 | Nellie | 428 | 6,837 | 16 | 508 | 1958 | Fife |
| 93 | Castlehill 6 | 422 | 2,948 | 7 | 465 | 1947 | Lanarkshire |
| 94 | Beoch 2, 3 & 4 | 408 | 8,561 | 21 | 460 | 1960 | Ayrshire |
| 95 | Woodend | 397 | 3,965 | 10 | 488 | 1954 | West Lothian |
| 96 | Blantyreferme 3 | 394 | 5,119 | 13 | 430 | 1956 | Lanarkshire |
| 97 | Whitehill 1 & 2 | 392 | 7,060 | 18 | 435 | 1949 | Ayrshire |
| 98 | Baads | 384 | 4,221 | 11 | 491 | 1955 | Midlothian |
| 99 | Canderrigg 4 & 5 and Canderside | 379 | 6,439 | 17 | 582 | 1947 | Lanarkshire |
| 100 | Minnivey | 378 | 6,420 | 17 | 470 | 1962 | Ayrshire |
| 101 | Highhouse 1 & 2 | 369 | 13,258 | 36 | 467 | 1947 | Ayrshire |
| 102 | Douglas Castle | 364 | 4,367 | 12 | 415 | 1952 | Lanarkshire |
| 103 | Hassockrig | 362 | 5,430 | 15 | 379 | 1958 | Lanarkshire |
| 104 | Twechar 1 | 355 | 6,023 | 17 | 413 | 1954 | Dunbartonshire |
| 105 | Forthbank | 353 | 2,118 | 6 | 992 | 1956 | Clackmannanshire |
| 106 | Lochgelly | 352 | 352 | 1 | 352 | 1947 | Fife |
| 107 | Ramsay | 346 | 6,228 | 18 | 385 | 1952 | Midlothian |
| 108 | Bothwell Castle 3 & 4 | 345 | 4,140 | 12 | 655 | 1947 | Lanarkshire |
| 109 | Gateside & Tower | 344 | 5,847 | 17 | 410 | 1952 | Dumfriesshire |
| 110 | Bannockburn | 344 | 5,833 | 17 | 563 | 1947 | Stirlingshire |
| 111 | Lady Helen | 342 | 5,807 | 17 | 455 | 1957 | Fife |
| 112 | Bellyford | 334 | 2,335 | 7 | 335 | 1954 | East Lothian |
| 113 | Bank 1, 2 & 6 | 332 | 7,291 | 22 | 400 | 1952 | Ayrshire |
| 114 | Blairmuckhill | 321 | 3,851 | 12 | 347 | 1952 | Lanarkshire |
| 115 | Carriden | 321 | 1,922 | 6 | 331 | 1948 | West Lothian |
| 116 | Branchall (Greenhead 2, 3) | 315 | 3,775 | 12 | 366 | 1952 | Lanarkshire |
| 117 | Gilmerton | 314 | 4,701 | 15 | 380 | 1950 | Midlothian |
| 118 | Wilsontown | 304 | 2,431 | 8 | 378 | 1948 | Lanarkshire |
| 119 | Wester Gartshore | 299 | 896 | 3 | 318 | 1947 | Dunbartonshire |
| 120 | Blackrigg 1 & 3 | 291 | 2,323 | 8 | 339 | 1951 | West Lothian |
| 121 | Polquhairn | 289 | 4,617 | 16 | 460 | 1960 | Ayrshire |
| 122 | Overtown | 284 | 4,260 | 15 | 466 | 1960 | Lanarkshire |
| 123 | Ardenrigg 6 | 282 | 4,513 | 16 | 304 | 1951 | Lanarkshire |
| 124 | Kepplehill & Stane | 279 | 1,392 | 5 | 297 | 1947 | Lanarkshire |

| Rank | Mine/Colliery | Mean workforce | Total workforce | Years open | Peak workforce | Peak year | County |
|---|---|---|---|---|---|---|---|
| 125 | Pennyvenie 4 | 278 | 3,887 | 14 | 335 | 1957 | Ayrshire |
| 126 | Fortissat | 278 | 555 | 2 | 283 | 1948 | Lanarkshire |
| 127 | Houldsworth | 276 | 4,970 | 18 | 346 | 1954 | Ayrshire |
| 128 | Tynemount & Oxenford | 273 | 3,003 | 11 | 370 | 1948 | East Lothian |
| 129 | Herbertshire | 272 | 3255 | 12 | 394 | 1948 | Stirlingshire |
| 130 | Limeylands | 269 | 1,877 | 7 | 335 | 1952 | East Lothian |
| 131 | Wellsgreen | 268 | 3,209 | 12 | 407 | 1949 | Fife |
| 132 | Dora (Little Raith) | 266 | 3,189 | 12 | 280 | 1957 | Fife |
| 133 | Hillhouserigg | 265 | 265 | 2 | 265 | 1947 | Lanarkshire |
| 134 | Chalmerston 4, 5, 6 & 7 | 256 | 3,329 | 13 | 300 | 1951 | Ayrshire |
| 135 | Woodmuir | 255 | 4,072 | 16 | 273 | 1963 | Midlothian |
| 136 | Rosie | 250 | 1,747 | 7 | 306 | 1947 | Fife |
| 137 | Killochan | 247 | 4,942 | 20 | 289 | 1952 | Ayrshire |
| 138 | Redding | 235 | 2,576 | 11 | 283 | 1948 | Stirlingshire |
| 139 | Dalzell & Broomside | 234 | 234 | 1 | 234 | 1947 | Lanarkshire |
| 140 | Sorn | 233 | 7,206 | 31 | 294 | 1970 | Ayrshire |
| 141 | Harvieston | 230 | 918 | 4 | 237 | 1960 | Clackmannanshire |
| 142 | Cairnhill | 221 | 4,194 | 19 | 250 | 1970 | Ayrshire |
| 143 | Brucefield | 216 | 3,017 | 14 | 265 | 1957 | Clackmannanshire |
| 144 | Wellwood | 215 | 429 | 2 | 215 | 1949 | Fife |
| 145 | Elgin & Wellwood | 214 | 214 | 1 | 214 | 1947 | Fife |
| 146 | Randolph | 208 | 4,351 | 21 | 284 | 1947 | Fife |
| 147 | Cuthill | 207 | 619 | 3 | 303 | 1959 | West Lothian |
| 148 | Garscube | 203 | 3,845 | 19 | 311 | 1963 | Lanarkshire |
| 149 | Benarty | 201 | 2,413 | 12 | 237 | 1957 | Fife |
| 150 | Zetland | 192 | 2,489 | 13 | 323 | 1957 | Clackmannanshire |
| 151 | Dullatur | 192 | 3,250 | 17 | 281 | 1963 | Dunbartonshire |
| 152 | Kennox 6 & 7 | 190 | 4,728 | 25 | 227 | 1960 | Lanarkshire |
| 153 | Meta (Devon 3) | 187 | 2,238 | 12 | 233 | 1956 | Clackmannanshire |
| 154 | Lumphinnans 1 | 182 | 1,635 | 9 | 202 | 1951 | Fife |
| 155 | Duntilland | 181 | 722 | 4 | 182 | 1947 | Lanarkshire |
| 156 | Shewalton 3 & 4 | 179 | 1,429 | 8 | 266 | 1947 | Ayrshire |
| 157 | Pirnhall | 174 | 2,780 | 16 | 241 | 1957 | Stirlingshire |
| 158 | Tillicoultry | 170 | 1,703 | 10 | 346 | 1950 | Clackmannanshire |
| 159 | Argyll | 168 | 3,363 | 20 | 245 | 1953 | Ayrshire |
| 160 | Lochlea | 164 | 3,933 | 24 | 239 | 1957 | Ayrshire |
| 161 | Cronberry Moor | 163 | 1,949 | 12 | 222 | 1947 | Ayrshire |
| 162 | Seaforth 1, 2 & 3 | 163 | 975 | 6 | 241 | 1947 | Ayrshire |
| 163 | Bankend | 161 | 1,767 | 11 | 205 | 1949 | Lanarkshire |
| 164 | Blair | 157 | 2,221 | 14 | 201 | 1960 | Ayrshire |
| 165 | Stane Mine | 155 | 464 | 3 | 210 | 1953 | Lanarkshire |
| 166 | Policy | 155 | 1,857 | 12 | 185 | 1948 | Stirlingshire |
| 167 | Shewalton 5 & 6 | 153 | 457 | 3 | 205 | 1949 | Ayrshire |
| 168 | Tofts 1 & 2 | 152 | 454 | 3 | 289 | 1947 | Ayrshire |

| Rank | Mine/Colliery | Mean workforce | Total workforce | Years open | Peak workforce | Peak year | County |
|------|---------------|----------------|-----------------|------------|----------------|-----------|--------|
| 169 | Oxenford 2 | 152 | 455 | 3 | 170 | 1948 | East Lothian |
| 170 | Knowehead | 152 | 1,663 | 11 | 203 | 1959 | Lanarkshire |
| 171 | Torry | 147 | 1,904 | 13 | 198 | 1962 | Fife |
| 172 | Dalquharran | 144 | 3,726 | 26 | 237 | 1958 | Ayrshire |
| 173 | Cameron | 143 | 1,717 | 12 | 207 | 1957 | Fife |
| 174 | East Benhar | 139 | 1,303 | 10 | 159 | 1951 | West Lothian |
| 175 | Barbeth | 137 | 682 | 5 | 149 | 1953 | Ayrshire |
| 176 | Dollar | 135 | 2,288 | 17 | 454 | 1963 | Perthshire |
| 177 | Criagrie | 134 | 669 | 5 | 136 | 1949 | Clackmannanshire |
| 178 | Boglea & Glentore | 134 | 1,733 | 13 | 221 | 1956 | Lanarkshire |
| 179 | Westoun (Coalburn) | 134 | 1,864 | 14 | 175 | 1956 | Lanarkshire |
| 180 | Blairenbathie | 132 | 1,836 | 14 | 151 | 1957 | Fife |
| 181 | Harwood | 132 | 1,580 | 12 | 196 | 1958 | Midlothian |
| 182 | Balmore | 131 | 1,694 | 13 | 173 | 1947 | Stirlingshire |
| 183 | Newfield | 126 | 1,135 | 9 | 144 | 1952 | Ayrshire |
| 184 | Rig | 122 | 1,454 | 12 | 93 | 1957 | Dumfriesshire |
| 185 | Meadowmill | 120 | 720 | 6 | 120 | 1954 | East Lothian |
| 186 | Glentaggart | 120 | 2,624 | 22 | 130 | 1952 | Lanarkshire |
| 187 | Glen Mine | 117 | 816 | 7 | 126 | 1948 | Lanarkshire |
| 188 | Lugar | 116 | 692 | 6 | 130 | 1949 | Ayrshire |
| 189 | Earlseat | 114 | 906 | 8 | 162 | 1956 | Fife |
| 190 | Glentore | 112 | 783 | 7 | 160 | 1957 | Lanarkshire |
| 191 | Warrix 1 & 2 | 111 | 332 | 3 | 130 | 1948 | Ayrshire |
| 192 | Rankin | 110 | 440 | 4 | 120 | 1947 | Lanarkshire |
| 193 | Whitehill 3 & 4 | 109 | 1,747 | 16 | 140 | 1957 | Ayrshire |
| 194 | Whitrigg 6 | 109 | 870 | 8 | 111 | 1952 | West Lothian |
| 195 | South Bantaskine | 107 | 1,277 | 12 | 141 | 1954 | Stirlingshire |
| 196 | Netherton | 106 | 318 | 3 | 106 | 1947 | Lanarkshire |
| 197 | Coatspark | 104 | 1,142 | 11 | 110 | 1952 | Lanarkshire |
| 198 | Barbauchlaw | 104 | 519 | 5 | 115 | 1947 | West Lothian |
| 199 | Castle Mine | 102 | 810 | 8 | 150 | 1950 | Midlothian |
| 200 | Maxwell | 95 | 2,466 | 26 | 133 | 1958 | Ayrshire |
| 201 | Quarter 1 (and mine) | 92 | 456 | 5 | 108 | 1948 | Lanarkshire |
| 202 | Southhook | 91 | 91 | 1 | 91 | 1947 | Ayrshire |
| 203 | Thinacres | 91 | 1,002 | 11 | 99 | 1956 | Lanarkshire |
| 204 | Skellyton 2 & 3 | 90 | 359 | 4 | 90 | 1948 | Lanarkshire |
| 205 | Bridgend | 89 | 1,337 | 15 | 146 | 1961 | Ayrshire |
| 206 | Lochend 5 | 89 | 89 | 1 | 89 | 1949 | Lanarkshire |
| 207 | Fortacres | 88 | 872 | 10 | 109 | 1952 | Ayrshire |
| 208 | Roger | 88 | 2,366 | 27 | 121 | 1975 | Dumfriesshire |
| 209 | Winton | 85 | 850 | 10 | 85 | 1952 | East Lothian |
| 210 | Auchmeddan 1, 2 & 3 | 85 | 1,686 | 20 | 101 | 1967 | Lanarkshire |
| 211 | Thornton | 84 | 586 | 7 | 113 | 1950 | Fife |
| 212 | Auldton | 84 | 584 | 7 | 96 | 1960 | Lanarkshire |

| Rank | Mine/Colliery | Mean workforce | Total workforce | Years open | Peak workforce | Peak year | County |
|------|---------------|----------------|-----------------|------------|----------------|-----------|--------|
| 213 | Lindsay 2 | 83 | 83 | 1 | 83 | 1947 | Fife |
| 214 | King O'Muirs | 82 | 812 | 10 | 153 | 1955 | Clackmannanshire |
| 215 | Hindsward | 78 | 233 | 3 | 83 | 1957 | Ayrshire |
| 216 | Gateside 2 | 78 | 233 | 3 | 78 | 1947 | Stirlingshire |
| 217 | Bogton | 76 | 602 | 8 | 99 | 1947 | Ayrshire |
| 218 | Quarter 2 | 74 | 74 | 1 | 74 | 1947 | Lanarkshire |
| 219 | Coalburn | 72 | 1,069 | 15 | 109 | 1952 | Ayrshire |
| 220 | Glencairn | 72 | 1,000 | 14 | 190 | 1948 | East Lothian |
| 221 | Gateside 1 | 72 | 362 | 5 | 78 | 1950 | Stirlingshire |
| 222 | Greenhill | 70 | 770 | 11 | 74 | 1954 | Ayrshire |
| 223 | Oakfield | 70 | 140 | 2 | 70 | 1947 | Fife |
| 224 | Pennyvenie 5 | 66 | 264 | 4 | 70 | 1947 | Ayrshire |
| 225 | Sundrum | 65 | 578 | 9 | 77 | 1958 | Ayrshire |
| 226 | Shieldmains 14 Drongan | 62 | 186 | 3 | 101 | 1947 | Ayrshire |
| 227 | Chapel | 61 | 122 | 2 | 61 | 1947 | Lanarkshire |
| 228 | Shieldmains, Barbeth 2 | 60 | 179 | 3 | 90 | 1949 | Ayrshire |
| 229 | Ashgill | 60 | 240 | 4 | 60 | 1947 | Lanarkshire |
| 230 | Longannet Mine | 56 | 1,063 | 19 | 60 | 1987 | Fife |
| 231 | Shewalton 8 & 9 | 50 | 50 | 1 | 50 | 1947 | Ayrshire |
| 232 | Headless Cross 1 | 50 | 100 | 2 | 50 | 1947 | Lanarkshire |
| 233 | Broomlands | 49 | 146 | 3 | 59 | 1952 | Ayrshire |
| 234 | Shieldmains 6 & 7 Drongan | 49 | 146 | 3 | 70 | 1948 | Ayrshire |
| 235 | Andershaw | 49 | 441 | 9 | 55 | 1952 | Lanarkshire |
| 236 | Headless Cross 2 | 49 | 290 | 6 | 91 | 1951 | Lanarkshire |
| 237 | Isle of Canty | 41 | 82 | 2 | 41 | 1947 | Fife |
| 238 | Carston | 39 | 271 | 7 | 44 | 1952 | Ayrshire |
| 239 | Overwood | 39 | 78 | 2 | 39 | 1952 | Lanarkshire |
| 240 | Edgehead | 39 | 388 | 10 | 40 | 1952 | Midlothian |
| 241 | Bankend 8 | 38 | 38 | 1 | 38 | 1947 | Lanarkshire |
| 242 | Woodside | 38 | 299 | 8 | 40 | 1954 | Lanarkshire |
| 243 | Bowhill (Patna) | 37 | 37 | 1 | 37 | 1947 | Ayrshire |
| 244 | Mortonmuir | 37 | 73 | 2 | 38 | 1952 | Ayrshire |
| 245 | Gateside | 37 | 37 | 1 | 37 | 1947 | Stirlingshire |
| 246 | Gillhead | 35 | 279 | 8 | 41 | 1947 | Stirlingshire |
| 247 | Fauldhouse 1 | 35 | 35 | 2 | 35 | 1948 | West Lothian |
| 248 | Avonbraes | 32 | 409 | 13 | 39 | 1952 | Lanarkshire |
| 249 | Knowetop | 31 | 453 | 15 | 42 | 1952 | Lanarkshire |
| 250 | Windyedge | 26 | 26 | 1 | 26 | 1950 | Fife |
| 251 | Dumback | 26 | 129 | 5 | 29 | 1958 | West Lothian |
| 252 | Bankend 15 & 16 | 24 | 24 | 1 | 24 | 1947 | Lanarkshire |
| 253 | Afton | 23 | 68 | 3 | 26 | 1959 | Ayrshire |
| 254 | Greenlees | 23 | 227 | 10 | 309 | 1957 | Lanarkshire |
| 255 | Bishop 3 | 21 | 21 | 1 | 21 | 1947 | Lanarkshire |
| 256 | Melloch | 19 | 38 | 2 | 29 | 1947 | Clackmannanshire |

| Rank | Mine/Colliery | Mean workforce | Total workforce | Years open | Peak workforce | Peak year | County |
|------|---------------|----------------|-----------------|------------|----------------|-----------|--------|
| 257 | Carnbroe | 18 | 18 | 1 | 18 | 1947 | Lanarkshire |
| 258 | Ayr 1 & 2 | 17 | 85 | 5 | 17 | 1947 | Ayrshire |
| 259 | Barleyside West Mine | 17 | 34 | 2 | 18 | 1947 | Lanarkshire |
| 260 | Langside | 17 | 17 | 1 | 17 | 1948 | Lanarkshire |
| 261 | Beaton's Lodge | 15 | 135 | 9 | 31 | 1952 | Lanarkshire |
| 262 | Blackston | 14 | 14 | 1 | 14 | 1947 | Stirlingshire |

# Appendix B
## Scottish NCB Collieries, ranked by total workforce, 1947–90

| Rank | Mine/Colliery | Total workforce | Years open | Mean workforce | Peak workforce | Peak year | County |
|---|---|---|---|---|---|---|---|
| 1 | Polkemmet | 59,836 | 40 | 1,496 | 1,896 | 1960 | West Lothian |
| 2 | Bilston Glen | 56,004 | 26 | 2,154 | 2,367 | 1970 | Midlothian |
| 3 | Cardowan | 55,240 | 37 | 1,493 | 1,870 | 1959 | Lanarkshire |
| 4 | Killoch | 54,357 | 27 | 2,014 | 2,305 | 1965 | Ayrshire |
| 5 | Michael | 51,962 | 20 | 2,598 | 3,353 | 1957 | Fife |
| 6 | Comrie | 49,778 | 40 | 1,245 | 1,498 | 1963 | Fife |
| 7 | Seafield | 47,927 | 22 | 2,178 | 2,466 | 1970 | Fife |
| 8 | Lady Victoria | 46,856 | 35 | 1,339 | 1,765 | 1953 | Midlothian |
| 9 | Barony 1, 2 & 3 | 45,257 | 42 | 1,078 | 1,695 | 1958 | Ayrshire |
| 10 | Frances | 45,241 | 41 | 1,104 | 1,482 | 1957 | Fife |
| 11 | Wellesley | 38,160 | 20 | 1,908 | 2,603 | 1957 | Fife |
| 12 | Monktonhall | 35,588 | 22 | 1,618 | 1,786 | 1971 | Midlothian |
| 13 | Kinneil | 33,150 | 28 | 1,184 | 1,268 | 1960 | West Lothian |
| 14 | Bedlay | 26,909 | 34 | 792 | 870 | 1959 | Lanarkshire |
| 15 | Polmaise 3 & 4 | 24,812 | 41 | 606 | 778 | 1957 | Stirlingshire |
| 16 | Bowhill | 23,954 | 18 | 1,331 | 1,490 | 1961 | Fife |
| 17 | Whitrigg 2, 4 & 5 | 23,824 | 25 | 953 | 1,208 | 1952 | West Lothian |
| 18 | Blairhall | 21,600 | 22 | 982 | 1,053 | 1947 | Fife |
| 19 | Fauldhead 1, 2, 3 & 4 | 21,361 | 21 | 1,018 | 1,155 | 1949 | Dumfriesshire |
| 20 | Manor Powis | 20,036 | 25 | 802 | 985 | 1963 | Stirlingshire |
| 21 | Aitken | 19,975 | 16 | 1,249 | 1,431 | 1956 | Fife |
| 22 | Glencraig | 18,786 | 17 | 1,105 | 1,316 | 1950 | Fife |
| 23 | Pennyvenie 2, 3 & 7 | 18,578 | 32 | 581 | 725 | 1961 | Ayrshire |
| 24 | Bogside | 18,561 | 30 | 618 | 875 | 1971 | Fife |
| 25 | Dalkieth | 18,506 | 30 | 617 | 898 | 1964 | Midlothian |
| 26 | Kingshill 1 | 18,381 | 21 | 876 | 1,381 | 1951 | Lanarkshire |
| 27 | Solsgirth (Longannet Complex) | 18,329 | 20 | 917 | 1,007 | 1975 | Clackmannanshire |
| 28 | Easthouses | 17,980 | 22 | 818 | 1,030 | 1963 | Midlothian |
| 29 | Auchencruive 4 & 5 | 17,112 | 26 | 659 | 958 | 1957 | Ayrshire |
| 30 | Douglas | 16,837 | 20 | 842 | 1,031 | 1952 | Lanarkshire |
| 31 | Valleyfield | 16,690 | 17 | 982 | 1,052 | 1959 | Fife |
| 32 | Woolmet | 16,397 | 19 | 863 | 960 | 1963 | Midlothian |
| 33 | Newcraighall | 15,952 | 21 | 760 | 810 | 1950 | Midlothian |
| 34 | Roslin | 15,822 | 21 | 754 | 770 | 1952 | Midlothian |
| 35 | Easton | 15,012 | 21 | 715 | 863 | 1958 | West Lothian |
| 36 | Mauchline 1, 2 & 3 | 14,879 | 19 | 783 | 876 | 1958 | Ayrshire |
| 37 | Lindsay | 14,714 | 18 | 818 | 970 | 1957 | Fife |
| 38 | Arniston (Emily & Gore) | 14,693 | 15 | 980 | 990 | 1951 | Midlothian |
| 39 | Lingerwood | 14,537 | 20 | 727 | 770 | 1951 | Midlothian |
| 40 | Littlemill 2 & 3 | 13,853 | 28 | 495 | 810 | 1960 | Ayrshire |

| Rank | Mine/Colliery | Total workforce | Years open | Mean workforce | Peak workforce | Peak year | County |
|---|---|---|---|---|---|---|---|
| 41 | Kingshill 3 | 13,795 | 23 | 600 | 769 | 1958 | Lanarkshire |
| 42 | Prestonlinks | 13,492 | 17 | 794 | 820 | 1950 | East Lothian |
| 43 | Castlehill | 13,480 | 18 | 749 | 770 | 1972 | Fife |
| 44 | Lochhead (inc. surface mine) | 13,324 | 23 | 580 | 841 | 1957 | Fife |
| 45 | Highhouse 1 & 2 | 13,258 | 36 | 369 | 467 | 1947 | Ayrshire |
| 46 | Burghlee | 13,049 | 17 | 768 | 795 | 1952 | Midlothian |
| 47 | Gartshore 9 & 11 | 12,919 | 21 | 616 | 740 | 1952 | Dunbartonshire |
| 48 | Minto | 12,784 | 20 | 640 | 730 | 1957 | Fife |
| 49 | Wester Auchengeich | 12,611 | 21 | 601 | 850 | 1959 | Lanarkshire |
| 50 | Kinglassie | 12,398 | 19 | 652 | 764 | 1956 | Fife |
| 51 | Auchencruive 1, 2 & 3 | 12,242 | 14 | 875 | 1,081 | 1947 | Ayrshire |
| 52 | Knockshinnoch Castle | 12,135 | 21 | 578 | 755 | 1956 | Ayrshire |
| 53 | Kames 1 & 2 | 11,936 | 21 | 569 | 634 | 1957 | Ayrshire |
| 54 | Lumphinnans XI & XII | 11,909 | 20 | 596 | 764 | 1947 | Fife |
| 55 | Auchengeich | 11,701 | 18 | 650 | 860 | 1956 | Lanarkshire |
| 56 | Riddochill | 11,358 | 21 | 541 | 687 | 1958 | West Lothian |
| 57 | Auchlochan 2, 6, 7, 9 & 10 | 11,250 | 21 | 534 | 687 | 1952 | Lanarkshire |
| 58 | Mary (Lochore) | 11,051 | 18 | 614 | 780 | 1957 | Fife |
| 59 | Kingshill 2 | 10,604 | 17 | 624 | 686 | 1954 | Lanarkshire |
| 60 | Prestongrange | 10,290 | 15 | 686 | 700 | 1952 | East Lothian |
| 61 | Fordell | 9,672 | 19 | 509 | 622 | 1959 | Fife |
| 62 | Dumbreck | 9,188 | 16 | 575 | 669 | 1947 | Stirlingshire |
| 63 | Beoch 2, 3 & 4 | 8,561 | 21 | 408 | 460 | 1960 | Ayrshire |
| 64 | Benhar | 8,173 | 15 | 545 | 772 | 1947 | Lanarkshire |
| 65 | Devon | 7,984 | 13 | 615 | 813 | 1954 | Clackmannanshire |
| 66 | Whitehill | 7,858 | 14 | 562 | 590 | 1950 | Midlothian |
| 67 | Carberry | 7,652 | 13 | 589 | 615 | 1952 | Midlothian |
| 68 | Blantyreferme 1 & 2 | 7,594 | 15 | 507 | 656 | 1950 | Lanarkshire |
| 69 | Bank 1, 2 & 6 | 7,291 | 22 | 332 | 400 | 1952 | Ayrshire |
| 70 | Greenrigg | 7,250 | 13 | 558 | 634 | 1952 | West Lothian |
| 71 | Polmaise 1 & 2 | 7,249 | 11 | 659 | 723 | 1951 | Stirlingshire |
| 72 | Sorn | 7,206 | 31 | 233 | 294 | 1970 | Ayrshire |
| 73 | Bardykes | 7,176 | 15 | 479 | 775 | 1947 | Lanarkshire |
| 74 | Whitehill 1 & 2 | 7,060 | 18 | 392 | 435 | 1949 | Ayrshire |
| 75 | Plean | 7,049 | 16 | 441 | 528 | 1958 | Stirlingshire |
| 76 | Nellie | 6,837 | 16 | 428 | 508 | 1958 | Fife |
| 77 | Fleets | 6,834 | 12 | 570 | 854 | 1947 | East Lothian |
| 78 | Hamilton Palace | 6,621 | 12 | 552 | 600 | 1948 | Lanarkshire |
| 79 | Canderrigg 4, 5, and Canderside | 6,439 | 17 | 379 | 582 | 1947 | Lanarkshire |
| 80 | Minnivey | 6,420 | 17 | 378 | 470 | 1962 | Ayrshire |
| 81 | Ramsay | 6,228 | 18 | 346 | 385 | 1952 | Midlothian |
| 82 | Northfield & Hall | 6,196 | 14 | 443 | 484 | 1958 | Lanarkshire |
| 83 | Loganlea | 6,118 | 12 | 510 | 674 | 1947 | West Lothian |
| 84 | Twechar 1 | 6,023 | 17 | 355 | 413 | 1954 | Dunbartonshire |
| 85 | Gateside & Tower | 5,847 | 17 | 344 | 410 | 1952 | Dumfriesshire |

| Rank | Mine/Colliery | Total workforce | Years open | Mean workforce | Peak workforce | Peak year | County |
|---|---|---|---|---|---|---|---|
| 86 | Bannockburn | 5,833 | 17 | 344 | 563 | 1947 | Stirlingshire |
| 87 | Balgonie | 5,815 | 13 | 448 | 490 | 1952 | Fife |
| 88 | Lady Helen | 5,807 | 17 | 342 | 455 | 1957 | Fife |
| 89 | Cowdenbeath 7 & 10 | 5,652 | 13 | 435 | 454 | 1950 | Fife |
| 90 | Jenny Gray | 5,603 | 12 | 467 | 492 | 1950 | Fife |
| 91 | Gartshore 3 & 12 | 5,465 | 10 | 547 | 611 | 1948 | Dunbartonshire |
| 92 | Hassockrig | 5,430 | 15 | 362 | 379 | 1958 | Lanarkshire |
| 93 | Southfield | 5,406 | 11 | 492 | 556 | 1947 | Lanarkshire |
| 94 | Ayr 9 & 10 (Enterkine) | 5,209 | 12 | 434 | 665 | 1948 | Ayrshire |
| 95 | Blantyreferme 3 | 5,119 | 13 | 394 | 430 | 1956 | Lanarkshire |
| 96 | Houldsworth | 4,970 | 18 | 276 | 346 | 1954 | Ayrshire |
| 97 | Killochan | 4,942 | 20 | 247 | 289 | 1952 | Ayrshire |
| 98 | Calderehead 3 & 4 | 4,807 | 11 | 437 | 617 | 1947 | Lanarkshire |
| 99 | Rothes | 4,800 | 5 | 960 | 1,235 | 1960 | Fife |
| 100 | Kennox 6 & 7 | 4,728 | 25 | 190 | 227 | 1960 | Lanarkshire |
| 101 | Gilmerton | 4,701 | 15 | 314 | 380 | 1950 | Midlothian |
| 102 | Polquhairn | 4,617 | 16 | 289 | 460 | 1960 | Ayrshire |
| 103 | Blantyre | 4,601 | 10 | 460 | 533 | 1947 | Lanarkshire |
| 104 | Ardenrigg 6 | 4,513 | 16 | 282 | 304 | 1951 | Lanarkshire |
| 105 | Foulshiels | 4,477 | 10 | 448 | 463 | 1951 | West Lothian |
| 106 | Douglas Castle | 4,367 | 12 | 364 | 415 | 1952 | Lanarkshire |
| 107 | Randolph | 4,351 | 21 | 208 | 284 | 1947 | Fife |
| 108 | Overtown | 4,260 | 15 | 284 | 466 | 1960 | Lanarkshire |
| 109 | Baads | 4,221 | 11 | 384 | 491 | 1955 | Midlothian |
| 110 | Cairnhill | 4,194 | 19 | 221 | 250 | 1970 | Ayrshire |
| 111 | Bothwell Castle 3 & 4 | 4,140 | 12 | 345 | 655 | 1947 | Lanarkshire |
| 112 | Woodmuir | 4,072 | 16 | 255 | 273 | 1963 | Midlothian |
| 113 | Woodend | 3,965 | 10 | 397 | 488 | 1954 | West Lothian |
| 114 | Lochlea | 3,933 | 24 | 164 | 239 | 1957 | Ayrshire |
| 115 | Pennyvenie 4 | 3,887 | 14 | 278 | 335 | 1957 | Ayrshire |
| 116 | Blairmuckhill | 3,851 | 12 | 321 | 347 | 1952 | Lanarkshire |
| 117 | Garscube | 3,845 | 19 | 203 | 311 | 1963 | Lanarkshire |
| 118 | Thankerton 1, 3 & 6 | 3,816 | 6 | 636 | 766 | 1947 | Lanarkshire |
| 119 | Branchall (Greenhead 2, 3) | 3,775 | 12 | 315 | 366 | 1952 | Lanarkshire |
| 120 | Dalquharran | 3,726 | 26 | 144 | 237 | 1958 | Ayrshire |
| 121 | Glenochil | 3,521 | 6 | 587 | 908 | 1960 | Clackmannanshire |
| 122 | Argyll | 3,363 | 20 | 168 | 245 | 1953 | Argyll |
| 123 | Chalmerston 4, 5, 6 & 7 | 3,329 | 13 | 256 | 300 | 1951 | Ayrshire |
| 124 | Herbertshire | 3,255 | 12 | 272 | 394 | 1948 | Stirlingshire |
| 125 | Dullatur | 3,250 | 17 | 192 | 281 | 1963 | Dunbartonshire |
| 126 | Wellsgreen | 3,209 | 12 | 268 | 407 | 1949 | Fife |
| 127 | Dora (Little Raith) | 3,189 | 12 | 266 | 280 | 1957 | Fife |
| 128 | Bothwell Castle 1 & 2 | 3,111 | 5 | 623 | 751 | 1947 | Lanarkshire |
| 129 | Brucefield | 3,017 | 14 | 216 | 265 | 1957 | Clackmannanshire |
| 130 | Tynemount & Oxenford | 3,003 | 11 | 273 | 370 | 1948 | East Lothian |

| Rank | Mine/Colliery | Total workforce | Years open | Mean workforce | Peak workforce | Peak year | County |
|------|---------------|-----------------|------------|----------------|----------------|-----------|--------|
| 131 | Castlehill 6 | 2,948 | 7 | 422 | 465 | 1947 | Lanarkshire |
| 132 | Pirnhall | 2,780 | 16 | 174 | 241 | 1957 | Stirlingshire |
| 133 | Hopetoun | 2,686 | 5 | 538 | 622 | 1951 | West Lothian |
| 134 | Glentaggart | 2,624 | 22 | 120 | 130 | 1952 | Lanarkshire |
| 135 | Redding | 2,576 | 11 | 235 | 283 | 1948 | Stirlingshire |
| 136 | Zetland | 2,489 | 13 | 192 | 323 | 1957 | Clackmannanshire |
| 137 | Maxwell | 2,466 | 26 | 95 | 133 | 1958 | Ayrshire |
| 138 | Wilsontown | 2,431 | 8 | 304 | 378 | 1948 | Lanarkshire |
| 139 | Benarty | 2,413 | 12 | 201 | 237 | 1957 | Fife |
| 140 | Roger | 2,366 | 27 | 88 | 121 | 1975 | Dumfriesshire |
| 141 | Bellyford | 2,335 | 7 | 334 | 335 | 1954 | East Lothian |
| 142 | Blackrigg 1 & 3 | 2,323 | 8 | 291 | 339 | 1951 | West Lothian |
| 143 | Dollar | 2,288 | 17 | 135 | 454 | 1963 | Perthshire |
| 144 | Meta (Devon 3) | 2,238 | 12 | 187 | 233 | 1956 | Clackmannanshire |
| 145 | Blair | 2,221 | 14 | 157 | 201 | 1960 | Ayrshire |
| 146 | Forthbank | 2,118 | 6 | 353 | 992 | 1956 | Clackmannanshire |
| 147 | Cronberry Moor | 1,949 | 12 | 163 | 222 | 1947 | Ayrshire |
| 148 | Carriden | 1,922 | 6 | 321 | 331 | 1948 | West Lothian |
| 149 | Torry | 1,904 | 13 | 147 | 198 | 1962 | Fife |
| 150 | Limeylands | 1,877 | 7 | 269 | 335 | 1952 | East Lothian |
| 151 | Westoun (Coalburn) | 1,864 | 14 | 134 | 175 | 1956 | Lanarkshire |
| 152 | Policy | 1,857 | 12 | 155 | 185 | 1948 | Stirlingshire |
| 153 | Blairenbathie | 1,836 | 14 | 132 | 151 | 1957 | Fife |
| 154 | Baton | 1,777 | 4 | 445 | 772 | 1950 | Lanarkshire |
| 155 | Bankend | 1,767 | 11 | 161 | 205 | 1949 | Lanarkshire |
| 156 | Rosie | 1,747 | 7 | 250 | 306 | 1947 | Fife |
| 157 | Whitehill 3 & 4 | 1,747 | 16 | 109 | 140 | 1957 | Ayrshire |
| 158 | Boglea & Glentore | 1,733 | 13 | 134 | 221 | 1956 | Lanarkshire |
| 159 | Cameron | 1,717 | 12 | 143 | 207 | 1957 | Fife |
| 160 | Tillicoultry | 1,703 | 10 | 170 | 346 | 1950 | Clackmannanshire |
| 161 | Balmore | 1,694 | 13 | 131 | 173 | 1947 | Stirlingshire |
| 162 | Auchmeddan 1, 2 & 3 | 1,686 | 20 | 85 | 101 | 1967 | Lanarkshire |
| 163 | Knowehead | 1,663 | 11 | 152 | 203 | 1959 | Lanarkshire |
| 164 | Lumphinnans 1 | 1,635 | 9 | 182 | 202 | 1951 | Fife |
| 165 | Harwood | 1,580 | 12 | 132 | 196 | 1958 | Midlothian |
| 166 | Rig | 1,454 | 12 | 122 | 93 | 1957 | Dumfriesshire |
| 167 | Shewalton 3 & 4 | 1,429 | 8 | 179 | 266 | 1947 | Ayrshire |
| 168 | Kepplehill & Stane | 1,392 | 5 | 279 | 297 | 1947 | Lanarkshire |
| 169 | Bridgend | 1,337 | 15 | 89 | 146 | 1961 | Ayrshire |
| 170 | East Benhar | 1,303 | 10 | 139 | 159 | 1951 | West Lothian |
| 171 | South Bantaskine | 1,277 | 12 | 107 | 141 | 1954 | Stirlingshire |
| 172 | Coatspark | 1,142 | 11 | 104 | 110 | 1952 | Lanarkshire |
| 173 | Newfield | 1,135 | 9 | 126 | 144 | 1952 | Ayrshire |
| 174 | Coalburn | 1,069 | 15 | 72 | 109 | 1952 | Ayrshire |
| 175 | Longannet Mine | 1,063 | 19 | 56 | 60 | 1987 | Fife |

| Rank | Mine/Colliery | Total workforce | Years open | Mean workforce | Peak workforce | Peak year | County |
|------|---------------|-----------------|------------|----------------|----------------|-----------|--------|
| 176 | Thinacres | 1,002 | 11 | 91 | 99 | 1956 | Lanarkshire |
| 177 | Glencairn | 1,000 | 14 | 72 | 190 | 1948 | East Lothian |
| 178 | Castlebridge | 985 | 6 | 985 | 985 | 1988 | Clackmannanshire |
| 179 | Seaforth 1, 2 & 3 | 975 | 6 | 163 | 241 | 1947 | Ayrshire |
| 180 | Harvieston | 918 | 4 | 230 | 237 | 1960 | Clackmannanshire |
| 181 | Earlseat | 906 | 8 | 114 | 162 | 1956 | Fife |
| 182 | Wester Gartshore | 896 | 3 | 299 | 318 | 1947 | Dunbartonshire |
| 183 | Fortacres | 872 | 10 | 88 | 109 | 1952 | Ayrshire |
| 184 | Whitrigg 6 | 870 | 8 | 109 | 111 | 1952 | West Lothian |
| 185 | Winton | 850 | 10 | 85 | 85 | 1952 | East Lothian |
| 186 | Glen Mine | 816 | 7 | 117 | 126 | 1948 | Lanarkshire |
| 187 | King O'Muirs | 812 | 10 | 82 | 153 | 1955 | Clackmannanshire |
| 188 | Castle Mine | 810 | 8 | 102 | 150 | 1950 | Midlothian |
| 189 | Glentore | 783 | 7 | 112 | 160 | 1957 | Lanarkshire |
| 190 | Greenhill | 770 | 11 | 70 | 74 | 1954 | Ayrshire |
| 191 | Duntilland | 722 | 4 | 181 | 182 | 1947 | Lanarkshire |
| 192 | Meadowmill | 720 | 6 | 120 | 120 | 1954 | East Lothian |
| 193 | Lugar | 692 | 6 | 116 | 130 | 1949 | Ayrshire |
| 194 | Barbeth | 682 | 5 | 137 | 149 | 1953 | Ayrshire |
| 195 | Criagrie | 669 | 5 | 134 | 136 | 1949 | Clackmannanshire |
| 196 | Cuthill | 619 | 3 | 207 | 303 | 1959 | West Lothian |
| 197 | Bogton | 602 | 8 | 76 | 99 | 1947 | Ayrshire |
| 198 | Thornton | 586 | 7 | 84 | 113 | 1950 | Fife |
| 199 | Auldton | 584 | 7 | 84 | 96 | 1960 | Lanarkshire |
| 200 | Sundrum | 578 | 9 | 65 | 77 | 1958 | Ayrshire |
| 201 | Fortissat | 555 | 2 | 278 | 283 | 1948 | Lanarkshire |
| 202 | Barbauchlaw | 519 | 5 | 104 | 115 | 1947 | West Lothian |
| 203 | Stane Mine | 464 | 3 | 155 | 210 | 1953 | Lanarkshire |
| 204 | Shewalton 5 & 6 | 457 | 3 | 153 | 205 | 1949 | Ayrshire |
| 205 | Quarter 1 (and mine) | 456 | 5 | 92 | 108 | 1948 | Lanarkshire |
| 206 | Oxenford 2 | 455 | 3 | 152 | 170 | 1948 | East Lothian |
| 207 | Tofts 1 & 2 | 454 | 3 | 152 | 289 | 1947 | Ayrshire |
| 208 | Knowetop | 453 | 15 | 31 | 42 | 1952 | Lanarkshire |
| 209 | Newbattle (surface) | 441 | 1 | 441 | 441 | 1947 | Midlothian |
| 210 | Andershaw | 441 | 9 | 49 | 55 | 1952 | Lanarkshire |
| 211 | Rankin | 440 | 4 | 110 | 120 | 1947 | Lanarkshire |
| 212 | Wellwood | 429 | 2 | 215 | 215 | 1949 | Fife |
| 213 | Avonbraes | 409 | 13 | 32 | 39 | 1952 | Lanarkshire |
| 214 | Edgehead | 388 | 10 | 39 | 40 | 1952 | Midlothian |
| 215 | Gateside 1 | 362 | 5 | 72 | 78 | 1950 | Stirlingshire |
| 216 | Skellyton 2 & 3 | 359 | 4 | 90 | 90 | 1948 | Lanarkshire |
| 217 | Lochgelly | 352 | 1 | 352 | 352 | 1947 | Fife |
| 218 | Warrix 1 & 2 | 332 | 3 | 111 | 130 | 1948 | Ayrshire |
| 219 | Netherton | 318 | 3 | 106 | 106 | 1947 | Lanarkshire |

| Rank | Mine/Colliery | Total workforce | Years open | Mean workforce | Peak workforce | Peak year | County |
|------|---------------|-----------------|------------|----------------|----------------|-----------|--------|
| 220 | Woodside | 299 | 8 | 38 | 40 | 1954 | Lanarkshire |
| 221 | Headless Cross 2 | 290 | 6 | 49 | 91 | 1951 | Lanarkshire |
| 222 | Gillhead | 279 | 8 | 35 | 41 | 1947 | Stirlingshire |
| 223 | Carston | 271 | 7 | 39 | 44 | 1952 | Ayrshire |
| 224 | Hillhouserigg | 265 | 2 | 265 | 265 | 1947 | Lanarkshire |
| 225 | Pennyvenie 5 | 264 | 4 | 66 | 70 | 1947 | Ayrshire |
| 226 | Ashgill | 240 | 4 | 60 | 60 | 1947 | Lanarkshire |
| 227 | Dalzell & Broomside | 234 | 1 | 234 | 234 | 1947 | Lanarkshire |
| 228 | Hindsward | 233 | 3 | 78 | 83 | 1957 | Ayrshire |
| 229 | Gateside 2 | 233 | 3 | 78 | 78 | 1947 | Stirlingshire |
| 230 | Greenlees | 227 | 10 | 23 | 309 | 1957 | Lanarkshire |
| 231 | Elgin & Wellwood | 214 | 1 | 214 | 214 | 1947 | Fife |
| 232 | Shieldmains 14 Drongan | 186 | 3 | 62 | 101 | 1947 | Ayrshire |
| 233 | Shieldmains, Barbeth 2 | 179 | 3 | 60 | 90 | 1949 | Ayrshire |
| 234 | Broomlands | 146 | 3 | 49 | 59 | 1952 | Ayrshire |
| 235 | Sheildmains 6 & 7 Drongan | 146 | 3 | 49 | 70 | 1948 | Ayrshire |
| 236 | Oakfield | 140 | 2 | 70 | 70 | 1947 | Fife |
| 237 | Beaton's Lodge | 135 | 9 | 15 | 31 | 1952 | Lanarkshire |
| 238 | Dumback | 129 | 5 | 26 | 29 | 1958 | West Lothian |
| 239 | Chapel | 122 | 2 | 61 | 61 | 1947 | Lanarkshire |
| 240 | Headless Cross 1 | 100 | 2 | 50 | 50 | 1947 | Lanarkshire |
| 241 | Southhook | 91 | 1 | 91 | 91 | 1947 | Ayrshire |
| 242 | Lochend 5 | 89 | 1 | 89 | 89 | 1949 | Lanarkshire |
| 243 | Ayr 1 & 2 | 85 | 5 | 17 | 17 | 1947 | Ayrshire |
| 244 | Lindsay 2 | 83 | 1 | 83 | 83 | 1947 | Fife |
| 245 | Isle of Canty | 82 | 2 | 41 | 41 | 1947 | Fife |
| 246 | Overwood | 78 | 2 | 39 | 39 | 1952 | Lanarkshire |
| 247 | Quarter 2 | 74 | 1 | 74 | 74 | 1947 | Lanarkshire |
| 248 | Mortonmuir | 73 | 2 | 37 | 38 | 1952 | Ayrshire |
| 249 | Afton | 68 | 3 | 23 | 26 | 1959 | Ayrshire |
| 250 | Shewalton 8 & 9 | 50 | 1 | 50 | 50 | 1947 | Ayrshire |
| 251 | Bankend 8 | 38 | 1 | 38 | 38 | 1947 | Lanarkshire |
| 252 | Melloch | 38 | 2 | 19 | 29 | 1947 | Clackmannanshire |
| 253 | Bowhill (Patna) | 37 | 1 | 37 | 37 | 1947 | Ayrshire |
| 254 | Gateside | 37 | 1 | 37 | 37 | 1947 | Stirlingshire |
| 255 | Fauldhouse 1 | 35 | 2 | 35 | 35 | 1948 | West Lothian |
| 256 | Barleyside West Mine | 34 | 2 | 17 | 18 | 1947 | Lanarkshire |
| 257 | Windyedge | 26 | 1 | 26 | 26 | 1950 | Fife |
| 258 | Bankend 15 & 16 | 24 | 1 | 24 | 24 | 1947 | Lanarkshire |
| 259 | Bishop 3 | 21 | 1 | 21 | 21 | 1947 | Lanarkshire |
| 260 | Carnbroe | 18 | 1 | 18 | 18 | 1947 | Lanarkshire |
| 261 | Langside | 17 | 1 | 17 | 17 | 1948 | Lanarkshire |
| 262 | Blackston | 14 | 1 | 14 | 14 | 1947 | Stirlingshire |

# Appendix C
## Previous owners of NCB Collieries

Key

*c.* – approximate date

*r.* – re-opened

*p.* – initially remained in private sector, nationalised later

| Company | Mine/Colliery | County | Open | Closed |
|---|---|---|---|---|
| A & G Anderson | Blairmuckhill | Lanarkshire | 1910 | 1959 |
| A G Moore & Co. | Barbeth 2 (Shieldmains) | Ayrshire | 1945 | 1955 |
| A G Moore & Co. | Shieldmains 6, 7 & 14 (Drongan) | Ayrshire | 1927 | 1950 |
| A G Moore & Co. | Blantyreferme 1, 2 | Lanarkshire | 1894 | 1962 |
| A G Moore & Co. | Blantyreferme 3 | Lanarkshire | 1850 | 1964 |
| A G Moore & Co. | Dalkeith | Midlothian | 1903 | 1978 |
| A Kenneth & Sons | Newfield | Ayrshire | 1940 | 1956 |
| A Kenneth & Sons | Shewalton 3 & 4 | Ayrshire | 1924 | 1955 |
| A Kenneth & Sons | Shewalton 5 & 6 | Ayrshire | 1933 | 1950 |
| A Kenneth & Sons | Warrix 1 & 2 | Ayrshire | 1944 | 1950 |
| Alloa Coal Co. | Craigrie | Clackmannanshire | *r.*1942 | 1952 |
| Alloa Coal Co. | Devon | Clackmannanshire | *r.*1879 | 1960 |
| Alloa Coal Co. | Dollar | Perthshire | 1943 | 1973 |
| Alloa Coal Co. | Forthbank Colliery 1 & 2 | Clackmannanshire | 1947 | 1958 |
| Alloa Coal Co. | King O' Muirs 1 | Clackmannanshire | 1938 | 1954 |
| Alloa Coal Co. | King O' Muirs 2 | Clackmannanshire | 1950 | 1957 |
| Alloa Coal Co. | Melloch | Clackmannanshire | *c.*1850 | 1948 |
| Alloa Coal Co. | Meta | Clackmannanshire | 1923 | 1959 |
| Alloa Coal Co. | Tillicoultry | Clackmannanshire | 1947 | 1957 |
| Alloa Coal Co. | Zetland | Clackmannanshire | 1935 | 1960 |
| Alloa Coal Co. | Isle of Canty | Fife | 1939 | 1948 |
| Archibald Russell | Fernigair | Lanarkshire | *c.*1850 | 1947 |
| Archibald Russell | Polmaise 1 & 2 | Stirlingshire | 1904 | 1958 |
| Archibald Russell | Polmaise 3 & 4 | Stirlingshire | 1904 | 1987 |
| Arden Coal Co. | Bankend (Westoun Mine) | Lanarkshire | 1948 | 1961 |
| Arden Coal Co. | Bankend Colliery | Lanarkshire | 1890 | 1958 |
| Ardenrigg Coal Co. | Ardenrigg 6 | Lanarkshire | 1926 | 1963 |
| Arniston Coal Co. | Arniston (Emily & Gore) | Midlothian | 1858 | 1962 |
| Auldton Colliery (Boyd Bros.) | Auldton 2 & 3 | Lanarkshire | 1941 | 1963 |
| Bairds & Dalmellington | Bogton Mine | Ayrshire | 1931 | 1954 |
| Bairds & Dalmellington | Littlemill 2, 3 & 5 | Ayrshire | 1860 | 1974 |
| Bairds & Dalmellington | Pennyvenie 2, 3 & 7 | Ayrshire | 1872 | 1978 |
| Bairds & Dalmellington | Tofts | Ayrshire | 1914 | 1948 |
| Bairds & Dalmellington | Fauldhead 1 & 3 | Dumfriesshire | 1896 | 1968 |

| Company | Mine/Colliery | County | Open | Closed |
|---|---|---|---|---|
| Bairds & Dalmellington | Tower Mine | Dumfriesshire | 1916 | 1964 |
| Bairds & Dalmellington | Auchincruive 1, 2 & 3 | Ayrshire | 1897 | 1960 |
| Bairds & Dalmellington | Auchincruive 4, 5, 6 & 7 | Ayrshire | 1912 | 1973 |
| Bairds & Dalmellington | Barony 1, 2 & 3 | Ayrshire | 1910 | 1989 |
| Bairds & Dalmellington | Beoch 3 | Ayrshire | 1866 | 1968 |
| Bairds & Dalmellington | Beoch 4 | Ayrshire | 1937 | 1968 |
| Bairds & Dalmellington | Chalmerston 4 & 5 | Ayrshire | 1925 | 1959 |
| Bairds & Dalmellington | Chalmerston 7 | Ayrshire | 1934 | 1952 |
| Bairds & Dalmellington | Cronberry Moor | Ayrshire | 1920 | 1957 |
| Bairds & Dalmellington | Enterkine 9 & 10 | Ayrshire | 1878 | 1959 |
| Bairds & Dalmellington | Highhouse | Ayrshire | 1894 | 1983 |
| Bairds & Dalmellington | Houldsworth | Ayrshire | 1901 | 1965 |
| Bairds & Dalmellington | Kames | Ayrshire | c.1870 | 1968 |
| Bairds & Dalmellington | Lugar Mine | Ayrshire | 1942 | 1953 |
| Bairds & Dalmellington | Mauchline 1, 2 & 4 | Ayrshire | 1925 | 1966 |
| Bairds & Dalmellington | Pennyvenie 4 | Ayrshire | 1911 | 1961 |
| Bairds & Dalmellington | Pennyvenie 5 | Ayrshire | 1911 | 1953 |
| Bairds & Dalmellington | Whitehill 1 & 2 | Ayrshire | c.1893 | 1965 |
| Bairds & Dalmellington | Whitehill 3 & 4 | Ayrshire | 1946 | 1965 |
| Bairds & Dalmellington | Gateside 4 & 5 | Dumfriesshire | 1891 | 1964 |
| Bairds & Scottish Steel | Gartshore 3 & 12 | Dunbartonshire | 1865 | 1959 |
| Bairds & Scottish Steel | Gartshore 9, 11, & Grayshill | Dunbartonshire | 1875 | 1968 |
| Bairds & Scottish Steel | Twechar No.1 & Gartshore 10 | Dunbartonshire | 1865 | 1964 |
| Bairds & Scottish Steel | Bedlay | Lanarkshire | 1905 | 1981 |
| Bairds & Scottish Steel | Bothwell Castle 1 & 2 | Lanarkshire | 1875 | 1950 |
| Bairds & Scottish Steel | Bothwell Castle 3 & 4 | Lanarkshire | 1889 | 1959 |
| Bairds & Scottish Steel | Dumbreck | Stirlingshire | 1887 | 1963 |
| Bairds & Scottish Steel | Easton | West Lothian | 1898 | 1973 |
| Bairds & Scottish Steel | Riddochhill | West Lothian | 1890 | 1968 |
| Balgonie Colliery Co. | Balgonie | Fife | 1883 | 1960 |
| Barr & Thornton | East Benhar Mine | West Lothian | 1940 | 1957 |
| Bent Colliery Co. | Hamilton Palace | Lanarkshire | 1884 | 1959 |
| Brownieside Coal Co. | Boglea | Lanarkshire | 1942 | 1962 |
| Brownieside Coal Co. | Lochend 5 | Lanarkshire | c.1880 | 1948 |
| Brownieside Coal Co. | Blackston Mine | Stirlingshire | 1928 | 1948 |
| Cadzow Coal Co. | Dullatur | Dunbartonshire | 1935 | 1964 |
| Cadzow Coal Co. | Wester Gartshore | Dunbartonshire | 1872 | 1950 |
| Callender Coal Co. | Policy | Stirlingshire | 1924 | 1959 |
| Callender Coal Co. | South Bantaskine | Stirlingshire | 1946 | 1959 |
| Carriden Coal Co. | Carriden 1 & 2 | West Lothian | 1914 | 1953 |
| Carron Co. | Bannockburn | Stirlingshire | 1894 | 1953 |
| Carron Co. | Pirnhall | Stirlingshire | 1933 | 1963 |
| Coltness Iron Co. | Blairhall | Fife | c.1870 | 1969 |
| Coltness Iron Co. | Branchal | Lanarkshire | 1924 | 1959 |

| Company | Mine/Colliery | County | Open | Closed |
|---|---|---|---|---|
| Coltness Iron Co. | Dewshill  (formerly Duntilland) | Lanarkshire | 1923 | 1943 |
| Coltness Iron Co. | Douglas | Lanarkshire | 1893 | 1968 |
| Coltness Iron Co. | Gillhead | Lanarkshire | 1945 | 1955 |
| Coltness Iron Co. | Hassockrig | Lanarkshire | 1885 | 1962 |
| Coltness Iron Co. | Kingshill 1 | Lanarkshire | 1919 | 1968 |
| Coltness Iron Co. | Kingshill 2 | Lanarkshire | 1931 | 1963 |
| Coltness Iron Co. | Overtown | Lanarkshire | 1931 | 1968 |
| Coltness Iron Co. | Woodend 5 | West Lothian | c. 1870 | 1965 |
| Coltness Iron Co. | Greenhead (Branchal) | Lanarkshire | 1924 | 1959 |
| Currie Bros. | Beaton's Lodge | Lanarkshire | 1940 | 1949 |
| Daniel Beattie & Co. | Auchmeddan 1, 2 & 3 | Lanarkshire | 1945 | 1968 |
| Daniel Beattie & Co. | Braehead | Lanarkshire | 1938 | 1948 |
| Earl of Buckinghamshire | Fordell | Fife | 1750 | 1966 |
| Edinburgh Collieries Co. | Fleets | East Lothian | 1866 | 1959 |
| Edinburgh Collieries Co. | Prestonlinks | East Lothian | 1899 | 1964 |
| Edinburgh Collieries Co. | Carberry | Midlothian | 1866 | 1960 |
| Fife Coal Co. | Aitken | Fife | 1895 | 1969 |
| Fife Coal Co. | Blairenbathie | Fife | 1945 | 1962 |
| Fife Coal Co. | Bowhill 1, 2 & 3 | Fife | 1895 | 1965 |
| Fife Coal Co. | Comrie | Fife | 1936 | 1986 |
| Fife Coal Co. | Cowdenbeath 7 | Fife | 1860 | 1960 |
| Fife Coal Co. | Frances | Fife | 1850 | 1988 |
| Fife Coal Co. | Kinglassie | Fife | 1908 | 1966 |
| Fife Coal Co. | Lindsay | Fife | 1873 | 1965 |
| Fife Coal Co. | Lumphinnans 1 | Fife | 1852 | 1957 |
| Fife Coal Co. | Lumphinnans 11 & 12 | Fife | c. 1895 | 1966 |
| Fife Coal Co. | Lumphinnans Mine | Fife | 1946 | 1966 |
| Fife Coal Co. | Mary (Lochore) | Fife | 1904 | 1966 |
| Fife Coal Co. | Oakfield | Fife | 1946 | 1949 |
| Fife Coal Co. | Randolph | Fife | c. 1850 | 1968 |
| Fife Coal Co. | Thornton Mine | Fife | 1945 | 1953 |
| Fife Coal Co. | Valleyfield 1 & 2 | Fife | 1908 | 1978 |
| Fife Coal Co. | Wellsgreen Pit | Fife | 1888 | 1959 |
| Flemington Coal Co. | Coatspark | Lanarkshire | 1937 | 1958 |
| Fordel Mains (Mid'thian) Colliery Co. | Edgehead | Midlothian | 1850 | 1959 |
| Fordell Mains Colliery Co. | Brucefield | Clackmannanshire | 1905 | 1961 |
| Gilmerton Colliery Co. | Gilmerton | Midlothian | 1928 | 1961 |
| Glasgow Iron & Steel Co. | Argyll (Machrihanish) | Argyll | 1946 | 1967 |
| Glasgow Iron & Steel Co. | Dalziel & Broomside | Lanarkshire | 1869 | 1948 |
| Glasgow Iron & Steel Co. | Harwood | Midlothian | 1946 | 1959 |
| Glencairn Coal Co. | Glencairn | East Lothian | 1936 | 1962 |
| Glentaggart Coal Co. | Glentaggart | Lanarkshire | 1943 | 1969 |
| Haywood Coal Co. | Chapel | Lanarkshire | 1935 | 1949 |
| Hugh Forrester & Co. | Balmore | Stirlingshire | 1943 | 1960 |

| Company | Mine/Colliery | County | Open | Closed |
|---|---|---|---|---|
| J Crawford & Co. | High Darngavil | Lanarkshire | *p.* 1947 | 1949 |
| J Hunter & Sons | Headlesscross 1 | Lanarkshire | 1938 | 1949 |
| J Hunter & Sons | Headlesscross 2 | Lanarkshire | 1946 | 1953 |
| James Nimmo & Co. | Auchengeich | Lanarkshire | 1908 | 1982 |
| James Nimmo & Co. | Canderigg 4 & 5 | Lanarkshire | 1902 | 1954 |
| James Nimmo & Co. | Canderside 6 & 7 | Lanarkshire | 1939 | 1964 |
| James Nimmo & Co. | Redding | Stirlingshire | *c.* 1894 | 1958 |
| John Hunter & Sons | Gateside 1 | Stirlingshire | 1938 | 1949 |
| John Hunter & Sons | Gateside 2 | Stirlingshire | 1939 | 1952 |
| John McAndrew & Co. | Thankerton | Lanarkshire | 1850 | 1949 |
| Kennox Coal Co. | Glespin | Lanarkshire | *c.* 1908 | 1964 |
| Kennox Coal Co. | Kennox 6 & 7 | Lanarkshire | 1908 | 1972 |
| Kinneil Cannel & Coking Coal Co. | Kinneil | West Lothian | 1880 | 1982 |
| Lochgelly Iron & Coal Co. | Dora (Little Raith) | Fife | 1875 | 1959 |
| Lochgelly Iron & Coal Co. | Jenny Gray | Fife | 1854 | 1959 |
| Lochgelly Iron & Coal Co. | Lady Helen (Dundonald) | Fife | 1892 | 1964 |
| Lochgelly Iron & Coal Co. | Minto | Fife | 1903 | 1967 |
| Lochgelly Iron & Coal Co. | Nellie | Fife | 1880 | 1965 |
| Logan & Co. | Garscube | Lanarkshire | *c.* 1850 | 1966 |
| Lothian Coal Co. | Easthouses | Midlothian | 1909 | 1969 |
| Lothian Coal Co. | Lady Victoria | Midlothian | 1895 | 1981 |
| Lothian Coal Co. | Lingerwood | Midlothian | 1798 | 1967 |
| Lothian Coal Co. | Whitehill | Midlothian | 1850 | 1961 |
| Manor Powis Coal Co. | Manor Powis | Stirlingshire | 1911 | 1972 |
| Mosside Coal & Fireclay Co. | Bishop No. 3 Mine | Lanarkshire | 1939 | 1948 |
| New Cumnock Collieries Ltd. | Afton No.1 | Ayrshire | 1871 | *c.* 1950 |
| New Cumnock Collieries Ltd. | Bank 1 | Ayrshire | *c.* 1850 | 1969 |
| New Cumnock Collieries Ltd. | Bank 2 | Ayrshire | 1946 | 1950 |
| New Cumnock Collieries Ltd. | Bank 6 | Ayrshire | 1925 | 1969 |
| New Cumnock Collieries Ltd. | Knockshinnoch Castle | Ayrshire | 1940 | 1968 |
| New Cumnock Collieries Ltd. | Seaforth 1, 2 & 3 | Ayrshire | 1940 | 1953 |
| Niddrie & Benhar Coal Co. | Newcraighall | Midlothian | 1897 | 1968 |
| Niddrie & Benhar Coal Co. | Woolmet | Midlothian | 1898 | 1966 |
| Nimmo & Dunlop | Cardowan | Lanarkshire | 1924 | 1983 |
| Nimmo & Dunlop | Wester Auchengeich | Lanarkshire | 1928 | 1968 |
| Ormiston Coal Co. | Limeylands | East Lothian | 1895 | 1954 |
| Ormiston Coal Co. | Tynemount | East Lothian | 1924 | 1962 |
| Ormiston Coal Co. | Oxenford 2 | Midlothian | 1926 | 1950 |
| Overwood Coal Co. | Overwood Mine | Lanarkshire | 1937 | 1955 |
| Overwood Coal Co. | Thinacre Mine | Lanarkshire | 1939 | 1963 |
| Plean Colliery Co. | Plean | Stirlingshire | 1932 | 1962 |
| Polquhairn Coal Co. | Greenhill | Ayrshire | 1936 | 1958 |
| Polquhairn Coal Co. | Polquhairn | Ayrshire | 1895 | 1962 |
| R Forrester & Co. | Whitrigg | West Lothian | *c.* 1900 | 1972 |

| Company | Mine/Colliery | County | Open | Closed |
|---|---|---|---|---|
| Robert Addie & Sons Collieries | Glen Mine | Lanarkshire | 1940 | 1954 |
| Robert Addie & Sons Collieries | Herbertshire | Stirlingshire | 1889 | 1959 |
| Robert Muir & Co. | Barbauchlaw | West Lothian | 1900 | 1952 |
| Robert Semple & Co. | Fortacre | Ayrshire | 1924 | 1957 |
| Shotts Iron Co. | Baton | Lanarkshire | c. 1850c | 1950 |
| Shotts Iron Co. | Calderhead 3 & 4 | Lanarkshire | c. 1850c | 1958 |
| Shotts Iron Co. | Castlehill No. 6 | Lanarkshire | 1916 | 1954 |
| Shotts Iron Co. | Fortissat | Lanarkshire | 1870 | 1949 |
| Shotts Iron Co. | Hillhouserigg | Lanarkshire | c. 1850 | 1949 |
| Shotts Iron Co. | Northfield | Lanarkshire | 1917 | 1961 |
| Shotts Iron Co. | Southfield | Lanarkshire | 1923 | 1959 |
| Shotts Iron Co. | Burghlee | Midlothian | c. 1860 | 1964 |
| Shotts Iron Co. | Ramsay | Midlothian | c. 1850 | 1965 |
| Shotts Iron Co. | Roslin | Midlothian | 1901 | 1969 |
| South Ayrshire Collieries (1928) | Killochan | Ayrshire | 1905 | 1967 |
| South Ayrshire Collieries (1928) | Maxwell 2 | Ayrshire | 1903 | 1973 |
| Summerlee Coal and Iron Co. | Bardykes | Lanarkshire | 1874 | 1962 |
| Summerlee Iron Co. | Prestongrange | East Lothian | c. 1820 | 1962 |
| Summerlee Iron Co. | Benhar | Lanarkshire | 1914 | 1962 |
| Thomas Spowart & Co. | Elgin & Wellwood (Leadside) | Fife | 1827 | 1950 |
| United Collieries | Netherton Mine | Lanarkshire | 1938 | 1950 |
| United Collieries | Quarter | Lanarkshire | 1815 | 1951 |
| United Collieries | Woodmuir | Midlothian | 1896 | 1963 |
| United Collieries | Blackrigg 1,2 & 3 | West Lothian | c. 1860 | 1955 |
| United Collieries | Foulshiels | West Lothian | c. 1900 | 1957 |
| United Collieries | Greenrigg | West Lothian | 1902 | 1960 |
| United Collieries | Loganlea | West Lothian | c. 1890 | 1959 |
| Wemyss Coal Co. | Cameron Mine | Fife | 1934 | 1959 |
| Wemyss Coal Co. | Lochhead (Lochhead/Victoria) | Fife | 1890 | 1970 |
| Wemyss Coal Co. | Michael | Fife | 1895 | 1967 |
| Wemyss Coal Co. | Rosie | Fife | 1880 | 1953 |
| Wemyss Coal Co. | Wellesley | Fife | 1883 | 1967 |
| William Dixon | Auchlochan 6 & 7 | Lanarkshire | 1894 | 1968 |
| William Dixon | Blantyre 1 & 2 | Lanarkshire | 1865 | 1957 |
| William Dixon | Polkemmet | West Lothian | 1913 | 1986 |
| William Dixon | Wilsontown | Lanarkshire | 1898 | 1955 |
| William Nicol | Coalburn | Ayrshire | 1924 | 1962 |
| Wilsons & Clyde Coal Co. | Glencraig | Fife | 1896 | 1966 |
| Wilsons & Clyde Coal Co. | Ashgill | Lanarkshire | 1945 | 1951 |
| Wilsons & Clyde Coal Co. | Skellyton 2 & 3 | Lanarkshire | 1944 | 1951 |
| Wilsons & Clyde Coal Co. | Douglas Castle | Lanarkshire | 1912 | 1959 |
| Wilsons & Clyde Coal Co. | Woodside | Lanarkshire | 1848 | 1955 |
| Young's Paraffin Light & Mineral Oil Co. | Baads | Midlothian | c. 1860 | 1964 |

# Appendix D
## Yearly closures of NCB Collieries, 1947–2002

| Year closed | Mine/Colliery | County | Opened | Annual total |
|---|---|---|---|---|
| 1947 | Fernigair | Lanarkshire | 1850 | 2 |
| | Rankin | Lanarkshire | 1947 | |
| 1948 | Afton 1 | Ayrshire | 1871 | 13 |
| | Bishop No. 3 Mine | Lanarkshire | 1939 | |
| | Blackston Mine | Stirlingshire | 1928 | |
| | Bowhill (Patna) | Ayrshire | 1947 | |
| | Braehead | Lanarkshire | 1949 | |
| | Dalkeith 1, 2 & 3 | Midlothian | 1903 | |
| | Dalziel/Broomside | Lanarkshire | 1869 | |
| | Isle of Canty | Fife | 1939 | |
| | Lochend 5 | Lanarkshire | 1880 | |
| | Melloch | Clackmannanshire | 1850 | |
| | Shewalton 8 & 9 | Ayrshire | 1947 | |
| | Southhook | Ayrshire | 1947 | |
| | Tofts 1 & 2 | Ayrshire | 1914 | |
| 1949 | Barleyside West | Stirlingshire | 1947 | 11 |
| | Beaton's Lodge | Lanarkshire | 1940 | |
| | Chapel 1 & 2 | Lanarkshire | 1935 | |
| | Fortissat | Lanarkshire | 1870 | |
| | Gateside 1 | Stirlingshire | 1938 | |
| | Headlesscross 1 | Lanarkshire | 1938 | |
| | High Darngavil | Lanarkshire | 1948 | |
| | Hillhouserigg | Lanarkshire | 1850 | |
| | Langside Mine | Lanarkshire | 1945 | |
| | Oakfield | Fife | 1946 | |
| | Thankerton 6 | Lanarkshire | 1850 | |
| 1950 | Bank 2 | Lanarkshire | 1946 | 11 |
| | Baton | Lanarkshire | 1850 | |
| | Bothwell Castle 1 & 2 | Lanarkshire | 1875 | |
| | Elgin & Wellwood | Fife | 1827 | |
| | Mount | Ayrshire | 1948 | |
| | Netherton Mine | Lanarkshire | 1938 | |
| | Oxenford 2 | Midlothian | 1926 | |
| | Shewalton 5 & 6 | Ayrshire | 1933 | |
| | Shieldmains 6, 7, 14 | Ayrshire | 1927 | |
| | Warrix 1 & 2 | Ayrshire | 1944 | |
| | Wester Gartshore | Dunbartonshire | 1872 | |
| 1951 | Ashgill | Lanarkshire | 1945 | 7 |
| | Duntilland | Lanarkshire | 1943 | |

| Year closed | Mine/Colliery | County | Opened | Annual total |
|---|---|---|---|---|
| | Kepplehill 1 & 2 | Lanarkshire | 1897 | |
| | Quarter 2 | Lanarkshire | 1815 | |
| | Quarter 5a | Lanarkshire | 1941 | |
| | Skellyton 2 & 3 | Lanarkshire | 1944 | |
| | Windyedge | Fife | 1950 | |
| 1952 | Barbauchlaw | West Lothian | 1900 | 6 |
| | Broomlands | Ayrshire | 1950 | |
| | Chalmerston 7 | Ayrshire | 1934 | |
| | Craigrie | Clackmannanshire | 1942 | |
| | Gateside 2 | Stirlingshire | 1939 | |
| | Tynemount | East Lothian | 1924 | |
| 1953 | Bannockburn | Stirlingshire | 1894 | 10 |
| | Carriden 1 & 2 | West Lothian | 1914 | |
| | Headlesscross 2 | Lanarkshire | 1946 | |
| | Lugar Mine | Ayrshire | 1942 | |
| | Mortonmuir 8 & 9 | Ayrshire | 1951 | |
| | Pennyvenie 5 | Ayrshire | 1911 | |
| | Rosie | Fife | 1880 | |
| | Seaforth 1, 2 & 3 | Ayrshire | 1940 | |
| | Thankerton 2 | Lanarkshire | 1850 | |
| | Thornton Mine | Fife | 1945 | |
| 1954 | Bogton Mine | Ayrshire | 1931 | 6 |
| | Canderigg 4 & 5 | Lanarkshire | 1902 | |
| | Castlehill No. 6 | Lanarkshire | 1916 | |
| | Glen Mine | Lanarkshire | 1940 | |
| | King O' Muirs 1 | Clackmannanshire | 1938 | |
| | Limeylands | East Lothian | 1895 | |
| 1955 | Barbeth | Ayrshire | 1950 | 8 |
| | Blackrigg 1, 2 & 3 | West Lothian | 1860 | |
| | Gillhead | Lanarkshire | 1945 | |
| | Overwood Mine | Lanarkshire | 1949 | |
| | Shewalton 3 & 4 | Ayrshire | 1924 | |
| | Stane Mine | Lanarkshire | 1948 | |
| | Wilsontown | Lanarkshire | 1898 | |
| | Woodside | Lanarkshire | 1848 | |
| 1956 | Carston | Ayrshire | 1950 | 2 |
| | Newfield | Ayrshire | 1940 | |
| 1957 | Blantyre 1 & 2 | Lanarkshire | 1865 | 9 |
| | Cronberry Moor | Ayrshire | 1920 | |
| | East Benhar Mine | West Lothian | 1940 | |
| | Fortacre | Ayrshire | 1949 | |
| | Foulshiels | West Lothian | 1900 | |
| | Greenlees | Lanarkshire | 1947 | |
| | King O' Muirs 2 | Clackmannanshire | 1950 | |

| Year closed | Mine/Colliery | County | Opened | Annual total |
|---|---|---|---|---|
| | Lumphinnans 1 | Fife | 1852 | |
| | Tillicoultry 1 & 2 | Clackmannanshire | 1876 | |
| 1958 | Bankend | Lanarkshire | 1890 | 9 |
| | Calderhead 3 & 4 | Lanarkshire | 1850 | |
| | Castle Mine | Midlothian | 1950 | |
| | Coatspark | Lanarkshire | 1937 | |
| | Earlseat | Fife | 1950 | |
| | Forthbank 1 & 2 | Clackmannanshire | 1949 | |
| | Greenhill | Ayrshire | 1936 | |
| | Polmaise 1 & 2 | Stirlingshire | 1904 | |
| | Redding | Stirlingshire | 1894 | |
| 1959 | Andershaw | Lanarkshire | 1949 | 27 |
| | Benarty | Fife | 1945 | |
| | Blairmuckhill | Lanarkshire | 1910 | |
| | Bothwell Castle 3 & 4 | Lanarkshire | 1889 | |
| | Branchal | Lanarkshire | 1947 | |
| | Cameron Mine | Fife | 1934 | |
| | Chalmerston 4 & 5 | Ayrshire | 1925 | |
| | Dora (Little Raith) | Fife | 1875 | |
| | Douglas Castle | Lanarkshire | 1912 | |
| | Dumback 1 | West Lothian | 1954 | |
| | Edgehead | Midlothian | 1949 | |
| | Enterkine 9 & 10 | Ayrshire | 1878 | |
| | Fleets | East Lothian | 1866 | |
| | Gartshore 3 & 12 | Dunbartonshire | 1865 | |
| | Hamilton Palace | Lanarkshire | 1884 | |
| | Harwood | Midlothian | 1946 | |
| | Herbertshire | Stirlingshire | 1889 | |
| | Hindsward 3 & 4 | Ayrshire | 1956 | |
| | Jenny Gray | Fife | 1890 | |
| | Loganlea | West Lothian | 1890 | |
| | Meta (Devon 3) | Clackmannanshire | 1946 | |
| | Oxenford 3 | Midlothian | 1952 | |
| | Policy | Stirlingshire | 1924 | |
| | Powharnal | Ayrshire | 1954 | |
| | South Bantaskine | Stirlingshire | 1946 | |
| | Southfield | Lanarkshire | 1923 | |
| | Wellsgreen Pit | Fife | 1888 | |
| 1960 | Auchincruive 1, 2 & 3 | Ayrshire | 1897 | 10 |
| | Balgonie | Fife | 1883 | |
| | Balmore | Stirlingshire | 1943 | |
| | Carberry | Midlothian | 1866 | |
| | Cowdenbeath 7 | Fife | 1860 | |
| | Cuthill | West Lothian | 1958 | |

| Year closed | Mine/Colliery | County | Opened | Annual total |
|---|---|---|---|---|
| | Devon 1 & 2 | Clackmannanshire | 1879 | |
| | Greenrigg | West Lothian | 1905 | |
| | Meadowmill Mine | East Lothian | 1954 | |
| | Zetland | Clackmannanshire | 1935 | |
| 1961 | Bankend (Westoun) | Lanarkshire | 1890 | 9 |
| | Bellyford | East Lothian | 1954 | |
| | Brucefield | Clackmannanshire | 1905 | |
| | Gilmerton | Midlothian | 1928 | |
| | Harvieston | Clackmannanshire | 1957 | |
| | Northfield | Lanarkshire | 1917 | |
| | Pennyvenie 4 | Ayrshire | 1911 | |
| | Sundrum | Ayrshire | 1953 | |
| | Whitehill | Midlothian | 1850 | |
| 1962 | Arniston | Midlothian | 1850 | 19 |
| | Baads | Midlothian | 1908 | |
| | Bardykes | Lanarkshire | 1874 | |
| | Benhar | Lanarkshire | 1914 | |
| | Blairenbathie Mine | Fife | 1949 | |
| | Blantyreferme 1, 2 | Lanarkshire | 1894 | |
| | Boglea | Lanarkshire | 1942 | |
| | Coalburn | Ayrshire | 1924 | |
| | Glencairn | East Lothian | 1949 | |
| | Glenochil 1 & 2 | Clackmannanshire | 1956 | |
| | Hassockrig | Lanarkshire | 1885 | |
| | Knowehead 2 | Lanarkshire | 1952 | |
| | Plean 3, 4 & 5 | Stirlingshire | 1932 | |
| | Polquhairn 1 & 4 | Ayrshire | 1895 | |
| | Polquhairn 5 & 6 | Ayrshire | 1955 | |
| | Prestongrange | East Lothian | 1874 | |
| | Rothes | Fife | 1957 | |
| | Westoun | Lanarkshire | 1948 | |
| | Winton | East Lothian | 1952 | |
| 1963 | Aitken | Fife | 1895 | 8 |
| | Ardenrigg 6 | Lanarkshire | 1926 | |
| | Auldton 2 & 3 | Lanarkshire | 1956 | |
| | Dumbreck | Stirlingshire | 1887 | |
| | Kingshill 2 | Lanarkshire | 1931 | |
| | Pirnhall | Stirlingshire | 1933 | |
| | Thinacre Mine | Lanarkshire | 1949 | |
| | Woodmuir | Midlothian | 1896 | |
| 1964 | Blantyreferme 3 | Lanarkshire | 1850 | 12 |
| | Bridgend | Ayrshire | 1949 | |
| | Burghlee | Midlothian | 1860 | |
| | Canderside 6 & 7 | Lanarkshire | 1939 | |

| Year closed | Mine/Colliery | County | Opened | Annual total |
|---|---|---|---|---|
| | Dullatur | Dunbartonshire | 1935 | |
| | Gateside 4 & 5 | Dumfriesshire | 1891 | |
| | Glentore | Lanarkshire | 1957 | |
| | Glespin | Lanarkshire | 1908 | |
| | Lady Helen (Dundonald) | Fife | 1895 | |
| | Prestonlinks | East Lothian | 1899 | |
| | Tower Mine | Dumfriesshire | 1916 | |
| | Twechar 1 Gartshore 10 | Dunbartonshire | 1865 | |
| 1965 | Auchengeich | Lanarkshire | 1908 | 12 |
| | Avonbraes | Lanarkshire | 1952 | |
| | Bowhill 1, 2 & 3 | Fife | 1895 | |
| | Cowdenfoot | Midlothian | 1956 | |
| | Houldsworth | Ayrshire | 1905 | |
| | Lindsay (Kelty 4 & 5) | Fife | 1873 | |
| | Nellie | Fife | 1880 | |
| | Ramsay | Midlothian | 1850 | |
| | Torry | Fife | 1952 | |
| | Whitehill 1 & 2 | Ayrshire | 1893 | |
| | Whitehill 3 & 4 | Ayrshire | 1946 | |
| | Woodend | West Lothian | 1870 | |
| 1966 | Fordell | Fife | 1750 | 11 |
| | Garscube | Lanarkshire | 1850 | |
| | Glencraig | Fife | 1896 | |
| | Kinglassie | Fife | 1908 | |
| | Knowetop | Lanarkshire | 1948 | |
| | Lumphinnans Mine | Fife | 1945 | |
| | Lumphinnans XI & XII | Fife | 1895 | |
| | Mary (Lochore) | Fife | 1904 | |
| | Mauchline 1, 2 & 4 | Ayrshire | 1925 | |
| | Rig | Dumfriesshire | 1949 | |
| | Woolmet | Midlothian | 1898 | |
| 1967 | Argyll | Argyll | 1946 | 8 |
| | Douglas | Lanarkshire | 1898 | |
| | Killochan | Ayrshire | 1905 | |
| | Lingerwood | Midlothian | 1798 | |
| | Manor Powis 1, 2 & 3 | Stirlingshire | 1914 | |
| | Michael | Fife | 1895 | |
| | Minto | Fife | 1903 | |
| | Wellesley | Fife | 1885 | |
| 1968 | Auchlochan 6 & 7 | Lanarkshire | 1894 | 16 |
| | Auchlochan 9 & 10 | Lanarkshire | 1894 | |
| | Auchmeddan 1, 2 & 3 | Lanarkshire | 1949 | |
| | Beoch 3 | Ayrshire | 1866 | |
| | Beoch 4 | Ayrshire | 1937 | |

| Year closed | Mine/Colliery | County | Opened | Annual total |
|---|---|---|---|---|
| | Fauldhead 1 & 3 | Dumfriesshire | 1896 | |
| | Gartshore 9, 11, Grayshill | Dunbartonshire | 1875 | |
| | Kames | Ayrshire | 1870 | |
| | Kingshill (1) | Lanarkshire | 1919 | |
| | Knockshinnoch Castle | Ayrshire | 1944 | |
| | Newcraighall | Midlothian | 1898 | |
| | Overtown | Lanarkshire | 1931 | |
| | Randolph | Fife | 1850 | |
| | Riddochhill | West Lothian | 1890 | |
| | Roger 3 & 4 | Dumfriesshire | 1956 | |
| | Wester Auchengeich | Lanarkshire | 1928 | |
| 1969 | Bank 1 | Lanarkshire | 1850 | 7 |
| | Bank 6 | Lanarkshire | 1925 | |
| | Blair 11 & 12 | Ayrshire | 1953 | |
| | Blairhall | Fife | 1870 | |
| | Easthouses | Midlothian | 1909 | |
| | Glentaggart | Lanarkshire | 1943 | |
| | Roslin | Midlothian | 1903 | |
| 1970 | Lochhead | Fife | 1890 | 1 |
| 1972 | Kennox 6 & 7 | Lanarkshire | 1908 | 2 |
| | Whitrigg 2, 4 & 5 | West Lothian | 1900 | |
| 1973 | Auchincruive 4, 5, 6, 7 | Ayrshire | 1912 | 6 |
| | Dollar | Perthshire | 1943 | |
| | Easton (Hopetoun) | West Lothian | 1898 | |
| | Lochlea 1 & 2 | Ayrshire | 1949 | |
| | Maxwell 2 | Ayrshire | 1903 | |
| | Maxwell 4 | Ayrshire | 1950 | |
| 1974 | Kingshill 3 | Lanarkshire | 1951 | 2 |
| | Littlemill 2, 3 & 5 | Ayrshire | 1860 | |
| 1975 | Minnivey 4 & 5 | Ayrshire | 1955 | 1 |
| 1976 | Cairnhill 1 & 2 | Ayrshire | 1957 | 1 |
| 1977 | Dalquharran 1 & 2 | Ayrshire | 1951 | 1 |
| 1978 | Dalkeith 5 & 9 | Midlothian | 1903 | 3 |
| | Pennyvenie 2, 3 & 7 | Ayrshire | 1945 | |
| | Valleyfield 1 & 2 | Fife | 1908 | |
| 1980 | Roger 1 & 2 | Dumfriesshire | 1952 | 1 |
| 1981 | Bedlay | Lanarkshire | 1905 | 2 |
| | Lady Victoria | Midlothian | 1895 | |
| 1982 | Kinneil | West Lothian | 1890 | 1 |
| 1983 | Cardowan | Lanarkshire | 1924 | 3 |
| | Highhouse | Ayrshire | 1894 | |
| | Sorn 1 & 2 | Ayrshire | 1953 | |
| 1984 | Polkemmet | West Lothian | 1916 | 1 |
| 1986 | Bogside 1, 2 & 3 | Fife | 1959 | 2 |

| Year closed | Mine/Colliery | County | Opened | Annual total |
|---|---|---|---|---|
| | Comrie | Fife | 1936 | |
| 1987 | Killoch | Ayrshire | 1960 | 2 |
| | Polmaise 3, 4 & 5 | Stirlingshire | 1904 | |
| 1988 | Frances | Fife | 1850 | 2 |
| | Seafield | Fife | 1966 | |
| 1989 | Barony 1, 2, 3 & 4 | Ayrshire | 1910 | 2 |
| | Bilston Glen | Midlothian | 1963 | |
| 1990 | Castlehill Mine | Fife | 1969 | 2 |
| | Solsgirth | Clackmannanshire | 1969 | |
| 1997 | Monktonhall | Midlothian | 1967 | 1 |
| 1999 | Castlebridge | Clackmannanshire | 1979 | 1 |
| 2002 | Longannet Mine | Fife | 1969 | 1 |

# Index of Scottish Deep Coal Mines, 1947–2002

Note

Further location data can be found in the Gazetteer.

pre – sometime close to, but before the stated date

c. – approximate date

In the case of private licensed mines, production may have ceased before the licences expired.

| Mine/Colliery | NCB/Private | County | Opened | Closed | Page |
|---|---|---|---|---|---|
| Abbey | M T & P Stark | Lanarkshire | 1954 | pre-1997 | 209 |
| Afton 1 | NCB | Ayrshire | 1871 | 1948 | 67/98 |
| Airngarth 1 | Dugald McKechnie | West Lothian | 1958 | 1961 | 268 |
| Airngarth 2 | Dugald McKechnie | West Lothian | 1961 | c. 1967 | 268 |
| Aitken | NCB | Fife | 1895 | 1963 | 135 |
| Aitkendean | W Walker | Midlothian | 1957 | 1973 | 233 |
| Alton | James Bell | Ayrshire | 1954 | 1956 | 98 |
| Amberly | Shankland & Co. | Lanarkshire | 1954 | 1957 | 209 |
| Amosknowes | J McCutcheon | Lanarkshire | 1955 | 1959 | 209 |
| Andershaw | NCB | Lanarkshire | 1949 | 1959 | 169 |
| Arbuckle | R & J Dow | Lanarkshire | pre-1947 | 1949 | 209 |
| Arbuckle 13/14 | R & J Dow | Lanarkshire | 1946 | c. 1949 | 209 |
| Arbuckle 15 | Charles Tobin | Lanarkshire | 1950 | c. 1967 | 209 |
| Arbuckle 16 | Charles Tobin | Lanarkshire | 1959 | c. 1962 | 209 |
| Arbuckle 17 | Charles Tobin | Lanarkshire | unknown | c. 1967 | 209 |
| Ardenrigg 6 | NCB | Lanarkshire | 1926 | 1963 | 169 |
| Argyll | NCB | Argyll | 1946 | 1967 | 63 |
| Arniston | NCB | Midlothian | 1850 | 1962 | 216 |
| Arnloss 2 | P & J Horne | Stirlingshire | 1954 | c. 1969 | 250 |
| Ashgill | NCB | Lanarkshire | 1945 | 1951 | 169 |
| Ashlea 4 | John Ferguson | Lanarkshire | pre-1947 | 1947 | 209 |
| Ashlea 5 | John Ferguson | Lanarkshire | 1946 | 1956 | 209 |
| Ashlea 6 | John Ferguson | Lanarkshire | 1956 | 1959 | 209 |
| Ashlea 7 | John Ferguson | Lanarkshire | 1958 | c. 1963 | 209 |
| Auchengeich | NCB | Lanarkshire | 1908 | 1965 | 170 |
| Auchenheath | Auchenheath Mining Co. | Lanarkshire | pre-1947 | 1951 | 209 |
| Auchincruive 1, 2 & 3 | NCB | Ayrshire | 1897 | 1960 | 67 |
| Auchincruive 4, 5, 6, 7 | NCB | Ayrshire | 1912 | 1973 | 68 |
| Auchlin | J R McLellan | Ayrshire | 1977 | 1991 | 98 |
| Auchlochan 6 & 7 | NCB | Lanarkshire | 1894 | 1968 | 171 |
| Auchlochan 9 & 10 | NCB | Lanarkshire | 1894 | 1968 | 171 |
| Auchmeddan 1, 2 & 3 | NCB | Lanarkshire | 1949 | 1968 | 171/209 |
| Auldton 2 & 3 | NCB | Lanarkshire | 1956 | 1963 | 172/209 |
| Avonbraes | NCB | Lanarkshire | 1952 | 1965 | 172 |

| Mine/Colliery | NCB/Private | County | Opened | Closed | Page |
|---|---|---|---|---|---|
| Avonhead | Headrigg Coal Co. | Lanarkshire | pre-1947 | 1948 | 209 |
| Avonhead 3 | Headrigg Coal Co. | Lanarkshire | pre-1947 | c. 1947 | 209 |
| Avonhead 4 | Headrigg Coal Co. | Lanarkshire | 1947 | 1948 | 209 |
| Avonhead 4a | Headrigg Coal Co. | Lanarkshire | 1947 | 1948 | 209 |
| Baads | NCB | Midlothian | 1908 | 1962 | 217 |
| Baads 42 | Young's Paraffin Light & Mineral Oil Co. | Midlothian | pre-1947 | c. 1947 | 233 |
| Backshot 4 | A Maxwell & Co. | Lanarkshire | 1964 | 1995 | 209 |
| Balgonie | NCB | Fife | 1883 | 1960 | 136 |
| Ballochney 1 | J & J Somerville | Lanarkshire | 1950 | 1951 | 209 |
| Ballochney 2 | J & J Somerville | Lanarkshire | 1951 | 1951 | 209 |
| Ballochney 3 | J & J Somerville | Lanarkshire | 1951 | 1951 | 209 |
| Balmoral | Palace Coal Co. | Lanarkshire | 1947 | 1948 | 209 |
| Balmore | NCB | Stirlingshire | 1943 | 1960 | 240 |
| Bank 1 | NCB | Lanarkshire | 1850 | 1969 | 68 |
| Bank 2 | NCB | Lanarkshire | 1946 | 1950 | 69 |
| Bank 6 | NCB | Lanarkshire | 1925 | 1969 | 69 |
| Bankend | NCB | Lanarkshire | 1890 | 1958 | 173 |
| Bankend (Westoun) | NCB | Lanarkshire | 1890 | 1961 | 207 |
| Banknock 3 | Joseph Bergin | Stirlingshire | 1937 | c. 1962 | 250 |
| Bannockburn | NCB | Stirlingshire | 1894 | 1953 | 241 |
| Barbauchlaw | NCB | West Lothian | 1900 | 1952 | 258 |
| Barbeth | NCB | Ayrshire | 1950 | 1955 | 70 |
| Bardykes | NCB | Lanarkshire | 1874 | 1962 | 173 |
| Barleydean | Temple Farm Coal Co. | Midlothian | 1967 | 1973 | 233 |
| Barleyside West | NCB | Stirlingshire | 1947 | 1949 | 242 |
| Barnsmuir | P & J Horne | Stirlingshire | 1949 | 1955 | 250 |
| Barony 1, 2, 3 & 4 | NCB | Ayrshire | 1910 | 1989 | 70 |
| Baton | NCB | Lanarkshire | 1850 | 1950 | 174 |
| Beaton's Lodge | NCB | Lanarkshire | 1940 | 1960 | 174/209 |
| Bedlay | NCB | Lanarkshire | 1905 | 1981 | 175 |
| Bellyford | NCB | East Lothian | 1954 | 1961 | 126 |
| Benarty | NCB | Fife | 1945 | 1959 | 136 |
| Benhar | NCB | Lanarkshire | 1914 | 1962 | 176 |
| Beoch 3 | NCB | Ayrshire | 1866 | 1968 | 71 |
| Beoch 4 | NCB | Ayrshire | 1937 | 1968 | 72 |
| Berryhill | T McDonach | Lanarkshire | 1962 | 1967 | 209 |
| Bethel | J G McCracken | Lanarkshire | 1967 | 1973 | 209 |
| Biggar 2 | J & J McCracken | Lanarkshire | 1953 | c. 1967 | 209 |
| Biggar Ford 2 | R Moffat & Sons | Lanarkshire | pre-1947 | 1948 | 209 |
| Biggar/Gorehill | J & J McCracken | Lanarkshire | 1947 | 1948 | 209 |
| Bilston Glen | NCB | Midlothian | 1963 | 1989 | 218 |
| Birkrigg | Jackson & Tweedie | Lanarkshire | 1952 | 1953 | 209 |
| Bishop No. 3 Mine | NCB | Lanarkshire | 1939 | 1948 | 177 |
| Blackbraes 8 | R Pringle & Co. | Stirlingshire | 1952 | 1952 | 250 |

| Mine/Colliery | NCB/Private | County | Opened | Closed | Page |
|---|---|---|---|---|---|
| Blackbraes 9 | R Pringle & Co. | Stirlingshire | 1952 | 1958 | 250 |
| Blackburnhall | David Allan & Co. | West Lothian | pre-1947 | 1947 | 268 |
| Blackhall | J B Milliken | Lanarkshire | 1957 | 1970 | 209 |
| Blackrigg 1, 2 & 3 | NCB | West Lothian | 1860 | 1955 | 258 |
| Blackston Mine | NCB | Stirlingshire | 1928 | 1948 | 242 |
| Blair 11 & 12 | NCB | Ayrshire | 1953 | 1969 | 72 |
| Blairburn | Flockhart & Buchan | Peebleshire | 1959 | 1959 | 235 |
| Blairenbathie Mine | NCB | Fife | 1949 | 1962 | 137 |
| Blairhall | NCB | Fife | 1870 | 1969 | 137 |
| Blairmuckhill | NCB | Lanarkshire | 1910 | 1959 | 177 |
| Blairmuckhole | Henderson Brothers | Lanarkshire | pre-1947 | 1947 | 209 |
| Blantyre 1 & 2 | NCB | Lanarkshire | 1865 | 1957 | 178 |
| Blantyreferme 1 & 2 | NCB | Lanarkshire | 1894 | 1962 | 178 |
| Blantyreferme 3 (Newton) | NCB | Lanarkshire | 1850 | 1964 | 179 |
| Blinkbonny | D Beattie | Midlothian | 1967 | 1995 | 233 |
| Blueknowe 1 | Shawfield Coal Co. | Lanarkshire | 1953 | 1953 | 209 |
| Blueknowe 2 | Shawfield Coal Co. | Lanarkshire | 1954 | 1955 | 209 |
| Boagstown | James Bell | Stirlingshire | 1951 | 1960 | 250 |
| Boglea | NCB | Lanarkshire | 1942 | 1962 | 179 |
| Bogside | Peter Hamil | Lanarkshire | 1953 | 1955 | 209 |
| Bogside 1 | A J & A Smillie | Lanarkshire | 1958 | c. 1959 | 209 |
| Bogside 1 ,2 & 3 | NCB | Fife | 1959 | 1986 | 138 |
| Bogton Mine | NCB | Ayrshire | 1931 | 1954 | 73 |
| Bonnyburn | Bonnyburn Fireclay Co. | Stirlingshire | 1950 | 1953 | 250 |
| Bonnyside | J Dougall & Sons | Stirlingshire | pre-1947 | c. 1960 | 250 |
| Bothwell Castle 1 & 2 | NCB | Lanarkshire | 1875 | 1950 | 180 |
| Bothwell Castle 3 & 4 | NCB | Lanarkshire | 1889 | 1959 | 180 |
| Bowhill (Patna) | NCB | Ayrshire | 1947 | 1948 | 73 |
| Bowhill 1, 2 & 3 | NCB | Fife | 1895 | 1965 | 139 |
| Braehead | NCB | Lanarkshire | 1949 | 1948 | 181/209 |
| Branchal | NCB | Lanarkshire | 1947 | 1959 | 181 |
| Bridgend | NCB | Ayrshire | 1949 | 1964 | 74/98 |
| Bridgend 1 | Deerpark Coal Co. | Lanarkshire | 1954 | c. 1963 | 209 |
| Broomhill | T Cook | Ayrshire | pre-1988 | 1991 | 98 |
| Broomknowe | R & W Henderson | Lanarkshire | pre-1947 | 1948 | 209 |
| Broomknowe 2 | R & W Henderson | Lanarkshire | 1949 | 1950 | 209 |
| Broomlands | NCB | Ayrshire | 1950 | 1952 | 74 |
| Brucefield | NCB | Clackmannanshire | 1905 | 1961 | 102 |
| Brunton | William McCracken | West Lothian | 1955 | 1966 | 268 |
| Buchanan Mine | N B Buchanan | Fife | 1979 | 1991 | 165 |
| Burghlee | NCB | Midlothian | 1860 | 1964 | 219 |
| Burnfoot | M T & P Stark | Lanarkshire | 1947a | Unknown | 210 |
| Burnside 3 | R Moffat & Sons | Lanarkshire | 1948 | 1951 | 210 |
| Burnside 4 | R Moffat & Sons | Lanarkshire | 1952 | 1961 | 210 |

| Mine/Colliery | NCB/Private | County | Opened | Closed | Page |
|---|---|---|---|---|---|
| Busbiehead 2b | J & R Howie | Ayrshire | pre-1947 | 1950 | 98 |
| Busbiehead 3 | J & R Howie | Ayrshire | 1947 | 1951 | 98 |
| Cairnhill 1 & 2 | NCB | Ayrshire | 1957 | 1976 | 74 |
| Calder 1 | Wester Moffat Coal Co. | Lanarkshire | 1957 | 1959 | 210 |
| Calder 2 | Wester Moffat Coal Co. | Lanarkshire | 1958 | 1960 | 210 |
| Calderhead 3 & 4 | NCB | Lanarkshire | 1850 | 1958 | 182 |
| Cameron Mine | NCB | Fife | 1934 | 1959 | 140 |
| Canderigg 4 & 5 | NCB | Lanarkshire | 1902 | 1954 | 182 |
| Canderside 6 & 7 | NCB | Lanarkshire | 1939 | 1964 | 183 |
| Carberry | NCB | Midlothian | 1866 | 1960 | 219 |
| Cardowan | NCB | Lanarkshire | 1924 | 1983 | 183 |
| Carriden 1 & 2 | NCB | West Lothian | 1914 | 1953 | 258 |
| Carse | Joseph Jack | Stirlingshire | 1953 | 1954 | 250 |
| Carston | NCB | Ayrshire | 1950 | 1956 | 75 |
| Cartmore | W S Rosland | Fife | 1981a | 1987 | 165 |
| Castle Mine | NCB | Midlothian | 1950 | 1958 | 220 |
| Castlebridge | NCB | Clackmannanshire | 1979 | 1999 | 102 |
| Castlehill Mine | NCB | Fife | 1969 | 1990 | 140 |
| Castlehill No. 6 | NCB | Lanarkshire | 1916 | 1954 | 184 |
| Castleview | J J Kane | Lanarkshire | 1960 | c. 1970 | 210 |
| Catcraig | Catcraig Coal Co. | Lanarkshire | 1955 | 1956 | 210 |
| Chalmerston 4 & 5 | NCB | Ayrshire | 1925 | 1959 | 75 |
| Chalmerston 7 | NCB | Ayrshire | 1934 | 1952 | 75 |
| Chancellorville | D Beattie | East Lothian | 1971 | 1979 | 131 |
| Chapel 1 & 2 | NCB | Lanarkshire | 1935 | 1949 | 185 |
| Chapelrigg | Bonnybridge Co. | Lanarkshire | pre-1947 | 1953 | 210 |
| Claremount | R D Poore | Lanarkshire | 1949 | 1950 | 210 |
| Clarkston 2 | T Reid & Co. | Lanarkshire | pre-1947 | 1947 | 210 |
| Coalburn | NCB | Ayrshire | 1924 | 1962 | 76 |
| Coalhall 1 | T & J Campbell | Lanarkshire | pre-1947 | 1951 | 210 |
| Coalhall 2 | T & J Campbell | Lanarkshire | pre-1947 | 1950 | 210 |
| Coatspark | NCB | Lanarkshire | 1937 | 1958 | 185 |
| Comrie | NCB | Fife | 1936 | 1986 | 141 |
| Cornton | Cornton Coal Co. | Midlothian | pre-1947 | c. 1964 | 233 |
| Couston | Couston Mining Co. | West Lothian | 1952 | 1972 | 268 |
| Cowdenbeath 7 | NCB | Fife | 1860 | 1960 | 142 |
| Cowdenfoot | NCB | Midlothian | 1956 | 1965 | 220 |
| Craigmad | James Penman | Stirlingshire | 1955 | 1955 | 250 |
| Craigman | Coleston Mining | Ayrshire | 1986 | 1991 | 98 |
| Craigrie | NCB | Clackmannanshire | 1942 | 1952 | 103 |
| Crindledyke | Auchinlea Quarries | Lanarkshire | pre-1947 | 1947 | 210 |
| Cronberry Moor | NCB | Ayrshire | 1920 | 1957 | 76 |
| Cuthill | NCB | West Lothian | 1958 | 1960 | 259 |
| Dalkeith 1, 2 & 3 | NCB | Midlothian | 1903 | 1948 | 220 |

| Mine/Colliery | NCB/Private | County | Opened | Closed | Page |
|---|---|---|---|---|---|
| Dalkeith 5 & 9 | NCB | Midlothian | 1903 | 1978 | 221 |
| Dalquharran 1 & 2 | NCB | Ayrshire | 1951 | 1977 | 77 |
| Dalziel/Broomside | NCB | Lanarkshire | 1869 | 1948 | 186 |
| Deerpark | Deerpark Coal Co. | Lanarkshire | 1941 | 1954 | 210 |
| Devon | Hillfarm Coal Co. | Clackmannanshire | 1983 | 1989 | 110 |
| Devon 1 & 2 | NCB | Clackmannanshire | 1879 | 1960 | 103 |
| Dollar | NCB | Clackmannanshire | 1943 | 1973 | 237 |
| Dora (Little Raith) | NCB | Fife | 1875 | 1959 | 142 |
| Douglas | NCB | Lanarkshire | 1898 | 1967 | 186 |
| Douglas Castle | NCB | Lanarkshire | 1912 | 1959 | 187 |
| Dowhail | T Cook | Ayrshire | 1978 | 1979 | 98 |
| Drossyhill | Armstrong Bros | Lanarkshire | 1958 | 1959 | 210 |
| Drum | Bonnybridge Silica & Fireclay Co. | Stirlingshire | c. 1947 | c. 1957 | 250 |
| Drumbreck | Beveridge Brose | Lanarkshire | 1946 | c. 1963 | 210 |
| Drumshangie 4 | Drumlangie Coal Co. | Lanarkshire | pre-1947 | 1947 | 210 |
| Drumshangie 5 | Drumlangie Coal Co. | Lanarkshire | pre-1947 | 1947 | 210 |
| Dryflats | Glenrigg Coal Co. | Lanarkshire | pre-1947 | 1950 | 210 |
| Dryflats 4 | M Wooton, Airdrie | Lanarkshire | pre-1947 | 1950 | 210 |
| Dullatur | NCB | Dunbartonshire | 1935 | 1964 | 120 |
| Dumback 1 | NCB | West Lothian | 1954 | 1959 | 259 |
| Dumback 1 | Whitburn Coal Co. | West Lothian | 1942 | 1954 | 268 |
| Dumback 2 | J & D McWilliam | West Lothian | pre-1947 | 1966 | 268 |
| Dumback 3 | J & D McWilliam | West Lothian | 1962 | 1968 | 268 |
| Dumbreck | NCB | Stirlingshire | 1887 | 1963 | 242 |
| Dunrobin | Browshot Coal Co. | Lanarkshire | pre-1947 | 1947 | 210 |
| Dunrobin 2 | Armstrong Coal Co. | Lanarkshire | 1948 | 1951 | 210 |
| Dunrobin 3 | Browshot Coal Co. | Lanarkshire | pre-1947 | 1947 | 210 |
| Dunrobin 4 | Browshot Coal Co. | Lanarkshire | 1948 | 1948 | 210 |
| Dunrobin 5 | Browshot Coal Co. | Lanarkshire | 1950 | 1950 | 210 |
| Duntilland | NCB | Lanarkshire | 1943 | 1951 | 187 |
| Dykhead | Hillfarm Coal Co. | Lanarkshire | 1968 | 1971 | 210 |
| Earlseat | NCB | Fife | 1950 | 1958 | 143 |
| East Benhar Mine | NCB | West Lothian | 1940 | 1957 | 260 |
| East Bonhard | Woodhead Coal Co. | West Lothian | 1968 | 1971 | 268 |
| Easter Clune | Easter Clune Coal Co. | Fife | 1957 | 1966 | 165 |
| Easter Jaw | J Broadby/J Drysdale | Stirlingshire | 1938 | 1953 | 250 |
| Easter Whin 1 & 2 | David Graham | Stirlingshire | 1953 | 1957 | 250 |
| Easter Windyedge | R Henderson & Co. | Lanarkshire | pre-1947 | 1948 | 210 |
| Easthouses | NCB | Midlothian | 1909 | 1969 | 221 |
| Easton (Hopetoun) | NCB | West Lothian | 1898 | 1973 | 260 |
| Edgehead | NCB | Midlothian | 1949 | 1959 | 222/233 |
| Elgin & Wellwood | NCB | Fife | 1827 | 1950 | 143 |
| Elrig | James Drysdale | Stirlingshire | 1941 | 1948 | 250 |
| Enterkine 9 & 10 | NCB | Ayrshire | 1878 | 1959 | 77 |

| Mine/Colliery | NCB/Private | County | Opened | Closed | Page |
|---|---|---|---|---|---|
| Fairview | M T & P Stark | Lanarkshire | 1953 | 1954 | 210 |
| Fairybank | Greenshields & Co. | Lanarkshire | 1950 | 1951 | 210 |
| Fardalehill/Newtonhead | Balgray Bauxite Co. | Ayrshire | pre- 1947 | 1950 | 98 |
| Fauldhead 1 & 3 | NCB | Dumfriesshire | 1896 | 1968 | 114 |
| Fence | Thomas Sinclair & Sons | Lanarkshire | 1944 | 1961 | 210 |
| Fernigair | NCB | Lanarkshire | 1850 | 1947 | 188 |
| Fleets | NCB | East Lothian | 1866 | 1959 | 126 |
| Foggermountain | Louia Moore, later Caledon Coal Co. | Stirlingshire | 1955 | 1958 | 250 |
| Fordell | NCB | Fife | 1750 | 1966 | 144 |
| Fortacre | NCB | Ayrshire | 1949 | 1957 | 78/98 |
| Forthbank 1 & 2 | NCB | Clackmannanshire | 1949 | 1958 | 104 |
| Fortissat | NCB | Lanarkshire | 1870 | 1949 | 188 |
| Foulshiels | NCB | West Lothian | 1900 | 1957 | 261 |
| Frances | NCB | Fife | 1850 | 1988 | 144 |
| Garallan | Garallan Brick & Tile Co. | Ayrshire | 1947 | 1961 | 98 |
| Garscube | NCB | Lanarkshire | 1850 | 1966 | 189 |
| Gartinkeir | J & W Miller | Clackmannanshire | 1967 | 1968 | 110 |
| Gartmillan 1 | W Wooton | Lanarkshire | 1951 | 1963 | 210 |
| Gartmorn | J Dawson | Clackmannanshire | 1967 | 1984 | 110 |
| Gartness | Barr & Sinclair | Lanarkshire | 1936 | 1960 | 210 |
| Gartshore 1,3 & 12 | NCB | Dunbartonshire | 1865 | 1959 | 120 |
| Gartshore 9 & 11, & Grayshill | NCB | Dunbartonshire | 1875 | 1968 | 121 |
| Gateside | Morton & Smith | Stirlingshire | 1970 | 1972 | 250 |
| Gateside 1 | NCB | Stirlingshire | 1938 | 1949 | 243 |
| Gateside 2 | NCB | Stirlingshire | 1939 | 1952 | 244 |
| Gateside 4 & 5 | NCB | Dumfriesshire | 1891 | 1964 | 115 |
| Gillfoot | Gillfoot Coal Co./Currie Bros | Lanarkshire | 1957 | 1969 | 210 |
| Gillhead | NCB | Lanarkshire | 1945 | 1955 | 190 |
| Gilmerton | NCB | Midlothian | 1928 | 1961 | 222 |
| Glen Mine | NCB | Lanarkshire | 1940 | 1954 | 190 |
| Glenbank | R Henderson | Lanarkshire | 1951 | 1964 | 210 |
| Glencairn | NCB | East Lothian | 1949 | 1962 | 127/131 |
| Glencraig | NCB | Fife | 1896 | 1966 | 145 |
| Glenellrigg | McNeill & Taylor | Stirlingshire | 1949 | 1952 | 250 |
| Glenend 1 | Thomas Heeps & Sons | Stirlingshire | 1948 | 1957 | 250 |
| Glenend 2 | Thomas Heeps & Sons | Stirlingshire | 1956 | c. 1973 | 250 |
| Glenend 3 & 4 | Thomas Heeps & Sons | Stirlingshire | 1958 | c. 1973 | 250 |
| Glenochil 1 & 2 | NCB | Clackmannanshire | 1956 | 1962 | 105 |
| Glentaggart | NCB | Lanarkshire | 1943 | 1969 | 191 |
| Glentore | NCB | Lanarkshire | 1957 | 1964 | 191 |
| Glenview/Law Mine | Wright & Moffat | Lanarkshire | 1970 | 1991 | 210 |
| Glespin | NCB | Lanarkshire | 1908 | 1964 | 191 |
| Goodockhill 3 | Greenshields & Co. | Lanarkshire | 1951 | 1958 | 210 |
| Grasshill | W Fisher & Sons | Ayrshire | 1961 | 1982 | 98 |

| Mine/Colliery | NCB/Private | County | Opened | Closed | Page |
|---|---|---|---|---|---|
| Grayrigg | Alex Mc Neill | Stirlingshire | 1954 | 1957 | 250 |
| Greencraig 1 | Greencraig Coal Co. | Stirlingshire | 1938 | 1950 | 250 |
| Greencraig 2 | Greencraig Coal Co. | Stirlingshire | 1950 | 1966 | 250 |
| Greenhill | NCB | Ayrshire | 1936 | 1958 | 79 |
| Greenlees | NCB | Lanarkshire | 1947 | 1957 | 192 |
| Greenrigg | NCB | West Lothian | 1905 | 1960 | 262 |
| Guildyhowers | W T Bathgate (Lime Works) | Midlothian | pre-1947 | c. 1950 | 233 |
| Hairstanes | A H Adams | Stirlingshire | 1959 | 1968 | 250 |
| Hairstanes 2 | A H Adams | Stirlingshire | 1958 | 1961 | 250 |
| Halkerston | Edward L Gray | Midlothian | 1958 | 1958 | 233 |
| Hall of Auchincross | BMC Mining Co. | Ayrshire | 1987 | 1991 | 98 |
| Hamilton Palace | NCB | Lanarkshire | 1884 | 1959 | 192 |
| Harestonehill | Blyth Bros | Lanarkshire | 1967 | 1983 | 210 |
| Harvieston | NCB | Clackmannanshire | 1957 | 1961 | 105 |
| Harviestoun | G Drysdale | Clackmannanshire | 1968 | 1991 | 110 |
| Harwood | NCB | Midlothian | 1946 | 1959 | 223 |
| Hassockrig | NCB | Lanarkshire | 1885 | 1962 | 193 |
| Hazelside Mine | Thomas Sinclair & Sons | Lanarkshire | 1966 | 1968 | 210 |
| Headlesscross 1 | NCB | Lanarkshire | 1938 | 1949 | 193 |
| Headlesscross 2 | NCB | Lanarkshire | 1946 | 1953 | 194 |
| Heads | Thomas Green & Sons | West Lothian | 1964 | 1965 | 268 |
| Heatheryknowe A | Glenrigg Coal Co. | Lanarkshire | 1950 | 1951 | 210 |
| Herbertshire | NCB | Stirlingshire | 1889 | 1959 | 244 |
| High Darngavil | NCB | Lanarkshire | 1948 | 1949 | 194/210 |
| Highhouse | NCB | Ayrshire | 1894 | 1983 | 79 |
| Hill | R Moffat & Sons | Lanarkshire | 1960 | 1962 | 210 |
| Hill of Drumgray 9 | Darngavil Brickworks | Lanarkshire | pre-1947 | 1951 | 210 |
| Hillfarm 1/2 | Hillfarm Coal Co. | Stirlingshire | pre-1947 | 1961 | 250 |
| Hillfarm 3 | Hillfarm Coal Co. | Stirlingshire | 1954 | 1961 | 250 |
| Hillfarm 4 | Hillfarm Coal Co. | Stirlingshire | 1960 | 1962 | 250 |
| Hillfarm 5 | Hillfarm Coal Co. | Stirlingshire | 1962 | 1968 | 250 |
| Hillfoot | Hillfoot Coal Co. | Stirlingshire | pre-1947 | 1956 | 250 |
| Hillhouserigg | NCB | Lanarkshire | 1850 | 1949 | 195 |
| Hills of Murdostoun 1 | Auchinlea Quarries | Lanarkshire | 1941c | 1951 | 210 |
| Hills of Murdostoun 3 | Auchinlea Quarries | Lanarkshire | 1950 | 1952 | 210 |
| Hillside 1 | Patrick McGrady | Lanarkshire | 1951 | 1956 | 211 |
| Hillside 2 | Patrick McGrady | Lanarkshire | 1956 | c. 1958 | 211 |
| Hillside 3 | Patrick McGrady | Lanarkshire | 1958 | 1968 | 211 |
| Hindsward 3 & 4 | NCB | Ayrshire | 1956 | 1959 | 80/98 |
| Holehousemuir | P & J Horne | Stirlingshire | pre-1947 | 1950 | 250 |
| Hopefield | Hopefield Coal Co. | Lanarkshire | pre-1947 | 1947 | 211 |
| Houldsworth | NCB | Ayrshire | 1905 | 1965 | 80 |
| Howierigg | Rumford Coal Co. | Stirlingshire | pre-1947 | 1950 | 250 |
| Hyndshaw 1 | Dr Arthur | Lanarkshire | 1947 | c. 1950 | 211 |

| Mine/Colliery | NCB/Private | County | Opened | Closed | Page |
|---|---|---|---|---|---|
| Hyndshaw 2 | Dr Arthur | Lanarkshire | 1948 | 1953 | 211 |
| Hyndshaw 3 | Hyndshaw Coal Co. | Lanarkshire | 1957 | 1963 | 211 |
| Isabella | William Fisher & Sons | Lanarkshire | 1959 | 1961 | 211 |
| Isle of Canty | NCB | Fife | 1939 | 1948 | 146 |
| Jawcraig Mid | John McKechnie | Stirlingshire | 1955 | 1957 | 250 |
| Jenny Gray | NCB | Fife | 1890 | 1959 | 146 |
| Junction | T McDonagh | Lanarkshire | 1973 | 1980 | 211 |
| Kames | NCB | Ayrshire | 1870 | 1968 | 81 |
| Kelvin View | J Graham | Dunbartonshire | 1963 | 1968 | 123 |
| Kendieshill | J Drysdale | Stirlingshire | 1950 | 1968 | 250 |
| Kennox Mines | NCB | Lanarkshire | 1908 | 1972 | 195 |
| Kepplehill 1 & 2 | NCB | Lanarkshire | 1897 | 1951 | 196 |
| Killernie | J Payne | Fife | 1966 | 1974 | 165 |
| Killoch | NCB | Ayrshire | 1960 | 1987 | 81 |
| Killochan | NCB | Ayrshire | 1905 | 1967 | 82 |
| King O' Muirs 1 | NCB | Clackmannanshire | 1938 | 1954 | 106 |
| King O' Muirs 2 | NCB | Clackmannanshire | 1950 | 1957 | 106 |
| Kinglassie | NCB | Fife | 1908 | 1966 | 147 |
| Kingshill 1 | NCB | Lanarkshire | 1919 | 1968 | 196 |
| Kingshill 2 | NCB | Lanarkshire | 1931 | 1963 | 197 |
| Kingshill 3 | NCB | Lanarkshire | 1951 | 1974 | 198 |
| Kinneil | NCB | West Lothian | 1890 | 1982 | 262 |
| Kippsbyre | Francis McLean & Co. | Lanarkshire | 1939 | 1960 | 211 |
| Kittymuir 3 | Matthew McCulloch | Lanarkshire | 1953 | 1953 | 211 |
| Kittymuir 4 | Matthew McCulloch | Lanarkshire | 1954 | 1954 | 211 |
| Kittymuirhill | Beattie Bros | Lanarkshire | 1951 | 1951 | 211 |
| Klondyke | Strathkelvin Mining Co. | Dunbartonshire | 1988 | 1991 | 123 |
| Knockshinnoch Castle | NCB | Ayrshire | 1944 | 1968 | 83 |
| Knowehead | J Bergin | Stirlingshire | pre-1947 | 1947 | 250 |
| Knowehead 2 | NCB | Lanarkshire | 1952 | 1962 | 198 |
| Knowetop | NCB | Lanarkshire | 1952 | 1966 | 199 |
| Lady Ann | D Thomson | Dumfriesshire | 1969 | c. 1977 | 177 |
| Lady Helen (Dundonald) | NCB | Fife | 1895 | 1964 | 147 |
| Lady Victoria | NCB | Midlothian | 1895 | 1981 | 224 |
| Langside Farm | Union Coal Co. | Lanarkshire | pre-1947 | 1948 | 211 |
| Langside Mine | NCB | Lanarkshire | 1945 | 1949 | 199/211 |
| Lassodie | John M Heeps | Fife | 1961 | 1991 | 165 |
| Lassodie 2 | J Methven & Sons | Fife | pre-1947 | c. 1950 | 165 |
| Lassodie 3 | J Methven & Sons | Fife | 1950 | c. 1952 | 165 |
| Lassodie 4 | J Methven & Sons | Fife | 1950 | 1952 | 165 |
| Lassodie 5 | J Methven & Sons | Fife | 1952 | 1953 | 165 |
| Lassodie 6 | J Methven & Sons | Fife | 1952 | c. 1959 | 165 |
| Lassodie 7 | J Methven & Sons | Fife | 1955 | c. 1957 | 165 |
| Lassodie 8 | J Methven & Sons | Fife | 1955 | c. 1956 | 165 |

| Mine/Colliery | NCB/Private | County | Opened | Closed | Page |
|---|---|---|---|---|---|
| Lassodie 9 | J Methven & Sons | Fife | 1956 | c. 1959 | 165 |
| Lethan's Mine 1 | John M Heeps | Fife | 1955 | 1958 | 165 |
| Lethan's Mine 2 | John M Heeps | Fife | 1957 | 1959 | 165 |
| Lime Road | J McCaig & Sons | Stirlingshire | 1948 | 1960 | 250 |
| Limeylands | NCB | East Lothian | 1895 | 1954 | 127 |
| Lindsay (Kelty 4 & 5) | NCB | Fife | 1873 | 1965 | 148 |
| Lingerwood | NCB | Midlothian | 1798 | 1967 | 225 |
| Lippie | J Walker | Stirlingshire | 1957 | c. 1959 | 250 |
| Little Whitehill 1 | Waddel Hawthorn | Lanarkshire | 1947 | 1950 | 211 |
| Little Whitehill 2 | Waddel Hawthorn | Lanarkshire | 1956 | 1956 | 211 |
| Littlemill 2, 3 & 5 | NCB | Ayrshire | 1860 | 1974 | 84 |
| Livingstone | Cloybank Minerals | Stirlingshire | c. 1933 | 1952 | 250 |
| Lochend | Alec McNeill | Stirlingshire | 1959 | c. 1965 | 250 |
| Lochend 5 | NCB | Lanarkshire | 1880 | 1948 | 199 |
| Lochhead | NCB | Fife | 1890 | 1970 | 149 |
| Lochhead | Lochside Coal & Fireclay Co. | Fife | c. 1957 | c. 1977 | 165 |
| Lochlea 1 & 2 | NCB | Ayrshire | 1949 | 1973 | 85 |
| Lochside 1, 2, & 3 | Lochside Coal & Fireclay Co. | Fife | pre-1947 | c. 1963 | 165 |
| Lochwood 1 | Lochwood Coal Co. | Ayrshire | pre-1947 | c. 1948 | 98 |
| Lochwood 2 | Lochwood Coal Co. | Ayrshire | 1942 | c. 1969 | 98 |
| Loganlea | NCB | West Lothian | 1890 | 1959 | 226 |
| Longannet Mine | NCB | Fife | 1969 | 2002 | 149/150 |
| Longriggend 1 | John Jenkins | Stirlingshire | 1953 | 1953 | 250 |
| Lugar Mine | NCB | Ayrshire | 1942 | 1953 | 85 |
| Lumphinnans 1 | NCB | Fife | 1852 | 1957 | 151 |
| Lumphinnans Mine | NCB | Fife | 1945 | 1966 | 151 |
| Lumphinnans XI & XII | NCB | Fife | 1895 | 1966 | 152 |
| Macbie Hill | William Potter | Peebleshire | 1950 | 1953 | 235 |
| Maddiston | Maddiston Coal Co. | Stirlingshire | pre-1947 | 1949 | 250 |
| Manor Powis 1, 2 & 3 | NCB | Stirlingshire | 1914 | 1967 | 245 |
| Manse | M T & P Stark | Lanarkshire | 1951 | 1951 | 211 |
| Marlage 1 | J & J McPhee | Lanarkshire | 1951 | 1953 | 211 |
| Marlage 2 | J & J McPhee | Lanarkshire | 1954 | 1959 | 211 |
| Marlage 3 | J & J McPhee | Lanarkshire | 1960 | c. 1963 | 211 |
| Mary (Lochore) | NCB | Fife | 1904 | 1966 | 152 |
| Mauchline 1, 2 & 4 | NCB | Ayrshire | 1925 | 1966 | 85 |
| Mauldslie | Andrew Maxwell & Co. | Lanarkshire | pre-1947 | 1949 | 211 |
| Mauldslie 7 | Andrew Maxwell & Co. | Lanarkshire | 1955 | c. 1968 | 211 |
| Mauldslie Upper Ell | Andrew Maxwell & Co. | Lanarkshire | 1949 | c. 1955 | 211 |
| Mavisbank / Pirleyhill | Callendar Brick & Fireclay Co. | Stirlingshire | 1962 | 1967 | 250 |
| Maxwell 2 | NCB | Ayrshire | 1903 | 1973 | 86 |
| Maxwell 4 | NCB | Ayrshire | 1950 | 1973 | 87 |
| Mayfield | Adams Pict Firebrick Co. | Ayrshire | 1956 | 1958 | 98 |
| Meadowhill | Meadowhill Coal Co. | Lanarkshire | 1951 | 1970 | 211 |

| Mine/Colliery | NCB/Private | County | Opened | Closed | Page |
|---|---|---|---|---|---|
| Meadowmill Mine | NCB | East Lothian | 1954 | 1960 | 128 |
| Meikle Drumgray 1 | Felix Travers | Lanarkshire | pre-1947 | 1958 | 211 |
| Melloch | NCB | Clackmannanshire | 1850 | 1948 | 107 |
| Meta (Devon 3) | NCB | Clackmannanshire | 1946 | 1959 | 107 |
| Michael | NCB | Fife | 1895 | 1967 | 153 |
| Milnquarter (Bonnybridge) | John G Stein & Co. | Stirlingshire | pre-1947 | c. 1957 | 251 |
| Minnivey 4 & 5 | NCB | Ayrshire | 1955 | 1975 | 87 |
| Minto | NCB | Fife | 1903 | 1967 | 154 |
| Monktonhall | NCB | Midlothian | 1967 | 1997 | 227 |
| Monktonhall | Monktonhhall Mineworkers | Midlothian | 1991 | 1997 | 233 |
| Moorside | William Potter | Peebleshire | 1955 | c. 1959 | 235 |
| Mortonmuir 8 & 9 | NCB | Ayrshire | 1951 | 1953 | 87 |
| Mossband 7 | A Allan & Sons | Lanarkshire | 1948 | 1950 | 211 |
| Mossband 8 | A Allan & Sons | Lanarkshire | 1947 | 1954 | 211 |
| Mossband 9 | A Allan & Sons | Lanarkshire | 1952 | 1953 | 211 |
| Mossband 10 | A Allan & Sons | Lanarkshire | 1957 | 1958 | 211 |
| Mount | NCB | Ayrshire | 1948 | 1950 | 88 |
| Muckraw | A Allan & Sons | West Lothian | pre-1947 | 1947 | 268 |
| Muirend | James S Burns | Ayrshire | 1954 | c. 1968 | 98 |
| Muirhouse | T M Thomson | Ayrshire | pre-1947 | 1949 | 98 |
| Muirside | J & R Howie | Ayrshire | pre-1947 | unknown | 98 |
| Murraysgate | Thomas Green & Co. | West Lothian | pre-1947 | 1965 | 268 |
| Nellie | NCB | Fife | 1880 | 1965 | 155 |
| Netherton Mine | NCB | Lanarkshire | 1938 | 1950 | 200 |
| New Biggarford | R Moffat & Sons | Lanarkshire | pre-1947 | 1948 | 211 |
| Newcraighall | NCB | Midlothian | 1898 | 1968 | 228 |
| Newfield | NCB | Ayrshire | 1940 | 1956 | 88 |
| North Dyke | James Penman | Stirlingshire | 1953 | 1954 | 251 |
| North Linrigg 1 | Greenshields & Co. | Lanarkshire | 1959 | c. 1963 | 211 |
| North Linrigg 3 | Greenshields & Co. | Lanarkshire | pre-1947 | 1950 | 211 |
| North Linrigg 4 & 5 | Greenshields & Co. | Lanarkshire | pre-1947 | 1947 | 211 |
| North Shaws | Auchinlea Quarries | Lanarkshire | 1941 | 1949 | 211 |
| North Steelend | J Payne | Fife | 1958 | c. 1965 | 165 |
| Northfield and Hall | NCB | Lanarkshire | 1917 | 1961 | 201 |
| Northrigg 1 | United Fireclay Co. | West Lothian | unknown | 1963 | 268 |
| Nunnery | Dalkieth Transport & Storage | Midlothian | 1955 | 1955 | 233 |
| O'Wood | John Dow jnr | Lanarkshire | pre-1947 | 1948 | 211 |
| Oakerdykes 2 | Mrs A White | Stirlingshire | pre-1947 | 1950 | 251 |
| Oakerdykes 3 | Mrs A White | Stirlingshire | pre-1947 | 1948 | 251 |
| Oakerdykes 4 | Mrs A White, White & Caine | Stirlingshire | 1950 | 1978 | 251 |
| Oakfield | NCB | Fife | 1946 | 1949 | 156 |
| Overtown | NCB | Lanarkshire | 1931 | 1968 | 201 |
| Overwood Mine | NCB | Lanarkshire | 1949 | 1955 | 202/211 |
| Oxenford 2 | NCB | Midlothian | 1926 | 1950 | 228 |

| Mine/Colliery | NCB/Private | County | Opened | Closed | Page |
|---|---|---|---|---|---|
| Oxenford 3 | NCB | Midlothian | 1952 | 1959 | 229 |
| Palace 2 | Palace Coal Co. | Lanarkshire | pre-1947 | 1950 | 211 |
| Palace 3 | Palace Coal Co. | Lanarkshire | 1948 | 1948 | 211 |
| Parkhead 1 | Auchinlea Quarries | Lanarkshire | 1952 | 1954 | 211 |
| Penkaet | Alex Gordon | East Lothian | pre-1947 | 1965 | 131 |
| Pennyvenie 2, 3 & 7 | NCB | Ayrshire | 1945 | 1978 | 89 |
| Pennyvenie 4 | NCB | Ayrshire | 1911 | 1961 | 89 |
| Pennyvenie 5 | NCB | Ayrshire | 1911 | 1953 | 90 |
| Pirleyhill 1 | Callender Brick & Fireclay Co. | Stirlingshire | 1947 | 1955 | 251 |
| Pirleyhill 2 | Callender Brick & Fireclay Co. | Stirlingshire | 1951 | c. 1967 | 251 |
| Pirleyhill 3 | Callender Brick & Fireclay Co. | Stirlingshire | 1954 | c. 1967 | 251 |
| Pirnhall | NCB | Stirlingshire | 1933 | 1963 | 246 |
| Plean 3, 4 & 5 | NCB | Stirlingshire | 1932 | 1962 | 246 |
| Policies | Daniel Beattie jnr | East Lothian | 1983 | 1987 | 131 |
| Policy | NCB | Stirlingshire | 1924 | 1959 | 247 |
| Polkemmet | NCB | West Lothian | 1916 | 1984 | 264 |
| Polmaise 1 & 2 | NCB | Stirlingshire | 1904 | 1958 | 247 |
| Polmaise 3, 4 & 5 | NCB | Stirlingshire | 1904 | 1987 | 248 |
| Polquhairn 1 & 4 | NCB | Ayrshire | 1895 | 1962 | 90 |
| Polquhairn 5 & 6 | NCB | Ayrshire | 1955 | 1962 | 91 |
| Powharnal | NCB | Ayrshire | 1954 | 1959 | 91 |
| Prestongrange | NCB | East Lothian | 1874 | 1962 | 128 |
| Prestonlinks | NCB | East Lothian | 1899 | 1964 | 129 |
| Printfield | Armstrong Bros | Lanarkshire | 1952 | 1959 | 211 |
| Purdielodge | J Sorbie | Lanarkshire | 1959 | 1959 | 211 |
| Quarter 2 | NCB | Lanarkshire | 1815 | 1951 | 203 |
| Quarter 5a | NCB | Lanarkshire | 1941 | 1951 | 203 |
| Raebog 6 | Drumslangie Coal Co. | Lanarkshire | pre-1947 | 1947 | 211 |
| Ramsay | NCB | Midlothian | 1850 | 1965 | 229 |
| Randolph | NCB | Fife | 1850 | 1968 | 156 |
| Rankin | NCB | Lanarkshire | 1947 | 1947 | 203 |
| Rashiehill | Hillfarm Coal Co. | Lanarkshire | 1987 | 1995 | 211 |
| Redding | NCB | Stirlingshire | 1894 | 1958 | 249 |
| Reddingmuir (Polmont) | James Penman | Stirlingshire | 1948 | 1949 | 251 |
| Redhall (Slamannan) | Hutchison Caine | Stirlingshire | 1957 | 1961 | 251 |
| Redhall 2 | Hutchison Caine | Stirlingshire | 1959 | unknown | 251 |
| Riddochhill | NCB | West Lothian | 1890 | 1968 | 265 |
| Rig | NCB | Dumfriesshire | 1949 | 1966 | 115 |
| Righead | Morton Bros | Stirlingshire | 1948 | c. 1967 | 251 |
| Roger 1 & 2 | NCB | Dumfriesshire | 1952 | 1980 | 115 |
| Roger 3 & 4 | NCB | Dumfriesshire | 1956 | 1968 | 116 |
| Rosie | NCB | Fife | 1880 | 1953 | 157 |
| Roslin | NCB | Midlothian | 1903 | 1969 | 230 |
| Ross 1 | Brora Coal Co., E E Pritchard | Sutherland | pre-1947 | 1969 | 254 |

| Mine/Colliery | NCB/Private | County | Opened | Closed | Page |
|---|---|---|---|---|---|
| Ross 2 | Brora Coal Co., E E Pritchard | Sutherland | 1969 | 1975 | 254 |
| Rothes | NCB | Fife | 1957 | 1962 | 157 |
| Saddler's Brae | Union Coal Co. | Dunbartonshire | 1939 | 1964 | 123 |
| Seafield | NCB | Fife | 1966 | 1988 | 158 |
| Seaforth 1, 2 & 3 | NCB | Ayrshire | 1940 | 1953 | 92 |
| Shewalton 3 & 4 | NCB | Ayrshire | 1924 | 1955 | 92 |
| Shewalton 5 & 6 | NCB | Ayrshire | 1933 | 1950 | 92 |
| Shewalton 8 & 9 | NCB | Ayrshire | pre-1947 | 1948 | 93 |
| Shieldmains 6, 7, & 14 | NCB | Ayrshire | 1927 | 1950 | 93 |
| Shotlinn | Overwood Coal Co. | Lanarkshire | pre-1947 | 1948 | 212 |
| Skellyton 2 & 3 | NCB | Lanarkshire | 1944 | 1951 | 204 |
| Slamannan | James Drysdale | Stirlingshire | 1954 | c. 1957 | 251 |
| Smithston | T Love | Ayrshire | 1966 | 1991 | 98 |
| Snabhead | A McNeil | Stirlingshire | 1953 | 1954 | 251 |
| Solsgirth | NCB | Clackmannanshire | 1969 | 1990 | 108 |
| Sorn 1 & 2 | NCB | Ayrshire | 1953 | 1983 | 94 |
| South Bantaskine | NCB | Stirlingshire | 1946 | 1959 | 249 |
| South Lanridge 2 | James Somerville | Lanarkshire | pre-1947 | 1948 | 212 |
| South Lanridge 3 | James Somerville | Lanarkshire | pre-1947 | 1948 | 212 |
| South Lanridge 6 | A Allan & Sons | Lanarkshire | pre-1947 | 1948 | 212 |
| Southfield | NCB | Lanarkshire | 1923 | 1959 | 204 |
| Southhook | NCB | Ayrshire | 1947 | 1948 | 94 |
| Southhook | Southhook Potteries | Ayrshire | pre-1947 | c. 1971 | 98 |
| Spalehall 3 | Whittagreen Coal Co. | Lanarkshire | 1951 | 1952 | 212 |
| Spalehall 5 | Whittagreen Coal Co. | Lanarkshire | pre-1947 | 1948 | 212 |
| Spalehall 6 | Whittagreen Coal Co. | Lanarkshire | pre-1947 | 1948 | 212 |
| Spalehall 7 | Whittagreen Coal Co. | Lanarkshire | 1948 | 1952 | 212 |
| Spoutcroft 2 | E & J Speirs | Lanarkshire | pre-1947 | 1975 | 212 |
| Spoutcroft 3 | E & J Speirs | Lanarkshire | 1958 | 1975 | 212 |
| Stane Mines | NCB | Lanarkshire | 1948 | 1955 | 205 |
| Stanley | Cloybank Minerals | Stirlingshire | 1947 | 1949 | 251 |
| Stanrigg 4 | R & J Dow | Lanarkshire | 1944 | 1946 | 212 |
| Stanrigg 5 | R & J Dow | Lanarkshire | 1945 | 1946 | 212 |
| Stanrigg 6 | R & J Dow | Lanarkshire | 1948 | c. 1969 | 212 |
| Staylea | J Jack | Lanarkshire | 1948 | 1955 | 212 |
| Stonehead | Kerr & Adams | West Lothian | 1951 | 1960 | 268 |
| Summerhouse | Maddiston Coal Co. | Stirlingshire | pre-1947 | 1947 | 251 |
| Sundrum | NCB | Ayrshire | 1953 | 1961 | 95 |
| Tannoch | Rumford Coal Co. | Dunbartonshire | 1952 | 1952 | 123 |
| Temple Braidwood | McNeill & Knox | Midlothian | 1955 | 1986 | 233 |
| Temple Farm/High Temple | Temple Farm Coal Co. | Midlothian | 1958 | 1968 | 233 |
| Thankerton 2 | NCB | Lanarkshire | 1850 | 1953 | 205 |
| Thankerton 6 | NCB | Lanarkshire | 1850 | 1949 | 205 |
| Thinacre Mine | NCB | Lanarkshire | 1949 | 1963 | 206/212 |
| Thorn | John Rankin | Lanarkshire | 1957 | 1957 | 212 |

| Mine/Colliery | NCB/Private | County | Opened | Closed | Page |
|---|---|---|---|---|---|
| Thornton Mine | NCB | Fife | 1945 | 1953 | 159 |
| Threaprig | J Frew | Stirlingshire | 1958 | 1970 | 251 |
| Tillicoultry 1 & 2 | NCB | Clackmannanshire | 1876 | 1957 | 108 |
| Tofts 1 & 2 | NCB | Ayrshire | 1914 | 1948 | 95 |
| Torrance | Torrance Coal Co. | Lanarkshire | pre-1947 | 1947 | 212 |
| Torry | NCB | Fife | 1952 | 1965 | 160 |
| Tower Mine | NCB | Dumfriesshire | 1916 | 1964 | 116 |
| Twechar 1 & Gartshore 10 | NCB | Dunbartonshire | 1865 | 1964 | 121 |
| Tynemount | NCB | East Lothian | 1924 | 1952 | 130 |
| Valleyfield 1 & 2 | NCB | Fife | 1908 | 1978 | 160 |
| Viaduct | Graham & Anderson | Ayrshire | 1959 | 1991 | 98 |
| Victor | J O Kane | Lanarkshire | 1954 | 1966 | 212 |
| Warrix 1 & 2 | NCB | Ayrshire | 1944 | 1950 | 96 |
| Wellesley | NCB | Fife | 1885 | 1967 | 161 |
| Wellsgreen Pit | NCB | Fife | 1888 | 1959 | 162 |
| West Machan | J Burns & Sons | Lanarkshire | pre-1947 | 1960 | 212 |
| Wester Auchengeich | NCB | Lanarkshire | 1928 | 1968 | 206 |
| Wester Burnhead 2 | William Walker & Sons | Stirlingshire | 1947 | 1952 | 251 |
| Wester Dunsyston 1 | James Shields | Lanarkshire | 1951 | 1952 | 212 |
| Wester Dunsyston 2 | James Shields | Lanarkshire | 1952 | 1954 | 212 |
| Wester Dunsyston 3 | William Hendrie | Lanarkshire | 1955 | 1956 | 212 |
| Wester Dunsyston 4 | William Hendrie | Lanarkshire | 1956 | 1970 | 212 |
| Wester Gartshore | NCB | Dunbartonshire | 1872 | 1950 | 122 |
| Westoun | NCB | Lanarkshire | 1948 | 1962 | 207 |
| Whitehill | NCB | Midlothian | 1850 | 1961 | 230 |
| Whitehill 1 & 2 | NCB | Ayrshire | 1893 | 1965 | 96 |
| Whitehill 3 & 4 | NCB | Ayrshire | 1946 | 1965 | 97 |
| Whiteside | Meadowhead Coal Co. | Lanarkshire | pre-1947 | 1948 | 212 |
| Whitrigg 2, 4 & 5 | NCB | West Lothian | 1900 | 1972 | 266 |
| Wilsontown | NCB | Lanarkshire | 1898 | 1955 | 207 |
| Windsor | Rigsmuir Coal Co. | Lanarkshire | pre-1947 | 1949 | 212 |
| Windsor 2 | Rigsmuir Coal Co. | Lanarkshire | pre-1947 | 1950 | 212 |
| Windsor 3 | Rigsmuir Coal Co. | Lanarkshire | 1950 | 1950 | 212 |
| Windsor 4 | Rigsmuir Coal Co. | Lanarkshire | 1950 | 1954 | 212 |
| Windyedge | NCB | Fife | 1950 | 1951 | 164 |
| Windyedge | J Summerville | Fife | pre-1947 | 1949 | 165 |
| Winton | NCB | East Lothian | 1952 | 1962 | 130 |
| Woodbank | Woodbank Mining Co./W Park | West Lothian | 1950 | 1965 | 268 |
| Woodend | NCB | West Lothian | 1870 | 1965 | 266 |
| Woodhall | William Crossan | Lanarkshire | 1951 | 1969 | 212 |
| Woodhead, Linlithgow | Savage & Moyser | West Lothian | 1957 | 1967 | 268 |
| Woodmuir | NCB | Midlothian | 1896 | 1963 | 231 |
| Woodside | NCB | Lanarkshire | 1848 | 1955 | 207 |
| Woolmet | NCB | Midlothian | 1898 | 1966 | 231 |
| Zetland | NCB | Clackmannanshire | 1935 | 1960 | 109 |

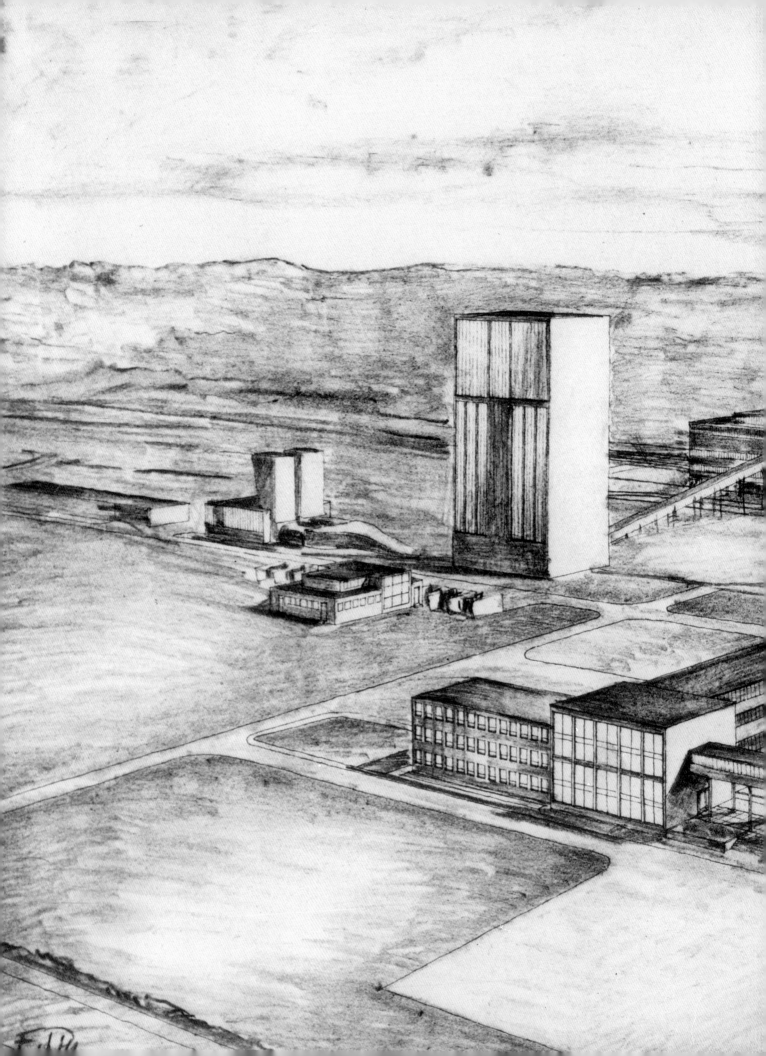